NONTRADITIONAL FEED SOURCES FOR USE IN SWINE PRODUCTION

P.A. Thacker and R.N. Kirkwood

Department of Animal & Poultry Science
University of Saskatchewan
Saskatoon, Saskatchewan

CRC Press is an imprint of the
Taylor & Francis Group, an **informa** business

First published 1990 by Butterworth Publishers
Taylor & Francis Group
6000 Broken Sound Parkway NW, Suite 300
Boca Raton, FL 33487-2742

Reissued 2018 by CRC Press

A Library of Congress record exists under LC control number: 90001588

Publisher's Note
The publisher has gone to great lengths to ensure the quality of this reprint but points out that some imperfections in the original copies may be apparent.

ISBN 13: 978-1-138-10583-6 (hbk)
ISBN 13: 978-1-138-56098-7 (pbk)
ISBN 13: 978-0-203-71124-8 (ebk)

Visit the Taylor & Francis Web site at http://www.taylorandfrancis.com and the
CRC Press Web site at http://www.crcpress.com

CONTENTS

CONTRIBUTORS

Dr. Frank X. Aherne
Department of Animal Science
University of Alberta
Edmonton, Alberta

Dr. Derek M. Anderson
Department of Animal Science
Nova Scotia Agricultural College
Truro, Nova Scotia

Dr. Henry S. Bayley
Department of Nutrition
University of Guelph
Guelph, Ontario

Mr. Ron B. Bazylo
Regional Swine Specialist
Alberta Agriculture
Vermillion, Alberta

Dr. Dick M. Beames
Department of Animal Science
University of British Columbia
Vancouver, B.C.

Dr. J. Milt Bell
Department of Animal Science
University of Saskatchewan
Saskatoon, Saskatchewan

Dr. John P. Bowland
Department of Animal Science
University of Alberta
Edmonton, Alberta

Dr. Adrian G. Castell
Swine Nutrition Section
Agric. Canada Research Station
Brandon, Manitoba

Dr. Des J.A. Cole
Department of Agriculture
University of Nottingham
Sutton Bonington, Leicester

Dr. James D. Chapman
Alltech Biotechnology Centre
3031 Catnip Hill
Nicholasville, Kentucky

Mr. Craig S. Darroch
Department of Animal Science
University of Saskatchewan
Saskatoon, Saskatchewan

Ms. Adrienne C. DeSchutter
Livestock Section
Ridgetown College of Agric.
Ridgetown, Ontario

Dr. William E. Dinusson
Department of Animal Science
North Dakota State University
Fargo, North Dakota

Dr. Sandra A. Edwards
Pig Lead Centre
N. Scotland College of Agric.
Aberdeen, Scotland

Dr. T.E. Ekpenyong
Department of Animal Science
University of Ibadan
Ibadan, Nigeria

Dr. David J. Farrell
Department of Animal Science
University of New England
Armidale, New South Wales

Dr. O.M. Hale
Department of Animal Science
Coastal Plain Station
Tifton, Georgia

Dr. Keith D. Haydon
Department of Animal Science
Coastal Plain Station
Tifton, Georgia

Dr. Palmer Holden
Department of Animal Science
Iowa State University
Ames, Iowa

Dr. Keith Hutton
Coprice Division
Rice Growers, Co-operative
Leeton, New South Wales

Dr. Ray H. King
Department of Agriculture
Animal Research Institute
Werribee, Victoria

Dr. Steve Leeson
Department of Animal Science
University of Guelph
Guelph, Ontario

Dr. R. Maitland Livingstone
Dept. of Animal Husbandry
Rowett Research Institute
Bucksburn, Aberdeen

Dr. Annette C. Longland
AFRC Institute
Shinfield, Reading
Berkshire, United Kingdom

Dr. A. Graham Low
AFRC Institute
Shinfield, Reading
Berkshire, United Kingdom

Dr. T. Pearse Lyons
Alltech Biotechnology Centre
3031 Catnip Hill
Nicholasville, Kentucky

Mr. W. Ian Magowan
Swine Nutritionist
Vigor Feeds Division
Edmonton, Alberta

Dr. Donald C. Mahan
Department of Animal Science
Ohio State University
Columbus, Ohio

Dr. Charles V. Maxwell
Department of Animal Science
Oklahoma State University
Stillwater, Oklahoma

Dr. Elwyn R. Miller
Department of Animal Science
Michigan State University
East Lansing, Michigan

Mr. Jim R. Morris
Department of Animal Science
Ridgetown College of Agric.
Ridgetown, Ontario

Dr. Surendra S. Negi
Indian Vet. Research Institute
Regional Station
Palampur, India

Dr. H. William Newland
Department of Animal Science
Ohio State University
Wooster, Ohio

Dr. G. Larry Newton
Department of Animal Science
University of Georgia
Tifton, Georgia

Dr. Emmanuel Nwokolo
Department of Animal Science
University of Alberta
Edmonton, Alberta

Dr. O.L. Oke
Chemistry Department
University of Ife
Ife-Ife, Nigeria

Ms. Daphine J. Peer
Animal Industry Branch
Ontario Ministry of Agriculture
Fergus, Ontario

Mr. Steven V. Radecki
Department of Animal Science
Michigan State University
East Lansing, Michigan

Dr. Velmurugu Ravindran
Department of Animal Science
University of Peradeniya
Peradeniya, Sri Lanka

Mr. J.P. Rodriquez
Nutro Products Incorporated
445 Wilson Way
Industry, California

Dr. Gerry C. Shurson
Department of Animal Science
Ohio State University
Columbus, Ohio

Dr. T.D. Tanksley, Jr.
Department of Animal Science
Texas A & M University
Betram, Texas

Dr. Phil A. Thacker
Department of Animal Science
University of Saskatchewan
Saskatoon, Saskatchewan

Dr. Peter J. Thorne
ODNRI
Culham, Abingdon
Oxford, United Kingdom

Mr. Ted A. Van Lunen
Swine Nutrition Section
Agric. Canada Research Station
Nappan, Nova Scotia

Dr. Richard C. Wahlstrom
Department of Animal Science
South Dakota State University
Brookings, South Dakota

Mr. Derek G. Waterworth
ICI Biological Products
Billingham, Cleveland
England

Dr. Julian Wiseman
Department of Agriculture
University of Nottingham
Sutton Bonington, Leicester

PREFACE

Feed represents the greatest single expense associated with raising pigs to market weight. Therefore, it may be possible to improve the economics of swine production if this cost can be reduced without detriment to pig performance. Pigs are "opportunity feeders" and have the ability to consume all manner of feedstuffs. However, in recent times, the ingredient list for swine rations has become fairly limited, and the majority of diets fed to pigs are based on a few staples, such as corn, wheat, barley and soybean meal.

There are many alternative sources of dietary energy and protein that may have potential for use in swine production. However, the successful incorporation of these unfamiliar nutrient sources into animal diets is limited by the availability of sufficient information on which to base feeding recommendations. The editors of this volume have attempted to fill this gap in the literature by presenting separate and self-contained chapters for over fifty nontraditional feedstuffs.

Each chapter was prepared by experts in their respective feed source areas and covers the background and history of the feed as well as providing information on its nutrient content and the presence of any anti-nutritional factors that might limit its nutritional value. Each chapter concludes with specific recommendations concerning the best use of the feedstuff in question and provides guidelines for optimum inclusion rates. As such, the book should provide a handy reference text for swine producers, feed industry nutritionists, extension personnel, university students and swine researchers.

We wish to express our sincere appreciation to Ms. Fran Teitge for her many hours spent preparing the text for publication and to the University of Saskatchewan for financial support during its preparation. Finally, we wish to thank each contributor, whose efforts made the volume possible.

CHAPTER 1

Alfalfa Meal

P.A. Thacker

INTRODUCTION

Alfalfa is of interest as a potential feedstuff for swine because it produces high yields of protein and is not consumed directly by humans. In the future, high quality protein sources such as soybean meal may become less available for livestock feeding because of greater direct human utilization. This may result in a greater dependency on more fibrous feedstuffs such as alfalfa for non-ruminant animals.

Unfortunately, there are problems associated with alfalfa which limit its usefulness as an ingredient in swine rations. The protein in alfalfa is poorly digested. Alfalfa has a low digestible energy content due to its high crude fiber content while toxins such as saponins and tannins reduce the growth rate of animals fed diets containing alfalfa. In addition, alfalfa is relatively unpalatable. Despite the negative factors present, there is still considerable interest in the use of alfalfa as a supplementary source of vitamins and as a component of gestation rations. Therefore, research is being conducted in an effort to overcome the problems associated with the feeding of alfalfa. If this work is successful, the use of alfalfa in swine rations may increase.

GROWING ALFALFA

Alfalfa (*Medicago species*) is one of the most popular forage crops grown throughout the world. It is adapted to a wide variety of soil and climatic conditions but grows particularly well on deep, well-drained, neutral soils (Goplen et al., 1982). Optimum temperatures for growth are 15 to 25°C during the day and 10 to 20°C at night. In addition, since alfalfa is a nitrogen fixing legume, proper inoculation eliminates the need for nitrogen fertilization.

Alfalfa is usually cut once or twice a year, although in very moist climates, three or four cuttings are possible. If the alfalfa is to be used as a protein supplement, cutting at the early bud stage is most appropriate, while for use as a roughage, cutting at the 10% bloom stage is recommended (Goplen et al., 1982). Under proper management, dry matter yields in excess of 7000 kg/ha are possible.

NUTRIENT CONTENT OF ALFALFA

The nutritional quality of alfalfa varies considerably due to stage of maturity, soil fertility, variety, physical handling and other factors. The most significant factor affecting the nutritional value of alfalfa is the growth stage at which it was cut. As the forage becomes more mature, it contains less protein and more fiber (National Academy of Science, 1972). The crude fiber content of alfalfa is extremely high compared to other commonly utilized feedstuffs. Since the pig has a simple stomach of relatively small capacity, it is less able to utilize crude fiber than are other types of farm livestock. Therefore, the digestible energy content of alfalfa is approximately half of that found in the common cereal grains (Table 1.1).

Table 1.1. Chemical Composition of Alfalfa (as fed)

	Alfalfa Hay[1]	Sun Cured Alfalfa Meal[1]	Dehydrated Alfalfa Meal[2]
Dry Matter (%)	91.4	90.7	92.0
Crude Protein (%)	15.5	17.6	17.4
Crude Fiber (%)	28.0	27.3	24.0
Ash (%)	9.0	9.6	----
Ether Extract (%)	1.7	2.1	2.8
D.E. (kcal/kg)	1419	1351	1880

[1]National Academy of Science, 1972.
[2]National Research Council, 1988.

Alfalfa ranges from 12 to 22% crude protein. Unfortunately, the protein in alfalfa is not very digestible. The high fiber content of alfalfa reduces the ability of digestive enzymes to gain access to the soluble cellular proteins. As a result, the protein in alfalfa is only about 60% digestible (Cheeke and Myer, 1975). Alfalfa contains a respectable amino acid balance and a reasonable level of lysine which is the limiting amino acid in most cereal grains (Table 1.2). However, because of the high fiber level, the availability of the lysine in alfalfa is low.

Table 1.2. Amino Acid Content of Alfalfa (% as fed)

	Alfalfa Hay[1]	Sun Cured Alfalfa Meal[1]	Dehydrated Alfalfa Meal[2]
Arginine	0.64	0.91	0.77
Histidine	0.27	0.27	0.33
Isoleucine	0.73	0.82	0.81
Leucine	0.91	1.27	1.28
Lysine	0.55	1.00	0.85
Methionine	0.09	0.18	0.27
Phenylalanine	0.55	0.82	0.80
Threonine	0.64	0.73	0.71
Tryptophan	0.09	0.27	0.34
Valine	0.64	0.82	0.88

[1]National Academy of Science, 1972.
[2]National Research Council, 1988.

Alfalfa is characteristically high in calcium (Table 1.3). However, it has only a fair phosphorus content. When grown on phosphorus-deficient soils, it may be very low in phosphorus. Therefore, rations containing high levels of alfalfa require supplemental phosphorus to meet the pigs phosphorus requirement and to narrow the rather wide calcium:phosphorus ratio present in this forage. Alfalfa is a reasonably good source of other minerals such as magnesium, potassium, copper, manganese, iron, chloride and zinc. The selenium content of alfalfa varies, depending on the area in which it is grown.

Alfalfa hay is a good source of most vitamins (Table 1.4) and is an excellent source of vitamins A, E and K. Field-cured alfalfa hay is also a good source of vitamin D. However, due to the fact that it is not exposed to sunlight in the curing process, the vitamin D content of artificially dehydrated alfalfa meal is only one-fourth to one-third as high as that found in sun-cured meal. Good alfalfa hay is also quite rich in riboflavin, pantothenic acid, biotin and niacin. However, the advent of relatively cheap sources of these nutrients has resulted in a reduction in the need for alfalfa in the diet as a source of these nutrients.

Table 1.3. Mineral Content of Alfalfa (% as fed)

	Alfalfa Hay[1]	Sun Cured Alfalfa Meal[1]	Dehydrated Alfalfa Meal[2]
Calcium (%)	1.29	0.87	1.40
Magnesium (%)	0.31	----	0.29
Phosphorus (%)	0.21	0.17	0.23
Potassium (%)	1.99	----	2.38
Sodium (%)	0.15	0.17	0.10
Sulfur (%)	0.29	0.24	0.23
Copper (mg/kg)	18.5	----	10.0
Zinc (mg/kg)	----	31.8	19.0
Manganese (mg/kg)	56.5	----	31.0
Selenium (mg/kg)	----	----	0.33

[1]National Academy of Science, 1972.
[2]National Research Council, 1988.

Table 1.4. Vitamin Content of Alfalfa (as fed)

	Alfalfa Hay[1]	Sun Cured Alfalfa Meal[1]	Dehydrated Alfalfa Meal[2]
Vitamin A (IU/g)	109	64	---
Vitamin D (IU/g)	1.5	---	---
Vitamin E (mg/kg)	102	386	111
Biotin (mg/kg)	0.16	0.32	0.33
Folic Acid (mg/kg)	3.10	5.72	4.40
Choline (mg/kg)	----	1021	1369
Niacin (mg/kg)	----	34	37
Pantothenic Acid (mg/kg)	----	28	30
Riboflavin (mg/kg)	----	10	13
Thiamine (mg/kg)	----	2.6	3.4

[1]National Academy of Science, 1972.
[2]National Research Council, 1988.

UNDESIRABLE CONSTITUENTS IN ALFALFA

Saponins are bitter tasting compounds present in alfalfa (Leamaster and Cheeke, 1979) which have been shown to impair growth rate in chicks and possibly in pigs. The mechanisms by which saponins cause growth depression have not been determined. However, Cheeke (1976) suggested several possibilities including a depressing effect on feed intake or an inhibitory effect on digestive enzymes. In addition, it was suggested that saponins may complex with nutrients to make them unavailable or that saponins may inhibit cellular metabolism (Cheeke, 1976).

It is possible through genetic selection to produce cultivars of alfalfa with a lower saponin content (Pedersen and Wang, 1971). It would appear that cultivars containing lower levels of saponins are more palatable and support higher levels of performance than do traditional alfalfa varieties. Future research, directed towards the development of high-yielding, low-saponin varieties, may allow for higher levels of alfalfa to be incorporated into swine diets.

Alfalfa also contains about 3.25% total tannins (Millic et al., 1972). Tannins are water-soluble polymeric phenolics that depress protein digestibility by binding dietary protein as well as by inhibiting digestive enzymes (Jung and Fahey, 1983; Mcleod, 1974). Tannins also depress feed intake due to astringency (Glick and Joslyn, 1970). In addition, a trypsin inhibitor has been identified in alfalfa (Ramirez and Mitchell, 1960) and there is also evidence that alfalfa contains a photosensitizing agent (Lohrey et al., 1974).

FEEDING ALFALFA TO SWINE

Starter Pigs

Alfalfa should not be used in diets fed to weanling pigs. Its high crude fiber content and low digestible energy level are likely to limit growth and reduce the efficiency of feed utilization when fed to pigs of this weight range. Other alternatives are available and producers would be wise to choose a higher energy feedstuff as the foundation for their starter diets.

Growing Pigs

Alfalfa should also be used sparingly in diets fed to grower pigs. The effects of the addition of 0, 20, 40, or 60% alfalfa meal on the performance of growing-finishing swine are shown in Table 1.5 (Powley et al., 1981). Pigs fed diets containing high levels of alfalfa meal gained significantly slower and with a reduced feed efficiency in comparison with those pigs fed the control diet. The reduction in gain occurring when high levels of dietary alfalfa are fed appears to be the result of insufficient dietary energy. Pigs fed high levels of alfalfa meal cannot consume sufficient feed to meet their energy requirements for maximum growth. Similar results were reported by Kass et al. (1980) and therefore the inclusion of alfalfa meal in the diet of grower pigs is not recommended.

Poor palatability is one factor accounting for the reduction in performance resulting from the inclusion of high levels of alfalfa in the diet of the growing-finishing pigs. Table 1.6 shows the results of a feeding trial in which pigs were given a choice between a standard corn-soybean meal diet and one in which pigs were given a choice between a standard corn-soybean meal diet and one in which 0.5, 1.0, 2.5, 5, 10, 20, or 30% alfalfa meal was added at the expense of corn (Leamaster and Cheeke, 1979). When only

0.5% alfalfa was present, pigs consumed approximately the same amount of each diet. However, at all other levels of inclusion, the pigs showed a distinct preference for the alfalfa free diet.

Table 1.5. Performance of Pigs Fed Alfalfa

	Level of Alfalfa (%)			
	0.0	20.0	40.0	60.0
Daily Gain (g/day)	860	730	630	410
Daily Intake (kg/day)	3.0	3.0	3.2	2.7
Feed Efficiency	3.6	4.1	5.0	6.7
Dressing Percentage	77.9	76.2	75.4	75.2
Backfat Thickness (cm)	3.9	3.5	3.2	2.9

Powley et al., 1981.

Table 1.6. Effect of Dietary Alfalfa on Diet Preference of Swine

Dietary Alfalfa (%)	Control	Alfalfa Diet
0.5	49	51
1.0	63	37
2.5	66	34
5.0	65	35
10.0	74	26
20.0	81	19
30.0	97	3

Leamaster and Cheeke, 1979.

Finishing Pigs

Backfat thickness has been shown to be reduced by feeding high levels of dietary alfalfa (Table 1.5). This may result in improvements in carcass grade. Unfortunately, the economic losses arising from the extended feeding period are unlikely to be offset by these improvements in carcass quality. However, if carcass quality is sufficiently poor it may be beneficial to include 10% alfalfa in the diet of finishing pigs.

Breeding Stock

Alfalfa has the greatest potential for use in diets fed to gestating sows. The use of alfalfa in gestation rations has been studied for over thirty years

(Danielson and Noonan, 1975; De Pape et al., 1953; Pollmann et al., 1979; Seerley and Wahlstrom, 1965; Teague, 1955) and most of this research has indicated little or no adverse effect on sow reproductive performance. In fact, studies conducted at Kansas State University have shown that a sow can successfully reproduce on a diet consisting of only good quality alfalfa hay supplemented with the necessary vitamins and minerals (Allee, 1977; Table 1.7). However, under stressful conditions, or when the sow's body maintenance requirement increases, an all alfalfa diet would not be adequate. The feeding of such a diet is not recommended or advocated.

Table 1.7. Reproductive Performance of Sows Fed Diets Containing 96.9% Dehydrated Alfalfa Meal Free Choice During Gestation

	Control	Alfalfa Meal
Percent Farrowing	90.3	90.3
Pigs Born Alive/Litter	8.4	8.3
Birth Weight (kg)	1.3	1.3
Pigs Weaned/Litter	7.1	6.5
Weaning Weight (kg)	6.0	5.9
Percent Mortality	15.4	21.6

Allee, 1977.

University of Nebraska workers investigated the use of 50% sun-cured alfalfa meal in pelleted corn-soybean rations for gestating sows (Pollmann et al., 1980). A corn-soybean meal control diet and the 50% alfalfa meal diet were both fed at a level to provide 6,000 kcal of metabolizable energy per day during the first 90 days of gestation. After this time, all sows were placed on a corn-soybean meal lactation diet. The experiment was conducted over three reproductive cycles. The results from this experiment showed that sows on the 50% alfalfa meal diet farrowed and weaned an additional pig per litter in comparison with those sows fed the control diet (Table 1.8). In addition, a higher percentage of the sows receiving the alfalfa treatment completed the three gestation-lactation cycles (88.6 vs. 65.9%).

Results from experiments such as the one described above have been widely touted as evidence for the necessity of inclusion of alfalfa in gestation diets. However, alfalfa is generally considered to be an excellent source of vitamin E and selenium, both of which have been implicated as being important in ensuring reproductive performance. When one looks at the results of experiments in which the control diet was adequately supplemented with these two nutrients, such as that conducted at Michigan State University, no benefit from the inclusion of alfalfa in the diet can be observed (Miller et al., 1981; Table 1.9). Since synthetic sources of both vitamin E and selenium are available, there does not appear to be any necessity for the inclusion of alfalfa in gestation diets.

Table 1.8. Reproductive Performance of Sows Fed Sun-Cured Alfalfa Meal for Three Successive Reproductive Cycles[1]

	Control	50% Alfalfa Meal
Percent Farrowing	77.9	92.5
Pigs Born Alive/Litter	10.8	11.8
Birth Weight (kg)	1.5	1.4
Pigs Weaned/Litter	7.7	8.9
Weaning Weight (kg)	3.9	3.7
Percent Mortality	28.7	24.5
Percent of Sows Culled[2]	34.1	11.4
Reason for Culling		
Poor Performance[3]	4	1
Death	3	1
Prolapse	2	1
Physical Abnormalities[4]	6	2

[1]Pollmann et al., 1980.
[2]Percentage of sows failing to complete three gestations.
[3]Saved three or fewer pigs from previous gestation.
[4]Feet, legs and vulva.

Table 1.9. Reproductive Performance of Sows Fed Dietary Alfalfa Meal in Diets Adequately Supplemented with Vitamins and Minerals

	Control	10% Alfalfa Meal
Pigs Born Alive/Litter	9.0	9.1
Birth Weight (kg)	1.5	1.5
Pigs Weaned/Litter	7.8	7.7
Weaning Weight (kg)	5.5	5.9
Percent Mortality	13.3	15.4

Miller et al., 1981.

Although alfalfa can be used successfully in sow gestation rations, it does not contain any magical properties. Provided the diet is properly supplemented with all the essential vitamins and minerals, sows will perform just as well on a diet devoid of alfalfa as they will on a diet containing alfalfa. Therefore, economics should dictate whether or not to include alfalfa in gestation diets.

As a word of caution, high levels of alfalfa in the gestation diet may reduce the birth weight of piglets. Since low birth weight piglets have been shown to have significantly greater mortality than do those piglets of a higher birth weight, it would seem wise to limit the levels of alfalfa fed during gestation. However, levels as high as 60% of the diet seem safe.

It has been suggested that constipation may predispose farrowing sows to post-farrowing fever and poor milk production. Therefore, many producers include bulky ingredients such as alfalfa or wheat bran in the diets of sows during prefarrowing and lactation. University of Florida workers conducted a study to determine the influence of adding 15% alfalfa meal to the diet three days before farrowing and during a two week lactation (Wallace et al., 1975). They found a reduction in the incidence of MMA in sows fed the alfalfa ration (Table 1.10). However, this reduced incidence was not reflected in a significant improvement in overall sow performance. Therefore, the use of alfalfa in prefarrowing diets to reduce constipation is not really necessary but can be included if desired (Baker et al., 1974).

Table 1.10. Reproductive Performance of Sows as Influenced by the Inclusion of Alfalfa Meal in the Farrowing and Lactation Diet

	Control	15% Alfalfa Meal
Pigs Born Alive/Litter	11.0	11.0
Birth Weight (kg)	1.4	1.4
Pigs Weaned/Litter	9.6	9.8
Weaning Weight (kg)	3.9	3.9
Percent Mortality	12.7	10.9
Percent Sows with MMA	56.7	37.1

Wallace et al., 1975.

Feeding of alfalfa meal during lactation will reduce the energy intake of sows and may result in a reduction in milk production. Therefore, it is recommended that alfalfa meal not be fed to sows during lactation.

SUMMARY

Alfalfa should not be used in diets fed to weaning pigs. Its high crude fiber content and low digestible energy level are likely to limit growth and reduce the efficiency of feed utilization when fed to pigs of this weight range. Similarly, grower pigs fed high levels of alfalfa meal cannot consume sufficient feed to meet their energy requirements for maximum growth and therefore alfalfa meal is not recommended in the diet of grower pigs. Backfat thickness has been shown to be reduced by feeding high levels of dietary alfalfa. This may result in improvements in carcass grade. Therefore, if carcass quality is sufficiently poor it may be beneficial to include 10% alfalfa in the diet of finishing pigs. Alfalfa has its greatest potential for use in gestation diets and can be used at levels as high as 60%, depending on economics. However, feeding of alfalfa meal during lactation will reduce the energy intake of sows and may result in a reduction in milk production. Therefore, it is recommended that alfalfa meal not be fed during lactation.

REFERENCES

Allee, G.L., 1977. Using dehydrated alfalfa to control intake of self-fed sows during gestation. Feedstuffs 49: 20-21.

Baker, D.H., Harmon, B.G. and Jensen, A.H., 1974. Value of alfalfa meal and wheat bran in diets for swine during prefarrowing and lactation. J. Anim. Sci. 39: 325-329.

Cheeke, P.R. and Myer, R.O., 1975. Protein digestibility and lysine availability in alfalfa meal and alfalfa protein concentrate. Nutr. Rep. Int. 12: 337-344.

Cheeke, P.R., 1976. Nutritional and physiological properties of saponins. Nutr. Rep. Int. 13: 315-324.

Danielson, D.M. and Noonan, J.J., 1975. Roughages in swine gestation diets. J. Anim. Sci. 41: 94-99.

De Pape, J.G., Burkitt, W.H. and Flower, A.E., 1953. Dehydrated alfalfa and antibiotic supplements in gestation-lactation rations for swine. J. Anim. Sci. 12: 77-83.

Glick, Z. and Joslyn, M.A., 1970. Food intake depression and other metabolic effects of tannic acid in the rat. J. Nutr. 100: 509-515.

Goplen, B.P., Baenziger, H., Bailey, L.D., Gross, A.T., Hanna, M.R., Michaud, R., Richards, K.W. and Waddington, J., 1982. Growing and managing alfalfa in Canada. Agriculture Canada, Ottawa, Publication No. 1705/E pp. 1-50.

Jung, H.G. and Fahey, G.C., 1983. Nutritional implications of phenolic monomers and lignin: A review. J. Anim. Sci. 57: 206-219.

Kass, M., Van Soest, P.J., Pond, W.G., Lewis, B. and McDowell, R.E., 1980. Utilization of dietary fiber from alfalfa by growing swine. I. Apparent digestibility of diet components in specific segments of the gastrointestinal tract. J. Anim. Sci. 50: 175-191.

Leamaster, B.R. and Cheeke, P.R., 1979. Feed preferences of swine: Alfalfa meal, high and low saponin alfalfa, and quinine sulfate. Can. J. Anim. Sci. 59: 467-469.

Lohrey, E., Tapper, B. and Hove, E.L., 1974. Photosensitization of albino rats fed on lucerne-protein concentrate. Brit. J. Nutr. 31: 159-166.

McLeod, M.N., 1974. Plant tannins: Their role in forage quality. Nutr. Abstr. Rev. 44: 803-815.

Miller, E.R., Hogberg, M.G. and Ku, P.K., 1981. Dietary laxatives for sows during gestation and lactation. Michigan State University, East Lansing, Report of Swine Research pp. 13-15.

Millic, B.L., Stojanovic, S., and Vucurevic, N., 1972. Lucerne tannins. II. Isolation of tannins from lucerne and their nature and influence on the digestive enzymes in vitro. J. Sci. Food Agric. 23: 1157-1162.

National Academy of Science, 1972. Atlas of Nutritional Data on United States and Canadian Feeds. NAS-NRC, Washington, D.C.

National Research Council, 1988. Nutrient Requirements of Domestic Animals No. 2. Nutrient Requirements of Swine. 9th ed. National Academy of Science, Washington, D.C.

Pedersen, M.W. and Wang, L., 1971. Modification of saponin content of alfalfa through selection. Crop Sci. 11: 833-835.

Pollmann, D.S., Danielson, D.M. and Peo, E.R., 1979. Value of high fiber diets for gravid swine. J. Anim. Sci. 48: 1385-1393.

Pollmann, D.S., Danielson, D.M., Crenshaw, M.A. and Peo, E.R., 1980. Long term effects of dietary additions of alfalfa and tallow on sow reproductive performance. J. Anim. Sci. 51: 294-299.

Powley, J.S., Cheeke, P.R., England, D.C., Davidson, T.P. and Kennick, W.H., 1981. Performance of growing-finishing swine fed high levels of alfalfa meal: Effects of alfalfa level, dietary additives and antibiotics. J. Anim. Sci. 53: 308-316.

Ramirez, J.S. and Mitchell, H.L., 1960. The trypsin inhibitor of alfalfa. J. Agric. Food Chem. 8: 393-395.

Seerley, R.W. and Wahlstrom, R.C., 1965. Dehydrated alfalfa meal in rations for confined brood sows. J. Anim. Sci. 24: 448-453.

Teague, H.S., 1955. The influence of alfalfa on ovulation rate and other reproductive phenomena in gilts. J. Anim. Sci. 14: 621-627.

Wallace, H.D., Thieu, D.D. and Combs, G.E., 1975. Alfalfa meal as a special bulky ingredient in the sow diet. Feedstuffs 47: 24.

CHAPTER 2

Bananas

V. Ravindran

INTRODUCTION

The scarcity of feed grains is one of the major constraints limiting the expansion of livestock industries in developing countries, particularly those in the tropical region. However, within the tropics, there exists a wide spectrum of resources that could form a feed base for increasing livestock production. One ingredient with particular potential is the banana (*Musa sapientum* L.). Plantain (*Musa paradisiaca* L.), or cooking bananas, are also included in the present discussion because many countries do not differentiate between these two species in their statistics and the appearance, composition and feed value of these species are somewhat similar.

In the majority of banana exporting countries, much of the banana crop is wasted due to rejection of the fruit in the fields or at buying points, and this material is generally left to rot. At packing plants, the bananas that are too small or too large, slightly bruised, off-colored or not in the optimal stage for shipping are rejected for export. On average, waste bananas constitute 10 to 20% of the total crop (Clavijo and Maner, 1975; Le Dividich et al., 1975) but wastage may be as high as 50% under unfavourable weather conditions (Gohl, 1981; Ravindran, 1987).

The reject or waste bananas, along with surplus quantities from small-farm production, represent a good carbohydrate source for livestock. Even if only 10% of the world production is available for livestock feeding, this would be equivalent to more than 6.85 million tonnes of fresh bananas or 1.36 million tonnes of air-dried material.

GROWING BANANAS

The banana is a large herb with a pseudostem built up from leaf sheaths with its true stem (rhizome) hidden underground. Bananas grow best in the humid tropics where bunches, containing up to 200 banana fingers each, are produced in 9 to 12 months. For favorable growth, the monthly rainfall must be at least 100 mm and can be as high as 250 mm. Where dry seasons occur, maturity is delayed unless the crop is irrigated. High temperature is the other requirement for banana growth and temperatures between 25 and 30°C are optimal. Bananas can be grown on any soil type provided the drainage is good. The recommended cultivation practices for banana production have been discussed in detail by Stover and Simmonds (1987).

The banana is generally harvested green and ripening is delayed by cool-storage (13°C) until the fruit is ready for sale (Salunkhe and Desai, 1984). The average banana yields in commercial plantations range from 40 to 55 tonnes/ha (Stover and Simmonds, 1987), although yields of up to 70 tonnes/ha have been recorded (Wilson, 1981). However, the world average is under 14 tonnes/ha due to the low productivity under small-farm situations. In 1986, the world production of bananas (including plantains) was 68.5 million tonnes.

There are about 20 countries producing more than one million tonnes of banana per year (FAO, 1986a). The major producers are Brazil, India, Uganda, Nigeria, Columbia, Mexico and Rwanda. Only 20% of the world production enters the international trade (FAO, 1986b), while the major portion is utilized within the producing countries.

NUTRIENT CONTENT OF BANANAS

The banana is characterized by its high moisture content, ranging from 70 to 80% (Table 2.1). In the green unripe state in which they are harvested, the dry matter is composed mainly of starch. This is converted to simple sugars (sucrose, glucose and fructose) during the ripening process (Desai and Deshpande, 1975; Forsyth, 1980). For swine, fresh banana is a bulky feed with a metabolizable energy value which is only 20% of that of corn.

The crude protein content of banana is only 1.2% on a fresh weight basis. Banana protein is deficient in most essential amino acids (Table 2.2) and of poor quality (Sharaf et al., 1979). Several authors have reported that green bananas have a negative protein digestibility, while the protein digestibility of ripe bananas is also reported to be low (Clavijo and Maner, 1975; Le Dividich and Canope, 1974), implying that inclusion of bananas would markedly lower the protein digestibility of the total diet. The negative protein digestibility in green bananas is due to the presence of free tannins (Von Loesecke, 1950), which are known to inhibit the action of proteolytic enzymes (Price and Butler, 1980). The free tannin levels in the peel of green bananas is five times greater than that in the pulp. Therefore, peeling of green bananas would greatly enhance their feeding value. However, this may not be practicable because of the labor cost involved and the loss of dry matter (20% of the total fruit weight) through peeling.

Table 2.1. Proximate Composition of Banana (as fed)

	Fresh Ripe Banana		Fresh Green Banana		Dried Banana Meal	
Dry Matter (%)	25.0	(21.0-30.0)	25.0	(19.0-31.0)	90.0	(88.0-90.0)
Protein (%)	1.2	(0.8-1.2)	1.2	(0.9-1.4)	7.5	(4.0-9.5)
Crude Fat (%)	0.1	(0.1-0.2)	0.1	(0.1-0.2)	1.5	(1.0-2.8)
Crude Fiber (%)	0.6	(0.4-1.2)	0.6	(0.3-0.8)	4.5	(3.0-6.2)
M.E. (kcal/kg)	740	(690-780)	700	(650-780)	2820	(2740-2850)

Van Loesecke, 1950; FAO, 1972; Oyenuga and Fetuga, 1974; Clavijo and Maner, 1975; Le Dividich et al., 1975; Gebhardt et al., 1982.

Table 2.2. Essential Amino Acid Profile of Banana and Corn (% as fed)

	Fresh Ripe Banana[1]	Corn[2]
Arginine	0.05 (4.21)[3]	0.43 (5.06)
Histidine	0.04 (3.36)	0.27 (3.18)
Isoleucine	0.03 (2.52)	0.35 (4.12)
Leucine	0.07 (5.88)	1.19 (13.99)
Lysine	0.04 (3.36)	0.25 (2.94)
Methionine	0.01 (0.84)	0.18 (2.12)
Cystine	0.02 (1.68)	0.22 (2.59)
Phenylalanine	0.04 (3.36)	0.46 (5.41)
Tyrosine	0.02 (1.68)	0.38 (4.47)
Threonine	0.03 (2.52)	0.36 (4.23)
Tryptophan	0.01 (0.84)	0.09 (1.06)
Valine	0.03 (2.52)	0.48 (5.65)

[1]FAO, 1972; Gebhardt et al., 1982.
[2]National Research Council, 1988.
[3]Values in parentheses refer to amino acid concentration as % of protein.

The banana is a poor source of most minerals, but is rich in potassium (Table 2.3). Fresh banana is a good source of ascorbic acid and vitamin B_6 (Gebhardt et al., 1982).

Table 2.3. Mineral Composition of Banana and Corn (as fed)

	Fresh Ripe Banana[1]	Corn[2]
Calcium (%)	0.01	0.03
Magnesium (%)	0.03	0.11
Phosphorus (%)	0.02	0.28
Potassium (%)	0.35	0.33
Copper (mg/kg)	3	4
Iron (mg/kg)	10	236
Manganese (mg/kg)	1	23
Zinc (mg/kg)	2	----

[1]FAO, 1972; Izonfuo and Omuaru, 1988.
[2]National Research Council, 1988.

USE OF BANANAS IN SWINE FEEDING

Fresh Bananas

In major production areas, waste bananas are often used to feed pigs. However, due to the high moisture and low nutrient content of bananas, supplementary sources of protein, energy, minerals and vitamins are necessary when fresh bananas are fed. Pigs given whole, ripe bananas will first consume the banana pulp leaving much of the peel. However, if the total quantity of bananas fed is limited, the pig will consume both the pulp and the peel (Clavijo and Maner, 1975).

Fresh, ripe bananas are usually fed to growing/finishing pigs ad lib. with about 1.0 to 1.5 kg supplement containing 30% crude protein. A 30% protein supplement appears optimal in terms of pig performance (Calles et al., 1970). Given bananas free choice, growing pigs will consume about 5 to 6 kg of ripe bananas per day and the finishing pigs up to 9 kg. Feeding growing/finishing pigs with more than these quantities of ripe bananas may cause severe diarrhea (Gohl, 1981; Ravindran, 1987). Banana feeding to finishing pigs has been reported to produce a leaner, more acceptable carcass (Butterworth and Houghton, 1963; Shillingford, 1971).

The superior feeding value of ripe over green bananas is well documented (Williamson and Payne, 1965; Shillingford, 1971; Clavijo and Maner, 1975; Gohl, 1981). The low voluntary consumption (Clavijo and Maner, 1975) and organic matter digestibility (Oyenuga and Fetuga, 1974; Le Dividich et al., 1975) of green, unripe bananas is related to the presence of free tannins. The free tannins cause a strongly astringent taste, thus lowering the palatability of green bananas. During ripening, these free tannins polymerize into a bound form which is insoluble and has no effect on palatability (Von Loesecke, 1950). Bananas should therefore be allowed to ripen before they are fed to pigs. However, it is relevant to note that, in some studies (Viteri et al., 1970; Le Dividich et al., 1975), little difference has been observed between the growth performance of pigs fed ripe or green bananas. This discrepancy appears related to variations in the levels of free tannins that may exist amongst banana cultivars (Clavijo, 1973).

When used in pig feeding, green bananas are usually chopped into thin slices or crushed. Cooking enhances the feeding value of green bananas, through its beneficial effects on starch digestion (Beroard-Cerning and Le Dividich, 1976) and in improving the apparent digestibilities of organic matter, energy and nitrogen (Le Dividich et al., 1975). In a study by Clavijo and Maner (1975), cooking green bananas significantly improved both banana consumption and pig growth rate, although not to the levels obtained with ripe bananas (Table 2.4). Furthermore, there seems to be little economic advantage in cooking because of the energy cost involved.

Table 2.4. Performance of Growing Pigs (30-90 kg) Fed Ripe, Green and Cooked-Green Bananas[1]

		Bananas + 30% Protein Supplement[2]		
	Control	Ripe	Green	Cooked-Green
Daily Gain (kg)	0.68	0.56	0.46	0.50
Daily Intake				
Bananas (kg)	----	8.85	4.25	6.20
Supplement (kg)	----	0.71	1.04	0.88
Dry Feed (kg)	2.31	2.48	1.89	2.13
Feed Efficiency	3.41	4.44	4.16	4.26

[1]Clavijo and Maner, 1975.
[2]Comprised of fish meal, cotton seed meal, corn, vitamins, minerals and antibiotic.

Fresh ripe bananas can be used efficiently for sows during gestation, with no adverse effects on reproductive performance (Table 2.5). In fact, the sows fed bananas plus the supplement gained more than those on the control diet.

Though fresh ripe bananas can be used to feed lactating sows (Table 2.6), they are not generally recommended as the major energy source because the quantity of bananas (at least 20 kg per day) that is needed to meet the high energy demands during lactation and to prevent body weight loss is greater than the sow's physical capacity to consume (Clavijo and Maner, 1975).

Table 2.5. Performance of Gestating Sows Fed Fresh Ripe Bananas Plus a 40% Protein Supplement

	Control	Banana + Supplement
Daily Feed		
Banana (kg)	----	5.00
Supplement (kg)	----	0.67
Total Dry Feed (kg)	1.66	1.67
Daily Sow Gain (kg)	0.26	0.37
Litter Size	8.90	8.40
Piglet Birth Weight (kg)	1.22	1.26

Clavijo and Maner, 1975.

Table 2.6. Performance of Lactating Sows Fed Fresh Ripe Bananas Plus a 40% Protein Supplement

	Control	Banana + Supplement
Daily Feed		
Banana (kg)	----	11.22
Supplement (kg)	----	1.02
Total Dry Feed (kg)	3.66	3.17
Weight Loss of Sows (kg)	9.53	11.30
Litter Size at Birth	8.50	8.70
Litter Size at Weaning	6.34	6.06
Preweaning Mortality (%)	26.40	30.30

Clavijo and Maner, 1975.

Banana Meal

Ripe bananas have poor drying characteristics. However, the green fruit can be readily dried and the dried slices can be ground to form a meal. On an as-fed basis, banana meal is a better source of energy than fresh bananas (Table 2.1), and may be used as a substitute for grains at levels of up to 50% in rations for growing/finishing pigs, gestating sows and lactating sows with little effect on performance (Clavijo and Maner, 1975). Lee et al. (1977) found the nutritional value of banana meal to be comparable to cassava root meal and sweet potato tuber meal. Use of higher levels of banana meal in pig diets is limited by its lower metabolizable energy content compared to corn (Table 2.1).

Banana pulp meal, prepared from peeled green fruits, is well tolerated by weaner pigs provided they weigh 5 kg or more. When it comprised 50% of the starter diet, growth performance and health were comparable to those obtained with cassava root meal (Le Dividich and Canope, 1974).

Banana Silage

When the supply of waste banana is abundant, preserving the fruit as a silage may be appropriate. Bananas can be successfully ensiled if normal rules of silage making are observed, ensuring adequate compaction and drainage. Ripe bananas compact easily, but green ones should be chopped. Alternatively, green bananas may be allowed to ripen prior to ensiling. The high content of fermentable carbohydrates in bananas makes the ensiling process easy. No additives are required. The silage stabilizes in 3 to 4 days and keeps for at least six months. The procedures for the preparation of quality banana silage have been described by Le Dividich et al. (1975, 1976).

Banana silage is highly palatable to pigs. Although protein digestibility of silage-based rations was poor, the carcass quality was good even when the pigs were slaughtered at weights of more than 100 kg (Seve et al., 1976). Banana silage can constitute a major energy feed for growing/finishing pigs (Seve et al., 1976) and gestating sows (Canope et al., 1975; Le Dividich et al., 1975), but not for lactating sows (Le Dividich et al., 1975).

SUMMARY

Ripe bananas can be used to supply up to 50 to 75% of the total dry matter intake of growing/finishing pigs and gestating sows when fed in combination with a protein supplement that provides adequate protein and moderate levels of energy. Due to their high moisture content, ripe bananas cannot be used as the major source of energy for lactating sows. Fresh bananas are not suitable as a feed for starter pigs. Green bananas should be ripened, cooked or ensiled prior to feeding to pigs.

The green fruit can be easily dried and the meal can be used as a substitute for up to 50% of grain in diets for pigs during all phases of their life cycle, except the starter phase. However, banana pulp meal, prepared from peeled green bananas, can form up to 50% of starter diets without any adverse effects. During periods of abundant supply, bananas can be ensiled and the banana silage can be fed in a manner similar to fresh, ripe bananas.

REFERENCES

Beroard-Cerning, J. and Le Dividich, J., 1976. Valeur alimentaire de quelques produits amylaces d'origine tropicale. Etude in vitro et in vivo de la palate douce, de l'igname, du malanga, du fruit a pain et de la banana. Annals Zootech. 25: 155-168.

Butterworth, M.H. and Houghton, T.R., 1963. The use of surplus bananas for pig feeding in the West Indies. Empire J. Exp. Agric. 31: 14-18.

Calles, A., Clavijo, H., Hervas, E. and Maner, J.H., 1970. Ripe bananas as energy source for growing finishing pigs. J. Anim. Sci. 31: 197 (Abstract).

Canope, I., Le Dividich, J., Hedreville, F. and Despois, E., 1975. Utilisation de la banane ensilee dans l'alimentation du porc en croissance et de la truie en reproduction. Nouvelles Agronomiques des Antilles et de la Guyane 1: 263-271.

Clavijo, H., 1973. Factors which affect the digestibility, nutritive value and energy value of bananas for rats and pigs. Revista Colombiano Agropecuario 10: 524-526.

Clavijo, H. and Maner, J., 1975. The use of waste bananas for swine feed. Proceedings of a Conference on Animal Feeds of Tropical and Subtropical Origin. Tropical Products Institute, London. pp. 99-106.

Desai, B.B. and Deshpande, P.B., 1975. Chemical transformations in three varieties of banana fruits stored at 20°C. Mysore J. Agric. Sci. 9: 634-638.

FAO, 1972. Food Composition Table for Use in East Asia. Food and Agriculture Organization of the United Nations, Rome and U.S. Dept. of Health, Education and Welfare, Washington, D.C.

FAO, 1986a. FAO Production Yearbook. Vol. 40. Food and Agriculture Organization, Rome.

FAO, 1986b. FAO Trade Yearbook. Vol. 40. Food and Agriculture Organization, Rome.

Forsyth, W.G., 1980. Banana and Plantain. In: S. Nagy and P.E. Shaw, eds. Tropical and Subtropical Fruits. AVI Publishing Company, Westport, Conneticut, pp. 258-278.

Gebhardt, S.E., Cutrufelli, R. and Matthews, R.H., 1982. Composition of Foods: Fruits and Fruit Juices. Agriculture Handbook No. 8-9, U.S. Department of Agriculture, Washington, D.C.

Gohl, B., 1981. Tropical Feeds. Food and Agriculture Organization, Rome.

Izonfuo, L.W. and Omuaru, V.O., 1988. Effect of ripening on the chemical composition of plantain peels and pulps. J. Sci. Food Agric. 45: 333-336.

Le Dividich, J. and Canope, I., 1974. Valeur alimentaire de la farine de banane et de manioc dans les regimes se sevrage du porcelet a 5 semaines: Influence du taux de proteines de la ration. Annals Zootech. 23: 161-169.

Le Dividich, J., Geoffroy, F., Canope, I. and Chenost, M., 1975. Using waste bananas as animal feed. World Anim. Rev. 3: 22-30.

Le Dividich, J., Seve, B. and Geoffroy, F., 1976. Preparation et utilisation de l'ensilage de banane en alimentation animale. I. Technologie, composition chimique et bilan des matieres nutritives. Annals Zootech. 25: 313-324.

Lee, P.K., Yang, Y.F. and Chen, F.N., 1977. Comparative study on the nutrient digestibility of sweet potato, cassava and banana meals by pigs. J. Taiwan Livestock Res. 10: 215-225.

National Research Council, 1988. Nutrient Requirements of Domestic Animals, No. 2. Nutrient Requirements of Swine. 9th ed. National Academy of Sciences, Washington, D.C.

Oyenuga, V.A. and Fetuga, B.L., 1974. The apparent digestibility of nutrients and energy value to pigs of plantains. Nigerian J. Anim. Prod. 1: 184-191.

Price, M.L. and Butler, L.G., 1980. Tannins and Nutrition. Purdue University Agriculture Experimental Station Bulletin No. 272, West Lafayette, Indiana.

Ravindran, V., 1987. Production and Utilization of Locally Available Resources for Animal Feeding in Cook Islands. Consultancy Report. Food and Agriculture Organization Regional Office, Bangkok, Thailand. 81 pp.

Salunkhe, D.K. and Desai, B.B., 1984. Postharvest Biotechnology. Volume I. CRC Press, Boca Raton, Florida.

Seve, B., Le Dividich, J. and Canope, I., 1976. Preparation et utilisation de l'ensilage de banane en alimentation animale. II. Incorporation dans la ration du porc en croissance-finition. Annals Zootech. 25: 325-336.

Sharaf, A., Sharaf, O.A., Hegazi, S.M. and Sedky, K., 1979. Chemical and biological studies on banana fruit. Zeitschrift fur Ernahrungwissenschaft 18: 8-15.

Shillingford, J.D., 1971. The economics of feeding bananas and coconut meal for pork production in Dominica, West Indies. Trop. Agric. (Trinidad). 48: 103-110.

Stover, R.H. and Simmonds, N.W., 1987. Bananas. John Wiley & Sons, Inc., New York.

Von Loesecke, H.W., 1950. Bananas. 2nd ed. Interscience Publishers. New York.

Viteri, J., Alvarado, P. and Hervas, E., 1970. Comparacion del valor nutritivo del banano verde V.S. banano maduro en iguales cantitades en alimentacion de cerdos en crecimiento. Instituto Nacional de Investigaciones Agropecuarias. Experimento SCEP 1.1.17.70. Quito, Ecuador.

Williamson, G. and Payne, W.J.A., 1965. An Introduction to Animal Husbandry in the Tropics. Longmans, Green, London.

Wilson, J.H., 1981. Phenology and yield of bananas in the Zimbabwe lowveld. Zimbabwe Agric. J. 78: 201-206.

CHAPTER 3

Barley: Hulless

F.X. Aherne

INTRODUCTION

Barley is frequently considered to be of poorer nutritive value for swine than wheat. However, a review of the literature by Pond and Manner (1984) indicated that in the majority of experiments there was no significant difference between the growth rate of pigs fed barley-based or wheat-based diets although feed efficiency was lower for pigs fed barley. The major factors responsible for the poorer feed efficiency of pigs fed barley-based diets are its higher crude fiber content and its lower digestible energy level (Larsen and Oldfield, 1961).

The hull of barley comprises about 13% of the kernel and is the major component contributing to the high crude fiber content of barley (Bhatty, 1986). Therefore, removal of the hull or production of hulless barleys would improve the energy and nutritional value of barley for monogastric animals. In barley, the hull consists of two glumes, the lemma and the palea, which completely enclose the seed. In traditional hulled varieties of barley, the glumes are fused together and are attached to the seed by a cementing substance produced by the caryopsis. This causes the hull to remain attached to the seed during harvest. Recently, cultivars of barley have been developed in which the fusion of the glumes does not occur, allowing the hull to be removed during the threshing process in a manner similar to that which occurs with wheat and other cereals. These so-called hulless varieties of barley would appear to have considerable potential for use in swine diets.

GROWING HULLESS BARLEY

Hulless barley is a naturally occurring barley belonging to the family Gramineae and tribe Hordeae which also includes wheat, rye, oats, millet, corn and sorghum (Bhatty, 1986). The hulless characteristic is controlled by a single recessive gene and is established during the development of the grain (Bhatty, 1986). Hulless barley is adapted to grow in traditional barley growing areas but is generally lower yielding than hulled cultivars of barley. Although yielding less on a weight of grain per hectare basis, hulless barley produces approximately the same or more total energy per hectare as hulled varieties of barley. In addition, the lower yield of hulless barley can be compensated for by increasing the seeding rate by 10 to 20% (Bhatty, 1986).

Hulless barley, like hulled barley, is susceptible to attack by a number of diseases. Among the more important of these diseases are root rot, net blotch, stem rust and smut (Alberta Agriculture, 1985; 1986). From an agronomic standpoint, the requirements for weed control and fertilization of hulless barley are the same as those for any other barley crop. However, conventional harvesting equipment may need adjusting in order to harvest hulless barley.

NUTRIENT CONTENT OF HULLESS BARLEY

The chemical composition of hulless barley, hulled barley and wheat are presented in Table 3.1. Hulless barley is higher in protein and digestible energy than hulled barley and is similar to wheat in both protein and digestible energy content (Mitchell et al., 1976; Bhatty, 1986). As with other cereal grains the composition of hulless barley varies considerably with variety, growing conditions and level of fertilizer application. Values of 11.1 to 18.9%

Table 3.1. Chemical Analysis of Hulless Barley, Hulled Barley and Wheat (as fed)

	Hulless Barley[1]	Hulled Barley[2]	Wheat[2]
Dry Matter (%)	86.5	89.0	88.0
D.E. (kcal/kg)	3367	3120	3402
Crude Protein (%)	13.5	11.5	12.6
Crude Fiber (%)	2.0	5.0	2.6
Neutral Detergent Fiber (%)	13.8	16.8	14.4
Acid Detergent Fiber (%)	2.4	2.6	2.8

[1]Bhatty, 1986.
[2]National Research Council, 1988.

crude protein in hulless barley have been reported in the literature (Newman et al., 1968; Mitchell et al., 1976).

Hulless barley also contains substantially more lysine than either barley or wheat. The dietary essential amino acid content of hulless barley, hulled barley and wheat is shown in Table 3.2. Because of the higher protein and lysine content of hulless barley, lower levels of protein supplementation are required in order to properly balance the grain source.

Table 3.2. Essential Amino Acid Content of Hulless Barley, Hulled Barley and Wheat (% as fed)

	Hulless Barley[1]	Hulled Barley[2]	Wheat[2]
Arginine	0.76 (0.69-0.85)[3]	0.52	0.65
Histidine	0.36 (0.29-0.40)	0.24	0.32
Isoleucine	0.59 (0.51-0.67)	0.46	0.45
Leucine	1.19 (1.06-1.31)	0.75	0.90
Lysine	0.49 (0.41-0.56)	0.40	0.36
Methionine	0.40 (0.38-0.41)	0.16	0.22
Phenylalanine	0.99 (0.84-1.13)	0.58	0.64
Threonine	0.59 (0.52-0.66)	0.36	0.39
Valine	0.81 (0.74-0.87)	0.57	0.58

[1]Mitchell et al., 1976; Bhatty, 1986.
[2]National Research Council, 1988.
[3]Values in parentheses are ranges reported in the literature.

There is no published data on the trace mineral and vitamin content of hulless barley, but it may be assumed to be similar to that of hulled barley (National Research Council, 1988). Cereal grains are not usually considered to be significant sources of trace minerals or vitamins for swine, but they do supply substantial amounts of thiamine, niacin and vitamin E to the pigs' diet.

UNDESIRABLE CONSTITUENTS IN HULLESS BARLEY

It has been suggested that the beta glucan content of barley is one factor which contributes to the reduction in performance of pigs fed barley-based diets (Honeyfield et al., 1983; Newman et al., 1983). Beta glucans are structural carbohydrates that constitute a substantial portion of the endosperm cell wall material of the barley grain (Fincher, 1975) and are similar to cellulose both chemically and structurally except that the glucose units are linked with beta, 1-4 and beta, 1-3 linkages (Bamforth, 1982).

Hulled barley contains 3 to 8% beta glucans while the beta glucan content of hulless barley may be even higher (Campbell et al., 1986). Although Wiseman and Cole (1984) reported that the beta glucan content of barley can be reduced by storing the grain for a few weeks post-harvest before inclusion in swine diets, Weltzien and Aherne (1987) reported that the beta glucan content of barley was not reduced by earlier harvest or by anaerobic storage for long periods of time.

Recent research has cast doubt on the role of beta glucans in reducing pig performance. Weltzien and Aherne (1987) reported a digestibility coefficient for beta glucans for growing pigs of 76.0 to 81.6. Graham et al. (1986) also reported similar digestibility coefficients for beta glucans in pig diets and suggested that the beta glucan digestion was due to the large intestinal flora of the pig. Davies and Radcliff (1984) reported no significant difference between the performance of pigs fed barley cultivars varying in beta glucan content. Further support for the conclusion that beta glucans in barley do not pose a significant nutritional problem for pigs is provided by experiments in which supplementation of barley-based diets with beta-glucanase enzyme did not improve pig performance (Table 3.3).

Table 3.3. Effect of Beta Glucanase Supplementation of Hulless Barley Diets on Pig Performance

	Control	Enzyme	
Starter Trial[1]			
Daily Gain (kg)	0.52	0.57	NS[3]
Daily Intake (kg)	0.74	0.77	NS
Feed Efficiency	1.44	1.36	NS
Grower Trial[2]			
Daily Gain (kg)	0.74	0.76	NS
Daily Intake (kg)	2.32	2.35	NS
Feed Efficiency	3.13	3.11	NS

[1]Aherne and Spicer, 1986.
[2]Thacker et al., 1988.
[3]NS=nonsignificant ($P>0.05$).

The effect of enzyme supplementation of hulless barley-based diets on nutrient digestibility has also been inconsistent. Thacker et al. (1988) reported that the addition of beta-glucanase to barley-based diets resulted in a significant improvement in the digestibility coefficients for protein and energy in growing pig diets. However, Spicer and Aherne (1988) observed no improvement in the digestibility coefficients for energy or protein in the diets of starter pigs fed diets based on hulless barley fed with or without beta-glucanase supplementation (Table 3.4).

Table 3.4. Digestibility of Crude Protein and Digestible Energy Content

	Digestibility Coefficient (%)		
	Hulled	Hulless	
Starter Trial[1]			
Crude Protein	80.6	79.8	NS[3]
Digestible Energy	83.6	87.4	NS
Grower Trial[2]			
Crude Protein	71.6	68.6	NS
Digestible Energy	72.1	76.2	NS

[1]Aherne and Spicer, 1986.
[2]Mitchall et al., 1976.
[3]NS=nonsignificant (P>0.05).

Barley contains several other antinutritional factors, the most important of which are the trypsin inhibitors and tannins. Tannins have been reported to depress the digestion and utilization of proteins (Chubb, 1982) and to reduce growth rate of rats and chickens (Featherstone and Rogler, 1975) possibly due to reduced protein digestibility. Trypsin inhibitors also affect protein digestibility.

FEEDING HULLESS BARLEY

On the basis of its chemical composition it would appear that hulless barley would offer a very attractive alternative to hulled barley or wheat in swine diets. However, the results of feeding trials involving the feeding of hulless barley to pigs have been inconsistent. In several experiments, pigs fed diets based on hulless barley had significantly better growth rates and/or feed to gain ratios than pigs fed diets supplemented with hulled barley (Gill et al., 1966; Newman et al., 1968; Thacker et al., 1987; Table 3.5).

In contrast, Spicer and Aherne (1988) reported no significant difference in the performance of pigs fed hulled or hulless barley-based diets (Table 3.6). Other experiments showing no significant difference in the performance of pigs fed either hulled or hulless barley have been reported by Mitchell et al. (1976) and Newman and Pepper (1984). The reasons for the differences in results are not obvious but could be related to the composition of the grain used in the experiments.

Table 3.5. Performance of Pigs Fed Hulless and Hulled Barley

	Hulled Barley	Hulless Barley
Starter Trial[1]		
Daily Gain (kg)[3]	0.36	0.42
Daily Intake (kg)[3]	0.67	0.77
Feed Efficiency	1.87	1.82
Grower Trial[2]		
Daily Gain (kg)	0.75	0.74
Daily Intake (kg)	2.46	2.32
Feed Efficiency[3]	3.30	3.13

[1]Newman et al., 1968.
[2]Thacker et al., 1987.
[3]Means are significantly different ($P<0.05$).

Table 3.6. Performance of Pigs Fed Condor Hulless or Samson Hulled Barley

	Hulless Barley (Condor)	Hulled Barley (Samson)
Starter Trial		
Daily Gain (kg)	0.58	0.59
Daily Intake (kg)	0.85	0.90
Feed Efficiency	1.47	1.57
Grower Trial		
Daily Gain (kg)	0.71	0.73
Daily Intake (kg)	1.95	2.05
Feed Efficiency	2.76	2.81

Spicer and Aherne, 1988.

One advantage of using hulless barley instead of hulled barley as an energy source in swine diets is that a reduction in feed costs may be achieved. The results of feeding trials conducted at the University of Alberta and the University of Saskatchewan clearly demonstrated that the level of soybean meal supplementation required to meet the essential amino acid requirements of the pig could be reduced by 2 to 11 kg per tonne (Thacker et al., 1987; Spicer and Aherne, 1988). Yet in both of these trials, pigs fed the hulless barley diets performed as well as those fed hulled barley or hulled barley plus wheat. Therefore, on a cost per unit of gain basis, the most cost effective grain to include in swine diets would be hulless barley, assuming a similar purchase price for hulless and hulled barley.

The effects of including hulless barley in swine diets on carcass characteristics has received little attention. However, Thacker et al. (1987) reported that there were no significant differences in dressing percent, backfat thickness or carcass grade for pigs fed hulled or hulless barley diets (Table 3.7).

Table 3.7. Carcass Characteristics of Pigs Fed Hulless and Hulled Barley Diets

	Hulless Barley	Hulled Barley
Slaughter Weight (kg)	97.8	98.0
Carcass Weight (kg)	78.2	79.0
Dressing Percentage	79.8	80.6
Backfat (cm)	3.5	3.5
Carcass Value Index	103.6	103.7

Thacker et al., 1987.

There is no data available on the nutritive value of hulless barley for the lactating sow. However, based on the current widespread interest in increasing the energy content of sow diets during lactation, it may be speculated that hulless barley could provide an excellant alternative grain source for use in lactating sow diets.

SUMMARY

Currently there is very little hulless barley being grown for use in swine diets. The reasons for this are that the yield of grain of hulless barley has been considerably lower than that of hulled barley and the nutritional quality of hulless barley has not always proven superior to that of hulled barley. However, the development of better yielding, higher protein cultivars and the fact that a reduction in feed costs may be achieved by including hulless barley in swine diets, may provide incentive for continued breeding programs to develop higher yielding cultivars of hulless barley.

REFERENCES

Aherne, F.X. and Spicer, H.M., 1986. An evaluation of the nutritive value of hulless barley for young pigs. 65th Annual Feeders' Day Report, University of Alberta, Edmonton. pp. 107-108.

Alberta Agriculture, 1985. Varietal Description of Tupper Hulless Six-Rowed Barley. Agdex 114/33-21.

Alberta Agriculture, 1986. Amended Varietal Description for Scout, Hulless Two-Row Spring Barley. Agdex 114/33-14.

Bamforth, C.W., 1982. Barley beta-glucans. Brewers Digest 57: 22-27.

Bhatty, R.S., 1986. The potential of hulless barley: A review. Cereal Chem. 63: 97-103.

Campbell, G.L., Classen, H.L., Thacker, P.A., Rossnagel, B.G., Grootwassink, J.W. and Salmon, R.E., 1986. Effect of enzyme supplementation on the nutritive value of feedstuffs. Proceedings 7th Western Nutrition Conference, Saskatoon. pp. 227-250.

Chubb, L.G., 1982. Anti-nutritional factors in animal feeds. In: W. Haresign, ed. Recent Advances in Animal Nutrition. Butterworths, London. pp. 21-37.

Davies, R.L. and Radcliff, B.C., 1984. Performance of growing pigs fed wheat, barley or triticale. Aust. J. Exp. Agric. Husb. 24: 501-506.

Featherstone, W.R. and Rogler, J.C., 1975. Influence of tannins on the utilization of sorghum grain by rats and chicks. Nutr. Rep. Int. 11: 491-497.

Fincher, G.B., 1975. Morphological and chemical composition of barley endosperm cell walls. J. Inst. Brew. 81: 116-122.

Gill, D.R., Oldfield, J.E. and England, D.C., 1966. Comparative values of hulless barley, regular barley, corn and wheat for growing pigs. J. Anim. Sci. 25: 34-36.

Graham, H., Hesselman, K., Jonsson, E. and Aman, P., 1986. Influence of B-glucanase supplementation on digestion of a barley based diet in the pig gastrointestinal tract. Nutr. Rep. Int. 34: 1089-1096.

Honeyfield, D.C., Froseth, J.A. and Lance, R., 1983. Components of barley that predict its nutritional value for finishing pigs. J. Anim. Sci. 57 (Suppl. 1): 250. (Abstract).

Larsen, L.M. and Oldfield, J.E., 1961. Improvement of barley rations for swine. II. Effect of fibre from barley hulls and purified cellulose in barley and corn rations. J. Anim. Sci. 20: 440-444.

Mitchall, K.G., Bell, J.M. and Sosulski, F.W., 1976. Digestibility and feeding value of hulless barley for pigs. Can. J. Anim. Sci. 56: 505-511.

National Research Council, 1988. Nutrient Requirements of Domestic Animals, No. 2. Nutrient Requirements of Swine. 9th ed. National Academy of Sciences, Washington, D.C.

Newman, C.W., Thomas, G.O. and Eslich, R.F., 1968. Hulless barley in diets for weanling pigs. J. Anim. Sci. 27: 981-984.

Newman, C.W., Eslich, R.F. and El-Negoumy, A.M., 1983. Bacterial diastase effect on the feed value of two hulless barleys for pigs. Nutr. Rep. Int. 28: 139-145.

Newman, C.W. and Pepper, J.W., 1984. Supplemental beta-glucanase for barley diets fed to growing swine. Montana State University, Bozeman, Montana Research Bulletin, pp. 15-17.

Pond, W.G. and Maner, J.H., 1984. Swine Production and Nutrition. AVI Publishing Company, Westport, Connecticut, pp. 273-276.

Spicer, H.M. and Aherne, F.X., 1988. An evaluation of a new variety of hulless barley for young pigs. 67th Annual Feeders' Day Report, University of Alberta, Edmonton, pp. 55-57.

Thacker, P.A., Bell, J.M., Classen, H.L., Campbell, G.L. and Rossnagel, B.G., 1987. The nutritive value of hulless barley for swine. Anim. Feed Sci. Technol. 19: 191-196.

Thacker, P.A., Campbell, G.L. and Grootwassink, J.D., 1988. The effect of beta-glucanase supplementation on the performance of pigs fed hulless barley. Nutr. Rep. Int. 38: 91-99.

Weltzien, E.M. and Aherne, F.X., 1987. The effect of anaerobic storage and processing of high moisture barley on its amino acid and B-glucan digestibility by growing swine. Can. J. Anim. Sci. 66: 829-840.

Wiseman, J. and Cole, D.J.A., 1984. Energy evaluation of cereals for pig diets. In: D.J.A. Cole and W. Haresign, eds. Pig Nutrition. Butterworths, London, pp. 246-261.

CHAPTER 4

Barley: Hydroponically Sprouted

D.J. Peer and S. Leeson

INTRODUCTION

Hydroponic culture is the technique of growing plants in nutrient enriched water rather than soil. Hydroponics were first used on a commerical scale in the early 1930's (Martin, 1939) but until recently, little research was conducted on the nutritional value of hydroponically sprouted grains for swine.

In the past few years, there has been renewed interest in feeding hydroponically sprouted grains to livestock. Many unfounded claims have been made by various manufacturers of hydroponic equipment and it is often difficult to separate fact from fiction regarding these products. Sprouted grains are widely believed to contain large amounts of vitamins as well as "unidentified growth factors". Despite this lack of candor, current research suggests that the use of hydroponically sprouted grains for swine may be feasible under certain conditions.

PRODUCTION OF HYDROPONICALLY SPROUTED BARLEY

A hydroponic sprouting chamber is a highly-insulated and thermostatically-controlled unit. Grain to be sprouted is pre-wetted with an equal volume of water and is then distributed evenly on a white plastic tray placed in the hydroponic chamber. The grain is sprayed with water, for 15 minutes every four hours, throughout the growth period. The sprouts are harvested after about seven days. Fluorescent lights are used to give an illumination of 600 lux and the temperature is maintained at 21°C.

Similar physiological changes occur in a germinating seed, regardless of the medium in which it is grown (soil or water). A dry seed is metabolically dormant; its respiration rate is extremely low and its enzymes are present but inactive. As soon as the seed imbibes water, respiration rate increases, enzyme systems are activated and protein synthesis is started. Starch stored in the embryo and endosperm is used as a source of energy for respiration. Amino acids required for plant growth are supplied by the stores in the endosperm. Thus, there is a net loss of dry matter at this time due to oxidation and the loss of stored materials from the seed (Mayer and Poljakoff-Mayber, 1975).

The hydroponic chambers used in research (Trubey et al., 1969; Peer and Leeson, 1985a,b) are labor intensive. Some manufacturers of the chambers recommend prewetting of the grain before placement in the chamber. The seed trays need to be washed and disinfected between batches of sprouts in order to prevent the growth of mold and yeasts. Washing and seeding the trays can take 2 to 5 minutes per tray; a 35 tray unit would require 1 to 3 hours for normal servicing. The water reservoir in the base of the chamber also requires emptying and disinfecting weekly in order to prevent algal growth.

Some hydroponic growth chambers are highly insulated, thermostatically controlled and electrically powered. Some of them have galvanized racks and stainless steel interiors in order to prevent rust. All these features are costly. The authors have seen quotes between $6,000 and $62,000 (Canadian dollars) for growth chambers, depending on their size. When the cost of the growth chamber is considered, as well as the cost of the original grain and the nutrient losses which occur during sprouting, hydroponic barley sprouts become a very costly feed indeed.

NUTRIENT CONTENT OF HYDROPONICALLY SPROUTED BARLEY

The physiological changes that take place during germination should influence the concentration of nutrients in the hydroponic barley sprouts. As starch is used as an energy source for the developing embryo, the concentration of starch (nitrogen free extract) in the sprouts should decrease. As starch accounts for 53 to 67% of the weight of a barley kernel (Kent-Jones and Amos, 1967), a corresponding decrease in dry matter concentration should also occur during sprouting. The concentration of crude protein, total lipids, and crude fiber should change as well (McCandlish and Struthers, 1938; Thomas and Reddy, 1962; Hillier and Perry, 1969; Trubey et al. 1969).

The nutrient concentration of hydroponic barley sprouts (Peer and Leeson, 1985a) and barley (National Academy of Sciences, 1982) are compared in Table 4.1. As predicted, day 2, 5 and 7 sprouts contain 75.4, 68.7 and 61.3% nitrogen free extract, respectively, as compared with barley grain which contains 76.6% nitrogen free extract. Correspondingly, barley grain contains 87.3% dry matter, and day, 2, 5 and 7 sprouts contain 52.7, 19.0 and 16.0% dry matter, respectively. This loss in dry matter and starch accounts, in part, for the lower digestible energy content of day 4 sprouts (3167 kcal/kg DM) (Peer and Leeson, 1985b). As the sprouts are lower in digestible energy than maize (4095 kcal/kg DM), or normal barley (3429 kcal/kg DM), they are limited in their usefulness as an energy source for pigs.

Table 4.1. Nutrient Concentrations (% DM) of Hydroponically Sprouted and Normal Barley

Feedstuffs	Dry Matter (%)	D.E. (kcal/kg)	NFE (%)	Crude Protein (%)	Crude Fiber (%)	Total Lipids (%)
Barley Grain	87.3	3429	76.6	12.1	5.4	3.3
Barley Sprouts						
Day 2	52.7	----	75.4	13.1	6.0	2.8
Day 4	29.2	3167	72.6	13.4	7.4	3.6
Day 5	19.0	----	68.7	14.1	9.8	4.4
Day 7	16.0	----	61.3	15.5	14.3	5.0

Peer and Leeson, 1985a.

Fiber concentration increases with increasing sprouting time (Alexander et al., 1984; Peer and Leeson, 1985a), which is probably a result of the increase in number and size of cell walls which occurs as the plant grows (James, 1940; Mayer and Poljakoff-Mayber, 1975). This increase in fiber content could also account for part of the decrease in digestibility of day 4 sprouts (Peer and Leeson, 1985b). Fiber is poorly digested by pigs (Church, 1977; Whittemore and Elsley, 1977; English et al., 1988), so that sprouts may only be useful in dry sow diets where some fiber is desirable.

The total lipid concentration of the barley sprouts increases with increasing sprout age. Day 7 sprouts contain 4.0% fat, which is higher than the fat concentration of barley grain. The increase in fat percentage could be due to an increase in structural lipids which occur during germination (Ching, 1972, as cited by Tluczkiewicz and Berendt, 1977) and plant growth (Bidwell, 1974).

The concentration of crude protein increases with increasing sprouting time. Day 4 and day 7 barley sprouts contain 13.4 and 15.5% crude protein, respectively. This increase in the percentage of crude protein is a proportional change due to the large decrease in starch content. As a consequence, when feeding sprouts to swine, the amount of crude protein supplied by the sprouts should be calculated on a grams per day basis. The balance of the pig's daily protein requirement can then be supplied by a balancer ration.

The amino acid profile of barley grain changes during germination (Folkes and Yemm, 1958). A comparison of the essential amino acid concentrations of hydroponic barley sprouts and normal barley is presented in Table 4.2. Proline and glutamic acid decrease as their nitrogen and carbon are used to synthesize other amino acids such as lysine, glycine and tryptophan (Smith, 1972; Dalby and Tsai, 1976). There is an increase in the lysine and tryptophan concentrations but a decrease in the concentrations of methionine, phenylalanine and threonine. As with other feedstuffs which are deficient in certain essential amino acids, diets containing barley sprouts should be supplemented with a source of methionine, tryptophan, threonine and lysine.

Table 4.2. Essential Amino Acid Concentration (% DM) of Hydroponically Sprouted and Normal Barley

	Barley[1] Grain	Hydroponic Barley Sprouts[1]			
		2 Day	4 Day	5 Day	7 Day
Isoleucine	0.39	0.47	0.38	0.37	0.44
Leucine	0.76	0.78	0.64	0.68	0.75
Lysine	0.39	0.28	0.47	0.44	0.54
Methionine	0.20	0.21	0.16	0.17	0.17
Phenylanaline	0.72	0.63	0.51	0.55	0.52
Threonine	0.44	0.32	0.25	0.32	0.43
Tryptophan	0.04[2]	----	----	----	0.06[2]
Valine	0.47	0.50	0.57	0.48	0.52

[1]Peer and Leeson, 1985a.
[2]Alexander et al., 1984.

Available data on the vitamin content of hydroponically sprouted barley is scarce. Table 4.3 shows that seven day old sprouts contain more ascorbic acid and riboflavin than unsprouted barley (Alexander et al., 1984). Ascorbic acid is not normally added to pig diets as pigs normally synthesize their own. The increase in riboflavin content is small.

The mineral concentration of hydroponic barley sprouts and normal barley are shown in Table 4.4. Similar to barley grain, the sprouts are a poor source of calcium and potassium. Phosphorus availability is improved with sprouting due to the decrease in phytic acid level (Alexander et al., 1984). However, this liberated phosphorus is then susceptible to leaching (Cook, 1962). Mineral content of the water and the type of metal used in the hydroponic growth chamber affects the mineral concentration of the sprouts. Peer and Leeson (1985a) noted an increase in the concentration of sodium and zinc which they attributed to the high level of sodium in the water supply and leaching from the galvanized racks in the chamber. Mineral analysis of the water supply and the sprouts should therefore be conducted before a supplemental mineral premix is developed.

Table 4.3. Ascorbic Acid and Riboflavin Content of Barley Grain and Barley Germinated For 84 Hours (DM Basis)

Vitamin (mg/100 g)	Barley Grain	Germinated Barley (7 Days)
Ascorbic Acid	6.99	10.98
Riboflavin	0.10	0.12

Alexander et al., 1984.

Table 4.4. Mineral Concentration of Hydroponically Sprouted Barley and Normal Barley (DM Basis)

	Barley[1] Grain	Hydroponic Barley Sprouts[2]			
		2 Day	4 Day	5 Day	7 Day
Calcium (%)	0.02	0.03	0.03	0.03	0.03
Magnesium (%)	0.14	0.13	0.14	0.13	0.14
Phosphorus (%)	0.52	0.51	0.48	0.46	0.47
Potassium (%)	0.50	0.45	0.48	0.46	0.47
Sodium (%)	0.02	0.04	0.07	0.14	0.20
Copper (mg/kg)	4.00	7.10	5.40	10.30	10.40
Iron (mg/kg)	61.40	71.10	82.40	68.70	73.20
Manganese (mg/kg)	18.50	17.30	17.30	17.40	18.30
Zinc (mg/kg)	40.70	48.90	47.70	55.80	65.90

[1]National Academy of Sciences, 1982.
[2]Peer and Leeson, 1985a.

FEEDING OF HYDROPONIC BARLEY SPROUTS

Starter Pigs

Young pigs digest fiber and whole grains very poorly (Whittemore and Elsley, 1977; Patience and Thacker, 1989). In addition, their daily intake capacity is small since intake is largely dependent on body size (Wiseman, 1987). As a consequence, barley sprouts are unsuitable as a feed for starter pigs due to their high moisture and fiber content and their large percentage of whole, ungerminated kernels.

Growing-Finishing Pigs

A few companies in the United Kingdom and Europe produced hydroponic systems for sprouting grain during the 1930's (Leitch, 1939). During this time period, several researchers investigated the effect of feeding sprouted barley on the growth rate of swine (Bostock, 1937; Fishwick, 1937; Schmidt et al., 1938, all as cited by Leitch, 1939). All three researchers reported no improvement in performance due to the sprouted grain. There was no difference in meat quality or in fat content (Schmidt et al., 1938, as cited by Leitch, 1939).

Little further interest in sprouting grain for livestock occurred until the early 1960's when new hydroponic growth chambers were developed. These new chambers were electronically powered, internally illuminated and thermostatically controlled. They used a seven day growth cycle instead of the ten day cycles used by the earlier systems.

Then, during the early 1970's, a new hydroponic chamber which used a four day growth cycle was developed and the interest in feeding barley sprouts to pigs was renewed. Peer and Leeson (1985b) found that the digestibility of dry matter, protein and energy of four day old barley sprouts was lower than for ground barley but superior to whole barley (Table 4.5).

Table 4.5. Digestibility of Dry Matter and Protein and Digestible Energy Content of 4-Day Barley Sprouts and Whole and Ground Barley

	Dry Matter Digestibility (%)	Protein Digestibility (%)	Digestible Energy (kcal/kg DM)
Ground Barley	85.0	83.5	3786
Sprouted Barley	68.8	66.0	3167
Whole Barley	32.3	31.0	1190

Peer and Leeson, 1985b.

Growing pigs fed four day old sprouts had significantly lower average daily gains and slightly poorer feed conversions than pigs fed ground barley (Table 4.6). The reduction in performance of pigs fed sprouts is likely due to their high fiber and low energy content and poor digestibility. These results suggest that hydroponic barley sprouts are unsuitable as a feedstuff for growing-finishing pigs in a commercial operation.

Table 4.6. Weight Gain and Feed Efficiency of Young Growing Pigs Fed on Diets Containing 4-Day Barley Sprouts or Ground Barley

	Daily Gain (g/day)	Feed Efficiency
Sprouted Barley	595	2.10
Ground Barley	643	1.99

Peer and Leeson, 1985b.

Drying and grinding the sprouts appears to improve their digestibility (Peer and Leeson, 1985b). Grinding the wet sprouts would likely increase the digestibility of both the ungerminated grain and the sprouts. Breaking the integrity of the whole kernel and reducing particle size increases the digestibility of grains for pigs (Whittemore and Elsley, 1977). However, grinding wet sprouts is likely to cause some loss of nutrients due to seepage and mechanical problems with automated feeding systems.

Breeding Stock

Dry sows have the ability to eat large quantities of low energy (high fiber) feeds. They can digest fiber to a greater extent than pigs at any other stage of production because the ability for pigs to digest fiber increases with age and size (Whittemore and Elsley, 1977). Since it was once common practice to pasture dry sows on alfalfa, the fiber content of hydroponic sprouts is unlikely to pose a problem with dry sows.

Very little research has been conducted on the effect of feeding sprouted barley to sows. One researcher, reported an improvement in reproductive performance when 30 gilts in dry lots were fed chopped, hydroponically grown oat grass (grown for 5 days; Perry et al., 1965). However, the reader should bear in mind that this trial used a very small number of gilts on each treatment and there is a lack of other research with which to compare this trial.

It is recommended that rations for dry sows using barley sprouts be balanced using the nutrient requirements of dry sows on a daily intake basis. The amount of nutrients supplied by the sprouts should be calculated first, then the amount and composition of a balancer ration can be calculated. One and a half to 2.5 kg of day 4 barley sprouts (29.2% dry matter) will supply 21 to 35% of a dry sow's daily requirements for energy and 16 to 24% of their daily lysine requirement (Table 4.7).

Table 4.7. Proportion of Daily Nutrient Requirements of Dry Sows Supplied by Different Amounts of Day 4 Hydroponic Barley Sprouts

	Daily Nutrient Requirement	Nutrients Supplied by Barley Sprouts			
		1.5 kg/day		2.5 kg/day	
D.E. (kcal)	6524	1381	(21.2)[1]	2310	(35.4)
Crude Protein (g)	262.0	59.0	(22.5)	97.0	(37.0)
Lysine (g)	12.0	2.0	(16.7)	3.0	(25.0)
Calcium (g)	15.5	0.1	(0.1)	0.2	(1.3)
Phosphorus (g)	12.3	2.0	(16.3)	4.0	(32.5)

[1]Percentage of requirements met by sprouts.

Sprouts can be top dressed to sows for a few days before and after farrowing as a laxative. They are highly palatable, and can also be used to encourage good levels of feed intake. However, because of their high moisture content and poor feed value they should not be used as a dry ration substitute during lactation.

SUMMARY

During sprouting, the nutrient profile of barley grain changes. There is a net loss of dry matter, mostly in the form of starch as the developing embryo uses starch stored in the endosperm as a source of energy. As starch is the major source of energy in barley, there is a corresponding decrease in the digestible energy content of the sprouts. Crude protein as a percentage of dry matter increases with increasing sprouting time because of the decrease in starch content. Lysine concentration increases slightly during sprouting, while concentrations of methionine, phenylalanine and threonine decrease. Fiber concentration increases markedly because of an increase in the number and size of cell walls during growth. Concentration of lipids also increases due mostly to the increase in structural lipids associated with growth. Mineral concentrations of the sprouts reflect the mineral content of the water used in the growth chamber. Some of the phytate phosphorus is liberated but is susceptible to leaching.

Several factors limit the usefulness of the barley sprouts as a swine feed. Germination rate affects the extent to which some nutrients (especially starch) change during germination. It also affects the digestibility of the sprouts as whole, unsprouted kernels are poorly digested by pigs. The sprouts' high fiber content also decreases digestibility for pigs, especially young pigs. The growth chamber is costly and labor intensive. Further processing of the sprouts (such as drying and grinding) would likely improve their digestibility but would add to their cost. Grinding the wet sprouts could cause mechanical problems with automated feeding systems and seepage of nutrients because of their high water content.

Sprouted barley is not recommended for starter pigs (10-20 kg) because of its high fiber and water content. Recent trials suggest that barley sprouts are too low in digestibility and result in a loss of growth rate and feed efficiency when fed to growing-finishing pigs; therefore the sprouts are also not recommended for these pigs. Barley sprouts can be top dressed to sows just prior and post farrowing as a laxative to prevent constipation. A small amount of sprouts can be fed during lactation in order to encourage feed consumption because they are quite palatable but should not be used as a substitute for the regular high energy lactation feed.

Dry sows are the only pigs which can be fed large quantities of sprouts. They have the ability to eat large quantities of high moisture, high fiber feeds. Rations using barley sprouts (day 4) should be balanced on a daily nutrient intake basis. Amounts of nutrients supplied by the sprouts should be calculated first; then a balancer ration can be made to compliment the nutrients supplied by the sprouts. Day 4 barley sprouts should be fed at the rate of 1.5 to 2.5 kg per head per day.

REFERENCES

Alexander, J.C., Gabriel, H.G., and Reichertz, J.L., 1984. Nutritional Value of Germinated Barley. Can. Inst. Food Sci. Technol. 17: 224-228.

Bidwell, R.G.S., 1974. Plant Physiology. MacMillan, New York.

Church, D.C., 1977. Livestock Feeds and Feeding. O. & B. Books, Corvallis, Oregon, 379 pp.

Cook, A.H., 1962. Barley and Malt. Academic Press, New York and London.

Dalby, A. and Tsai, C.Y., 1976. Lysine and tryptophan increases during germination of cereal grains. Cereal Chem. 53: 222-226.

English, P.R., Fowler, V.R., Baxter, S. and Hart, B., 1988. The Growing and Finishing Pig: Improving Efficiency. Farming Press, Ipswich, U.K. 311 pp.

Folkes, B.F. and Yemm, E.W., 1958. Respiration of barley plants. X. Respiration and the metabolism of amino acids and proteins in germinating grains. New Phytol. 57: 106-131.

Hillier, R.J. and Perry, T.W., 1969. Effect of hydroponically produced oat grass on ration digestibility of cattle. J. Anim. Sci. 29: 783-785.

James, A.L., 1940. The carbohydrate metabolism of germinating barley. New Phytol. 39: 133-144.

Kent-Jones, D.W. and Amos, A.J., 1967. Modern Cereal Chemistry. 6th ed. Food Trade Press, London.

Leitch, I., 1939. Sprouted fodder and germinated grain in stock feeding. Imperial Bureau of Animal Nutrition Technical Communication 11: 3-63.

Martin, D.R., 1939. Growing Plants Without Soil. Chemical Publishing Company, New York, New York.

Mayer, A.M. and Poljakoff-Mayber, A., 1975. The Germination of Seeds. 2nd ed. Pergamon Press, Toronto.

McCandlish, A.C. and Struthers, J.P., 1938. The influence of sprouted grain on digestibility. Scottish J. Agric. 21: 141-142.

National Academy of Sciences, 1982. United States-Canadian Tables of Feed Composition, 3rd ed., National Academy Press, Washington, D.C.

Patience, J.F. and Thacker, P.A. 1989. Swine Nutrition Guide. Prairie Swine Centre, University of Saskatchewan, 260 pp.

Peer, D.J. and Leeson, S., 1985a. Nutrient content of hydroponically sprouted barley. Anim. Feed Sci. Technol. 13: 191-202.

Peer, D.J. and Leeson, S., 1985b. Feeding value of hydroponically sprouted barley for poultry and pigs. Anim. Feed Sci. Technol. 13: 183-190.

Perry, T.W., Pickett, R.A. and Hillier, R.J., 1965. Comparative value of dehydrated alfalfa meal and hydroponically grown oat grass in the ration of pregnant gilts in dry lot. Purdue University Research Progress Report 212: 1-3.

Smith, D.B., 1972. The amino acid composition of barley grain during development and germination. J. Agric. Sci. 78: 265-273.

Thomas, J.W. and Reddy, B.S., 1962. Sprouted oats as a feed for dairy cows. Quarterly Bull. Michigan State University, East Lansing, Agricultural Experimental Station Report No. 44: 654-665.

Tluczkiewicz, J. and Berendt, W., 1977. Dynamics of phosphorus compounds in ripening and germinating cereal grains. II. Changes in phosphorus compounds content during germination of wheat, barley and rye grains. Acta Soc. Bot. Poland 46: 15-30.

Trubey, C.R.M., Rhykerd, C.L., Noller, C.H., Ford, D.R. and George, J.R., 1969. Effect of light, culture solution, and growth period on growth and chemical composition of hydroponically produced oat seedlings. Agron. J. 61: 663-665.

Whittemore, C.T. and Elsley, F.W.H., 1977. Practical Pig Nutrition. Farming Press, Suffolk, 190 pp.

Wiseman, J., 1987. Feeding of Non-Ruminant Livestock. Butterworths, London.

CHAPTER 5

Beans: Culled

J.P. Rodriguez and H.S. Bayley

INTRODUCTION

The common dry bean, *Phaseolus vulgaris*, is the legume consumed in greatest quantity around the world. It is particularly popular in Central and South America, where the dietary staples have traditionally been corn and beans (Carpenter, 1981). Cull beans are dry beans that have been damaged by either weather, abrasion, storage, mold or insects. They are no longer considered fit for human consumption. As a consequence, they occasionally find their way into the livestock feed market and may find a place in swine production.

GROWING BEANS

All beans are annuals belonging to the family *Leguminosae* and are grown from seeds. Several varieties differing in size, shape and color are available including navy, pinto, great northern, small white, white marrow, red kidney, pink, small red, cranberry, yellow eye, flat small white and black turtle soup. Most beans are temperate season crops and grow in areas with mean summer temperatures of 21 to 23°C. The onset of flowering is affected by the length of the day and there are long-day, day-neutral and short-day requiring cultivars (Kay, 1979; Duke, 1981).

Beans grow well in well drained, sandy loams, silt loams or clay loams (Duke, 1981). Optimum soil pH is between 6.0 and 6.8. With a pH below 5.2 manganese toxicity is likely and above pH 7.0 manganese deficiency may occur (Kay, 1979). Beans respond variably to nitrogen fertilization but require good supplies of phosphorus and potassium. Zinc and manganese are the most likely limiting micro-minerals, particularly in lake bed or outwash soils (McGraw Hill, 1987).

Beans are planted when the danger of frost has passed and soil temperatures are about 18°C. The seeds are placed at a distance of 5 to 10 cm in rows 50 to 70 cm apart, depending on cultivar and type of soil (Kay, 1979). Maturity is reached in 80 to 120 days, but can take up to 150 days at higher elevations or in cooler seasons. The dry beans are harvested and threshed when their moisture is lower than 20%, but higher than 14%. Higher moisture levels result in mold development while lower moisture levels make the seeds split or break when threshed (McGraw Hill, 1987). Yields of 1500 kg/ha can be expected.

Annual bean production in Europe (excluding the USSR) averaged three quarters of a million metric tons between 1981 and 1985 (Todorov, 1988). Dry bean production in the USA was one million metric tons in 1986 with Michigan, California, North Dakota, Colorado and Nebraska accounting for nearly 75% of that production (USDA, 1987). Dry bean production in Canada is concentrated in Ontario with 60,000 metric tons being produced in 1985 (Statistics Canada, 1987). The proportion of cull beans was estimated as 5% of the total crop in Ontario (Rodriguez and Bayley, 1987).

NUTRIENT CONTENT OF BEANS

The available information on the proximal components and amino acid content of cull beans compared to values reported for non-cull or regular beans is shown in Table 5.1. The white cull beans (Rodriguez and Bayley, 1987) were not recleaned and their low dry matter content (84.2%) is typical of cull beans in Ontario. The digestible energy value shown for the white cull beans is lower than that reported by the National Research Council (1979) for navy beans and for other commonly used protein supplements. However, Ensminger and Olentine (1978) indicated a metabolizable energy content of 2910 kcal/kg (12 kJ/g) for seed navy beans. Beans are low in fat and have moderate fiber levels and both of these factors contribute to their low energy content.

The protein content of regular dry beans is reported to range between 22 and 30% with a median of 23.4% (Tobin and Carpenter, 1978). The crude protein values shown in Table 5.1 for the cull beans are within that range. Dry beans are a good source of lysine but are relatively deficient in methionine and cystine and this has to be considered when formulating pig diets. There are varietal differences in amino acid content, with kidney beans having lower lysine and methionine levels than black or red beans (Evans and Bandemer, 1967).

Table 5.1. Proximal and Amino Acid Composition of Regular and Cull Dry Beans

	Regular Beans		Cull Beans	
	Dry Beans[1]	Navy Beans[2]	White Beans[3]	Red Beans[4]
Dry Matter (%)	90.0	89.0	84.2	88.2
D.E. (kcal/kg)	----	----	3024	----
Crude Protein (%)	23.4	22.6	22.2	24.2
Crude Fat (%)	1.5	1.3	----	1.2
Crude Fiber (%)	4.3	4.5	----	----
Arginine (g/16g N)	5.3	----	----	8.0
Lysine (g/16g N)	5.7	7.1	----	7.4
Methionine (g/16g N)	1.1	1.2	----	1.5
Cystine (g/16g N)	1.0	0.9	----	1.0
Tryptophan (g/16g N)	1.1	1.1	----	----

[1]National Research Council, 1979.
[2]Tobin and Carpenter, 1978.
[3]Rodriguez and Bayley, 1987.
[4]Myer, 1981.

Up to 13.4% of the crude protein in beans consists of non-protein amino acids such as homoserine and S-methylcysteine (Evans and Bandemer, 1967; Lucas et al., 1988). Meiners et al. (1976) and Lucas et al. (1988) found that both the crude and true protein contents of beans differed between varieties, with pinto beans having a lower protein content and black beans having greater levels of non-protein nitrogen.

No detailed information is available on the mineral and vitamin content of cull beans and the data shown in Table 5.2 for these nutrients pertains to regular dry beans. Beans are low in calcium but are high in potassium and phosphorus. However, about 70% of the phosphorus is in the form of phytates (Lolas and Markakis, 1975). Beans are not a significant source of fat-soluble vitamins but they are a good source of niacin (Table 5.2), although their content of this and other B vitamins is lower than that of other commonly used protein supplements (National Research Council, 1979). Therefore, the vitamin premix used would have to contain higher levels of these vitamins when feeding bean-containing diets to pigs.

Table 5.2. Mineral and Vitamin Content of Dry and Navy Beans (as fed)

	Dry Beans[1]	Navy Beans[2]
Calcium (%)	0.13	0.16
Phosphorus (%)	0.42	0.52
Potassium (%)	1.24	1.31
Magnesium (%)	0.13	----
Copper (mg/kg)	10.0	----
Iron (mg/kg)	99.0	149.0
Manganese (mg/kg)	21.0	----
Biotin (mg/kg)	0.1	----
Niacin (mg/kg)	25.0	21.0
Riboflavin (mg/kg)	1.8	1.6
Thiamine (mg/kg)	6.3	6.7

[1]National Research Council, 1979.
[2]Tobin and Carpenter, 1978.

UNDESIRABLE CONSTITUENTS IN BEANS

The main anti-nutritional factors in raw dry beans are trypsin inhibitors, hemagglutinins and tannins (Liener and Kakade, 1980; Jaffe, 1980; Myer, 1981). The trypsin inhibitors bind the intestinal trypsin which is rich in cystine, thus decreasing the free levels of this enzyme in the intestine and stimulating its secretion by the pancreas. The result is that protein digestibility is lowered and the dietary requirement for sulfur amino acids is increased (Myer, 1981). White cull beans had 13.6 trypsin inhibitor units per gram (raw SBM had 18.6; Rodriguez and Bayley, 1987) while small red cull beans had 50 trypsin inhibitor units per mg protein (raw soybeans had 193; Myer, 1981).

The hemagglutinins are proteins which are resistant to proteolysis in the stomach. Once in the small intestine, they bind to the mucosal surfaces causing extensive disruption of the microvilli, interfering with the secretory and absorptive functions of the intestine. Absorption of these proteins causes toxic reactions (King et al., 1983). The bean hemagglutinins are located in the cotyledon and there is little hemagglutinating activity in the seed coat (Elias et al., 1979). There are four types of hemagglutinins, depending on their specificity for agglutination for blood from different animal species, but only those that display agglutinating activity towards trypsinized cow's blood have been shown to be toxic (Jaffe, 1980). Therefore, assays for hemagglutinating activity should be conducted using cow's blood. White cull beans had 810 hemagglutinin units per gram (Rodriguez and Bayley, 1987) while small red cull beans had 213 hemagglutinin units per mg protein.

Tannins are polyphenolic compounds which interfere with protein digestibility by binding to proteins. This decreases the digestion of ingested proteins and inhibits the activities of digestive enzymes directly (Norton et al.,

1985). Tannins are concentrated in the seed coat (Elias et al., 1979) and are present at higher levels in the deeply colored beans (reds and blacks) than in the lighter color varieties (Elias et al., 1979; Adams et al., 1985; Lowgren, 1988). Small red cull beans had 9.1 mg of tannins (as catechin equivalents) per mg of protein (Myer, 1981).

FEEDING BEANS TO SWINE

Young pigs receiving a diet containing 15% raw small red cull beans had depressed protein digestibility and nitrogen retention and decreased rate and efficiency of gain (Myer, 1981). Growing pigs fed a diet containing 40% raw beans showed loss of appetite, growth depression, intermittent scouring and extensive disruption of the intestinal microvilli (King et al., 1983). As a consequence, cull beans must be cooked or otherwise heat treated to inactivate their heat labile toxic constituents (trypsin inhibitors and hemagglutinins) and thereby avoid deleterious effects on swine performance.

Autoclaving navy beans for five minutes at 121°C and two atmospheres of pressure denaturated the lectins and trypsin inhibitors (Kakade and Evans, 1965). However, autoclaving may not be a practical means of processing cull beans on a large scale for animal feed. Miller (1972) roasted cull beans under conditions that were adequate to prepare soybeans in a "Roastatron" but found that the roasted cull beans depressed the performance of finishing pigs even when included at only 10% of the diet.

Myer (1981) conducted an exhaustive investigation on the effects of the extrusion and autoclaving of cull beans on their nutritive value for pigs (Table 5.3). He compared the performance of young pigs (9.9 kg) fed a basal diet or diets containing either 15% cull beans (raw, steamed and autoclaved) or 40% cull beans (extruded, autoclaved and autoclaved with 0.1% added methionine). Autoclaving was conducted at 120°C for 15 minutes at 2

Table 5.3. Effect of Feeding Diets Containing Raw or Heat Processed Small Red Cull Beans on Pig Performance

	Feed Intake (kg)	Daily Gain (g)	Feed Efficiency
Corn-Soybean (Basal)	1.08	564	1.93
15% Raw Beans	0.32	100	3.21
15% Extruded Beans	1.04	558	1.88
15% Autoclaved Beans	1.06	554	1.92
40% Extruded Beans	0.88	443	2.00
40% Autoclaved Beans	0.96	402	2.20
40% Autoclaved Beans + 0.1% Methionine	1.01	503	1.98

Myer, 1981.

atmospheres while extrusion was at 150°C for 16 seconds (InstaPro model 500). Pigs fed the 15% raw bean diet had decreased intake and gain but this was eliminated by autoclaving or extruding the beans. Incorporating beans as 40% of the diet depressed weight gains. Methionine supplementation of the autoclaved beans resulted in performance similar to that obtained with the basal diet.

In the same experiment, Myer (1981) also studied the effects of bean extrusion and autoclaving on protein and amino acid digestibilities and nitrogen retention (Table 5.4). Feeding young pigs diets in which all the protein originated from autoclaved or extruded beans resulted in decreases in crude protein and amino acid digestibilities and in nitrogen retention. The decreases were smaller for the extruded beans and he concluded that extrusion may be a practical method of heat processing cull beans for swine feeding. He also indicated that the beans were difficult to dry-extrude due to their low fat content.

Table 5.4. Effect of Feeding Diets Containing Autoclaved or Extruded Small Red Beans on Protein and Amino Acid Digestibility and on Nitrogen Retention

	Soybean Meal	Autoclaved Beans	Extruded Beans
Nitrogen Retained (g/day)	55	31	39
Protein Digestibility (%)	87	68	79
Amino Acid Digestibility (%)			
Arginine	93	79	87
Lysine	89	71	80
Methionine	81	54	56
Cystine	86	57	50
Threonine	86	67	76

Myer, 1981.

Rodriguez and Bayley (1987) studied the effects of steam heating the cull beans because this was a simpler process than autoclaving or extruding. They found that steaming beans at 100°C for 30 minutes reduced the trypsin inhibitory activity from 14.2 to 1.5 units/g and the hemagglutinins from 620 to 0 units/g. However, incorporation of beans steamed for either 30 or 45 minutes into pig diets at a level of 34.5% resulted in a high incidence of scours and depressed dry matter, protein and energy digestibilities and a much lower nitrogen retention than for pigs fed a bean-free control diet. These data indicate that the processing conditions were inadequate to prepare the beans for feeding to pigs.

In a second experiment, the beans were steamed for 75 minutes. These beans had 0.9 trypsin inhibitory units and 0 hemagglutinins per gram. The beans were included at 0, 11.5, 23.0 or 34.5% and the animals readily consumed the diets (Table 5.5). However, nutrient digestibility decreased linearly as the level of beans in the diet increased. The results of these two experiments indicate that toxic factors other than the trypsin inhibitors and hemagglutinins are limiting pig performance because these toxins had been inactivated by steam processing for 75 minutes. Tannins could have been responsible for the lowered digestibility of the protein in the second experiment (Table 5.5) but there are no reports that tannins cause scours such as those observed in the earlier experiment.

The linear declines in the digestibilities of energy and crude protein of the diets as the level of inclusion of the beans increased allowed calculation of the digestibilities of these nutrients in the beans themselves. The steamed beans provided 3.08 Mcal/kg (12.7 MJ/kg) of digestible energy and 103 g/kg of digestible protein. These values for the nutrient parameters of the steam-processed cull beans indicate that because the processing conditions must be severe enough to destroy the trypsin inhibitor and the hemagglutinins, beans cannot be considered as protein rich feeds, but they are useful because of the energy which they contribute to a diet. The monetary worth of raw culled beans should be based upon the cost of cereal grains, making an allowance for the costs of processing to destroy the toxins.

Table 5.5. Effect of Dietary Level of Inclusion of Beans Steamed at 100°C for 75 Minutes on Nutrient Utilization by Pigs

	Level of Bean in Diet (%)			
	0	11.5	23.0	34.5
Feed Intake (kg/d)	1.91	1.95	1.91	1.89
Nitrogen Retention	0.60	0.54	0.55	0.46
Scours (Animal Days)	2	0	0	2
Digestibility (%)				
Dry Matter	0.92	0.91	0.90	0.88
Energy	0.92	0.90	0.88	0.86
Crude Protein	0.90	0.83	0.78	0.74

Rodriguez and Bayley, 1987.

SUMMARY

The toxins in beans make them an unsatisfactory feed ingredient for pig production. However, heat processing, such as steaming at 100°C for at least 75 minutes, reduces trypsin inhibitor activity and destroys hemagglutinins. In spite of the virtual elimination of these toxins by processing, the palatability of diets containing beans remains low. However, if the price of processed culled beans is appropriate, they can be used to supply up to one-half of the supplementary protein for pigs of greater than 40 kg liveweight.

REFERENCES

Adams, M.W., Coyne, D.P., Davis, J.H.C., Graham, P.H. and Francis, C.A., 1985. Common bean (*Phaseolus vulgaris* L.). In: R.J. Summerfield and E.H. Roberts, eds. Grain Legume Crops. William Collins & Sons and Company, London, pp. 433-476.

Carpenter, K.J., 1981. The nutritional contribution of dry bean (*Phaseolus vulgaris*) in perspective. Food Technol. 35 (March): 77-78.

Duke, J.A., 1981. Handbook of Legumes of World Economic Importance. Plenum Press, New York, pp. 195-200.

Elias, L.G., de Fernandez, D.G. and Bressani, R., 1979. Possible effects of seed coat polyphenolics on the nutritional quality of bean protein. J. Food Sci. 44: 524-527.

Ensminger, M.E. and Olentine, C.G., 1978. Feeds and Nutrition. The Ensminger Publishing Company. Clovis, California, 1417 pp.

Evans, R.J. and Bandemer, S.L., 1967. Nutritive value of legume seed proteins. J. Agric. Food Chem. 15: 439-443.

Jaffe, W.G., 1980. Hemagglutinins (lectins). In: I.E. Liener, ed. Toxic Constituents of Plant Foodstuffs. Academic Press, New York, pp. 73-102.

Kakade, M.L. and Evans, R.J., 1965. Nutritive value of navy beans (*Phaseolus vulgaris*). Brit. J. Nutr. 19: 269-276.

Kay, D.E., 1979. Crops and Product Digest Number 3: Food Legumes. Tropical Products Institute, London, pp. 125-176.

King, T.P., Begbie, R. and Cadenhead, A., 1983. Nutritional toxicity of raw kidney beans in pigs. Immunocytochemical and cytopathological studies on the gut and pancreas. J. Sci. Food Agric. 34: 1404-1412.

Liener, I.E. and Kakade, M.L., 1980. Protease inhibitors. In: I.E. Liener, ed. Toxic Constituents of Plant Feedstuffs. Academic Press, New York, pp. 7-71.

Lolas, G.M. and Markakis, P., 1975. Phytic acid and other phosphorus compounds of beans (*Phaseolus vulgaris* L.). J. Agric. Food Chem. 23: 13-15.

Lucas, B., Guerrero, A., Sigales, L. and Sotelo, A., 1988. True protein content and non-protein amino acids present in legume seeds. Nutr. Rep. Int. 37: 545-553.

Lowgren, M., 1988. The effect of hull on the protein quality of brown beans (*Phaseolus vulgaris* L.). Nutr. Rep. Int. 38: 873-883.

McGraw Hill, 1987. McGraw Hill Encyclopedia of Science and Technology, 6th ed. McGraw Hill Book Company, New York, vol. 12: 434-437.

Meiners, C.R., Derise, N.L., Lau, H.C., Ritchey, H.J. and Murphy, E.W., 1976. Proximate composition and yield of raw and cooked mature dry legumes. J. Agric. Food Chem. 24: 1122-1126.

Miller, E.R., 1972. Roasted soybeans and roasted cull beans for swine. Michigan State University, East Lansing, Swine Report, pp. 69-75.

Myer, R.O., 1981. Protein utilization and toxic effects of raw beans (*Phaseolus vulgaris*) in diets for young pigs and evaluation of heat processed cull beans including extruded cull bean-soybean mixtures in diets for young pigs and chicks. Ph.D. Thesis. Washington State University, Pullman, Washington.

National Research Council, 1979. Nutrient Requirements of Domestic Animals, No. 2. Nutrient Requirements of Swine. 8th ed. National Academy of Sciences, Washington, D.C.

Norton, G., Bliss, F.A. and Bressani, R., 1985. Biochemical and nutritional attributes of grain legumes. In: R.J. Summerfield and E.H. Roberts, eds. Grain Legume Crops. William Collins & Sons and Company, London, pp. 88-103.

Rodriguez, J.P. and Bayley, H.S., 1987. Steam-heated cull beans: Nutritional value and digestibility for swine. Can. J. Anim. Sci. 67: 803-810.

Statistics Canada, 1987. Canada Yearbook 1988. Ministry of Supply and Services, Ottawa, Canada, pp. 9-36.

Tobin, G. and Carpenter, K.J., 1978. The nutritional value of the dry bean (*Phaseolus vulgaris*): A literature review. Nutr. Abstr. Rev. 48: 919-936.

Todorov, N.A., 1988. Livestock feed resources and feed evaluation in Europe. II. Cereals, Pulses and Oilseeds. Livest. Prod. Sci. 19: 47-95.

United States Department of Agriculture, 1987. Agricultural Statistics 1987. U.S. Government Printing Office, Washington, D.C., pp. 251-254.

CHAPTER 6

Blood Meal: Flash-Dried

E.R. Miller

INTRODUCTION

The feed industry makes good use of by-products of many human food production processes. Blood is a valuable by-product of the animal slaughter and meat processing industry. Historically, blood has been cooked and dried as blood meal for use in livestock feeds using the traditional vat cooking process. Blood meal produced by conventional vat cooking and drying processes has been recognized for a long time as a high protein ingredient but has found limited use in pig rations because of poor palatability and low availability of lysine. Spray drying and newer flash-drying procedures have significantly improved both the palatability and lysine availability of blood meal, giving it potential for use in cereal grain-based swine diets (Doty, 1973; Waibel et al., 1977; Miller, 1977; Wahlstrom and Libal, 1977; Miller and Parsons, 1981).

PRODUCTION OF FLASH-DRIED BLOOD MEAL

In order to produce blood meal, blood is dried using much the same process as that used in producing evaporated milk (Hall and Hedrick, 1966). In the spray drying system, blood is evaporated to 40 to 50% solids in a vacuum at low heat (49°C). Then the material is sprayed into a hot air stream (316°) and drying is completed. For the flash-drying procedure, blood is coagulated, the fluid is pressed out and the solid matter enters an elliptical hot air stream (400°C) as small particles and is rapidly dried. The dried blood (less than 10% moisture) is removed from the ring by a cyclone harvester. The blood is in the hot air stream for less than one minute. Thus, it is called flash-drying. Another flash-drying procedure is one in which liquid blood is applied in a thin film to the ascending side of a rotating steam-heated drum and scraped from the descending side as a dried sheet which is then flaked and pulverized to form a meal.

COMPOSITION OF FLASH-DRIED BLOOD MEAL

The nutrient composition and energy density of spray-dried and flash-dried blood meals is presented in Table 6.1. The high protein concentration and the high concentration of lysine make flash-dried blood meal particularly valuable in cereal grain-based swine diets. Vat-dried blood is about 20% lower in lysine concentration than flash-dried blood (Kramer et al., 1978) and the lysine from vat-dried blood is very poorly available (Kratzer and Green, 1957).

Table 6.1. Nutrient Composition of Spray-Dried and Flash-Dried Blood Meal (as fed)

	Spray-Dried[1]	Flash-Dried[2]
Dry Matter (%)	93	90
Crude Protein (%)	86	83
Ether Extract (%)	1.2	1.5
Crude Fiber (%)	1.0	1.0
D.E. (kcal/kg)	2980	3600
M.E. (kcal/kg)	2330	3400

[1]National Research Council, 1988.
[2]Miller and Parsons, 1981.

Blood meal is very rich in the amino acid leucine. However, there is some suggestion that high levels of leucine in the diet may increase the requirement for isoleucine and isoleucine has been shown to be the first limiting amino acid in a corn, soybean meal and flash-dried blood meal diet for swine (Table 6.2). Incorporation of up to 6% of flash-dried blood into corn-soybean meal starter diets containing adequate lysine is about the limit before a deficiency of isoleucine occurs (Parsons et al., 1985).

Table 6.2. Amino Acids Content (% as fed) of Spray-Dried and Flash-Dried Blood Meal

	Spray-Dried[1]	Flash-Dried[2]
Arginine	3.6	4.0
Histidine	5.2	5.3
Isoleucine	0.9	1.0
Leucine	11.0	12.5
Lysine	7.4	9.7
Methionine	1.0	1.0
Cystine	1.0	1.0
Phenylalanine	5.9	7.5
Tyrosine	2.3	3.0
Threonine	3.6	4.4
Tryptophan	1.0	1.1
Valine	7.5	9.0

[1]National Research Council, 1988.
[2]Miller and Parsons, 1981.

The high concentration of lysine in dried blood makes it a valuable feedstuff for incorporation into cereal grain-based diets of swine and poultry. However, if the drying process lowers the availability of the lysine in blood, the value of this potential feedstuff is reduced accordingly. Waibel et al. (1977) reported lysine availabilities of 19.2, 0 and 14.4% for vat-cooker-dried blood meal and 86.6, 90.7 and 81.5% for flash-dried blood meal using rat, chick and turkey poult growth assays, respectively. Thus, vat-cooker-dried blood meal is of little value in swine diets. On the other hand, cattle or pig blood, dried by the flash-drying processes (ring or steam drum drying), contains highly available lysine as determined by pig growth assays (Parsons et al., 1985).

In these assays, 5-week old pigs were fed a basal diet which was adequate in all nutrients except lysine. Then in a 4-week trial, pigs were fed this basal diet supplemented with 0, 0.1 or 0.2% of L-lysine or the basal diet supplemented with 0, 1.5 or 3.0% of flash-dried blood. Pigs were weighed and feed intake was recorded weekly. From these data, equations of the regression of weight gain on L-lysine or flash-dried blood intake were calculated. The slope ratio of the two regression lines gave the measure of bioavailability of the lysine in flash-dried blood. This assay demonstrated that the bioavailability of lysine in flash-dried blood was about 70%. Thus, with the total lysine in flash-dried blood of 9.7% (Table 6.1), the available lysine is 7% and this is a very safe value to assign to flash-dried blood meal in diet formulations.

Mineral elements are in quite low concentration in flash-dried blood with the exception of iron (Table 6.3). The iron in flash-dried blood (over 2000 mg/kg) is quite highly available to swine (Miller, 1978). In fact, the iron requirement of young pigs (100 mg/kg) is met by incorporating 5% of flash-dried blood into the diet.

Table 6.3. Mineral Composition of Spray-Dried and Flash-Dried Blood Meal (as fed)

	Spray-Dried[1]	Flash-Dried[2]
Calcium (%)	0.41	0.30
Phosphorus (%)	0.30	0.25
Sodium (%)	0.38	0.33
Chloride (%)	0.25	0.27
Magnesium (%)	0.15	0.22
Potassium (%)	0.15	0.90
Sulfur (%)	0.34	0.32
Iron (mg/kg)	2769	2000
Copper (mg/kg)	8.2	8.0
Manganese (mg/kg)	6.4	5.0
Zinc (mg/kg)	----	50.0

[1]National Research Council, 1988.
[2]Miller and Parsons, 1981.

Flash-dried blood is not a notably rich source of any of the vitamins required by swine and with an upper limit of 6% flash-dried blood in the diet would not make a substantial contribution to meeting the requirements of any of the vitamins (Table 6.4).

Table 6.4. Vitamin Composition of Spray-Dried Blood Meal (as fed)

Thiamin (mg/kg)	0.3
Riboflavin (mg/kg)	2.9
Niacin (mg/kg)	22
Pantothenic Acid (mg/kg)	3.2
Pyridoxine (mg/kg)	4.4
Choline (mg/kg)	600
Folacin (mg/kg)	0.4
Biotin (mg/kg)	0.3

National Research Council, 1988.

FEEDING FLASH-DRIED BLOOD MEAL

Parsons et al. (1979) conducted a feeding trial in which corn-soybean meal starter, grower, and finisher diets were supplemented with 0 or 5% of tallow and 0 or 5% flash-dried blood in a 2 x 2 factorial design. Diets met National Research Council (1988) requirements for lysine, tryptophan, methionine and cystine, isoleucine, calcium and phosphorus. Ratios of these nutrients to M.E. concentration were maintained when tallow was added. In all cases, pig performance was as good or better when flash-dried blood meal was incorporated into the diet (Table 6.5).

Table 6.5. Effects of Incorporation of 5% of Flash-Dried Blood Meal and 5% Tallow into Corn-Soybean Meal Diets on Pig Performance

	Diets			
Tallow (%)	0	0	5	5
Flash-Dried Blood (%)	0	5	0	5
Starter Phase (11 to 23 kg)				
Daily Gain (kg)	0.35	0.43	0.40	0.49
Daily Intake (kg)	0.87	0.96	0.87	0.95
Feed Efficiency	2.48	2.23	2.17	1.93
Grower Phase (23 to 63 kg)				
Daily Gain (kg)	0.76	0.77	0.77	0.90
Daily Intake (kg)	1.86	1.87	1.68	1.80
Feed Efficiency	2.44	2.42	2.18	2.00
Finisher Phase (63 to 104 kg)				
Daily Gain (kg)	0.86	0.87	0.87	0.89
Daily Intake (kg)	3.11	2.96	2.81	2.83
Feed Efficiency	3.61	3.40	3.22	3.18

Parsons et al., 1979.

In another study, Ilori et al. (1984) substituted combinations of peanut meal and flash-dried blood meal for soybean meal in corn-based, growing-finishing pig diets. Diets were calculated to be nutritionally adequate and isolysinic during the grower (0.75% lysine for 8 weeks) and finisher (0.60% lysine for 4 weeks) phases. Over the entire 12-week period, rate and efficiency of gain were similar in pigs consuming the basal corn-soybean meal diet or the corn plus 15 to 20% peanut meal plus 3 to 4% flash-dried blood meal diets. Apparent biological value of protein and net protein utilization of these diets were similar and carcass measures did not differ. Wahlstrom and Libal (1977) reported that replacing soybean meal with flash-dried blood on an equal-weight basis up to 6% of the diet did not affect the performance of growing-finishing pigs.

ECONOMIC EVALUATION

The economic value of flash-dried blood meal in pig diets is calculated primarily on its available lysine and metabolizable energy content and hence its ability to replace soybean meal in grain-soybean meal diets. The low isoleucine concentration (1.0%) in flash-dried blood meal limits the extent of soybean meal replacement. Nevertheless, when used up at to 5% in corn-soybean meal diets, 40 kg of flash-dried blood meal plus 60 kg of corn will replace 100 kg of dehulled soybean meal. Thus, with corn priced currently at 11 U.S. cents per kg and dehulled soybean meal priced at 22 cents per kg, flash-dried blood meal is worth about 38 cents per kg.

SUMMARY

Flash-dried blood meal is a viable source of essential amino acids in swine diets and because of its high concentration of available lysine is ideal for incorporation into cereal grain-based diets. However, the low isoleucine concentration (1.0%) in flash-dried blood meal limits the extent of soybean meal replacement. If proper formulation values are employed, flash-dried blood meal can be used with no difficulty at up to 2% of the diet for starting pigs and up to 5% for growing, finishing and adult pigs. Higher levels are possible if care is taken to ensure proper diet formulation.

REFERENCES

Doty, D.M., 1973. Developments in processing meat and blood by-products. In: Alternative Sources of Protein for Animal Production. National Academy of Sciences, Washington, D.C.

Hall, C.W. and Hedrick, T.I., 1966. Drying Milk and Milk Products. The AVI Publishing Company, Westport, Connecticut.

Ilori, J.O., Miller, E.R., Ullrey, D.E., Ku, P.K. and Hogberg, M.G., 1984. Combinations of peanut meal and blood meal as substitutes for soybean meal in corn-based, growing-finishing pig diets. J. Anim. Sci. 59: 394-399.

Kramer, S.L., Waibel, P.E., Behrends, B.R. and El Kandelgy, S.M., 1978. Amino acids in commercially produced blood meals. J. Agric. Food Chem. 26: 979-981.

Kratzer, F.H. and Green, N., 1957. The availability of lysine in blood meal for chicks and poults. Poult. Sci. 36: 562-565.

Miller, E.R., 1977. Formulating swine, poultry rations using flash-dried blood meal. Feedstuffs 49 (16): 22-23.

Miller, E.R., 1978. Bioavailability of iron in iron supplements. Feedstuffs 59 (30): 20-21.

Miller, E.R. and Parsons, M.J., 1981. Flash-dried blood meal (FDBM) as an ingredient for pig diets. Pig News and Information 2: 407-409.

National Research Council, 1988. Nutrient Requirements of Domestic Animals, No. 2. Nutrient Requirements of Swine. 9th ed. National Academy of Sciences, Washington, D.C.

Parsons, M.J., Miller, E.R., Bebiak, D.M., and Hill, G.M., 1979. Effects of incorporating 5% flash-dried blood meal and 5% tallow into starter, grower and finisher diets. Michigan Agricultural Experimental Station Report No. 386: 110-114.

Parsons, M.J., Ku, P.K. and Miller, E.R., 1985. Lysine availability in flash-dried blood meals for swine. J. Anim. Sci. 60: 1447-1453.

Wahlstrom, R.C. and Libal, G.W., 1977. Dried blood meal as a protein source in diets of growing-finishing swine. J. Anim. Sci. 44: 778-783.

Waibel, P.E., Cuperlovic, M., Hurrell, R.F. and Carpenter, K.J., 1977. Processing damage to lysine and other amino acids in the manufacture of blood meal. J. Agric. Food Chem. 25: 171-175.

CHAPTER 7

Buckwheat

P.A. Thacker

INTRODUCTION

Common buckwheat (*Fagopyrum esculentum*) is a summer annual widely grown throughout the world. At present, it is most commonly grown for human consumption with small amounts used in pancake mixes, breakfast cereals and in certain breads and ethnic dishes. Only small quantities of buckwheat are used as livestock feed.

The term buckwheat can be confusing. In this review, buckwheat refers to the common, cultivated buckwheat (*Fagopyrum esculentum*). A related species, Tartary buckwheat (*Fagopyrum tataricum*), is cultivated in some parts of the world but it is generally classed as a noxious weed. Wild buckwheat (*Polygonum convolvulus* L.), another common weed, is more distantly related.

GROWING BUCKWHEAT

Buckwheat is not a true cereal but it is often classed as one because the handling of the crop and the type of seed are similar to the cereal grains. Total world production of buckwheat is about 50 million bushels, most of which is produced by the USSR. Under normal conditions, buckwheat yields averaging 800 to 1000 kg/ha are not uncommon but under excellent conditions, yields may reach 2000 kg/ha (Campbell and Gubbels, 1977).

Traditionally, buckwheat has been considered as a catch crop, grown when seeding is delayed or when an earlier seeding of cereals fails to become established. It has also been used as a smother crop to reduce weed infestations as it germinates rapidly and is a good competitor. Buckwheat has a short growing season of 10 to 12 weeks and as a result, it can be seeded later than cereal grains and still produce a crop (Ali-Khan, 1972). It appears to have a special capacity to grow in cool climates under a wide variety of soil conditions. In addition, buckwheat has been shown to outproduce any of the commonly grown cereals under infertile or acidic soil conditions (Farrell, 1978). Therefore, based on its agronomic characteristics, it may become more attractive to grow buckwheat in the future, particularly in areas where the land is of marginal quality.

A self-incompatible, sexually-propagated crop such as buckwheat is not well adapted for improvement through breeding (De Jong, 1972). Nevertheless, several cultivars of common buckwheat have been developed. The prime objective of plant breeders in developing these cultivars has been to increase seed size and yield. The history of their development has been reported previously (Campbell and Gubbels, 1977). Some agronomic characteristics of the more commonly grown buckwheat cultivars are given in Table 7.1 (Alberta Agriculture, 1983).

Table 7.1. Agronomic Characteristics of Buckwheat Cultivars

Variety	Relative Yield	Days to Maturity	Seed Color	Seed Weight (g/1000)
Mancan	100	75	Black	31
Manor	108	75	Black	29
Tokyo	104	75	Silver	22

Alberta Agriculture, 1983.

NUTRIENT COMPOSITION OF BUCKWHEAT

A chemical analysis of buckwheat and some of the commonly grown cereal grains is presented in Table 7.2. Buckwheat is similar to oats in chemical composition with a protein content of approximately 12.0%, a crude fiber level of 12.8% and a digestible energy level of 2999 kcal/kg. The calcium content of buckwheat is slightly higher than that found in the cereal grains while the phosphorus level is approximately the same as the level found in barley and wheat.

The protein quality of buckwheat is reported to be among the highest in the plant kingdom (Sure, 1955; Biely and Pomeranz, 1975; Lyman et al., 1956; Pomeranz and Robbins, 1972). The concentration of several of the essential amino acids is higher in buckwheat compared with the cereal grains (Table 7.3). Of particular importance are the levels of lysine and threonine which are the first and second limiting amino acids in cereal protein. Buckwheat contains significantly higher levels of both of these amino acids (Thacker et al., 1983a). Isoleucine has been suggested to be the only amino acid not present in adequate amounts to support optimum growth in rats (Farrell, 1976).

Table 7.2. Chemical Analysis of Buckwheat (as fed)

	Wheat	Corn	Barley	Oats	Buckwheat
Dry Matter (%)	87.0	89.0	89.0	89.0	90.3
Crude Protein (%)	14.1	8.8	11.6	11.4	12.0
Crude Fiber (%)	2.4	2.2	5.1	10.8	12.8
Ether Extract (%)	1.9	3.8	1.8	4.2	2.1
Calcium (%)	0.05	0.02	0.05	0.06	0.10
Phosphorus (%)	0.37	0.28	0.26	0.27	0.35
D.E. (kcal/kg)	3483	3525	3086	2866	2999

National Research Council, 1988.

Table 7.3. Essential Amino Acid Composition (% as fed) of Buckwheat

	Wheat	Corn	Barley	Oats	Buckwheat
Arginine	0.52	0.40	0.40	0.65	0.90
Histidine	0.32	0.25	0.24	0.22	0.33
Isoleucine	0.46	0.33	0.40	0.37	0.46
Leucine	0.91	1.20	0.79	0.73	0.84
Lysine	0.39	0.26	0.42	0.40	0.77
Methionine	0.18	0.18	0.15	0.15	0.19
Phenylalanine	0.64	0.47	0.59	0.49	0.56
Threonine	0.40	0.33	0.39	0.35	0.49
Valine	0.56	0.44	0.55	0.50	0.60

National Research Council, 1988.

Based on its amino acid composition, buckwheat would appear to have considerable potential for use as a supplement to the cereal grains, as its high lysine content may compensate for the limiting lysine level in diets consisting predominately of cereal grains. This would allow less protein concentrate to be utilized in the formulation of balanced diets.

UNDESIRABLE CONSTITUENTS IN BUCKWHEAT

When exposed to sunlight, pigs fed high levels of buckwheat develop peculiar eruptions and intense itching of the skin (Morrison, 1956). This condition is know as Fagopyrism or buckwheat poisoning (Bruce, 1917). It is caused by a photosensitizing agent in buckwheat know as Fagopyrin (Wender,

1946). Only white or light-colored areas of the skin are affected but they must be exposed to direct sunlight. If animals are kept indoors, away from sunlight, they remain normal. Therefore, under modern systems of confinement, Fagopyrism is unlikely to be a problem for pigs fed buckwheat.

Buckwheat is also reported to contain a trypsin inhibitor which may decrease the digestibility of buckwheat protein (Ikeda et al., 1983; Ikeda and Kusano, 1983). Condensed tannins are also present in buckwheat but at a level considerably lower than the level present in sorghum or fababeans (Eggum et al., 1981).

FEEDING BUCKWHEAT TO SWINE

Starter Pigs

Despite its high quality protein, buckwheat should not be used in diets fed to starter pigs. Its high crude fiber content and low digestible energy level are likely to limit growth and reduce the efficiency of feed utilization when fed to pigs of this weight range. Other alternatives are available and producers would be wise to choose a higher energy feedstuff as the foundation of their starter diets.

Growing-Finishing Pigs

Buckwheat has much more potential for use in the growing-finishing phase. Some of the initial research conducted with buckwheat suggested that it was somewhat unpalatable and therefore, should not make up more than one-third of the total swine ration (Morrison, 1956; van Wyk et al., 1952). However, these studies were conducted with unselected varieties of buckwheat. More recent work conducted with rats, has shown that the newer varieties of buckwheat (Mancan, Tempest or Tokyo) will support improved animal performance in comparison with the unselected varieties of buckwheat (Thacker et al., 1983b). These varieties appear to be more palatable in comparison with the unselected varieties.

The results of two feeding trials in which buckwheat was used to replace 0, 25, 50, 75, or 100% of the cereal portion of swine diets containing either barley or wheat are shown in Tables 7.4 and 7.5. Substitution of buckwheat for either barley or wheat had no significant effect on feed intake, daily grain or feed efficiency. In addition, pigs fed diets containing higher levels of buckwheat tended to have leaner carcasses compared with pigs fed barley or wheat.

The results of these experiment indicate that it may be possible to include higher levels of buckwheat in swine diets than earlier research had indicated. Use of the newer cultivars of buckwheat would appear to be beneficial. However, at the present time, the price of buckwheat is so much higher than that of the common cereal grains, that its inclusion in swine diets would not be economical. However, if the cost of protein supplementation increases substantially, there may be economic benefits from the inclusion of buckwheat in swine rations.

Table 7.4. Performance of Pigs Fed Diets Containing Various Levels of Buckwheat in Combination with Barley

	Percent of Cereal as Buckwheat				
	0	25	50	75	100
Daily Gain (kg)	0.70	0.70	0.72	0.66	0.70
Daily Intake (kg)	1.95	2.12	2.27	1.94	2.01
Feed Efficiency	2.79	3.02	3.16	2.93	2.91
Backfat (mm)	30.9	31.1	29.0	27.3	27.5

Anderson and Bowland, 1981.

Table 7.5. Performance of Pigs Fed Diets Containing Various Levels of Buckwheat in Combination with Wheat

	Percent of Cereal as Buckwheat				
	0	25	50	75	100
Daily Gain (kg)	0.67	0.72	0.69	0.69	0.70
Daily Intake (kg)	1.89	2.02	2.18	2.04	2.01
Feed Efficiency	2.85	2.96	3.18	2.82	2.91
Backfat (mm)	30.9	32.6	32.0	31.9	27.5

Anderson and Bowland, 1981.

Breeding Stock

There has been very little research conducted on the feeding value of buckwheat for the breeding herd. Based on its nutrient content, it is likely that buckwheat could be used in gestation diets. However, buckwheat should not be used if the gestating sows are housed outdoors. In addition, because of its low energy content, buckwheat should not be fed to sows during lactation.

SUMMARY

Buckwheat contains a high quality protein which may compensate for the limiting lysine content in swine diets consisting predominantly of cereal grains.

This may allow for the use of smaller quantities of protein supplements in the formulation of balanced diets. The overall performance of growing-finishing pigs fed buckwheat has been shown to be comparable to those of pigs fed cereal grains. However, at the present time, the price of buckwheat is so much higher than that of the cereal grains that its inclusion in the diet would not be economical. Despite its high quality protein, buckwheat should not be used in diets fed to starter pigs or to sows during lactation.

REFERENCES

Alberta Agriculture, 1983. Varieties of special crops for Alberta. Edmonton, Agdex 140/32-1.

Ali-Khan, S.T., 1972. Growing buckwheat. Canadian Department of Agriculture, Ottawa, Publication No. 1468.

Anderson, D.M. and Bowland J.P., 1981. Evaluation of buckwheat in diets for growing pigs. Proc. Western Section Amer. Soc. Anim. Sci. 32: 422-425.

Biely, J. and Pomeranz, Y., 1975. The amino acid composition of wild buckwheat and No. 1 wheat feed screening. Poult. Sci. 54: 761-766.

Bruce, E.A., 1917. Fagopyrismus (buckwheat poisoning) and similar affections. Amer. Vet. Med. Assoc. J. 52: 189-194.

Campbell, C.G. and Gubbels, G.H., 1977. Buckwheat for profit. Canadian Department of Agriculture, Ottawa, Contribution No. M-204.

De Jong, H., 1972. Buckwheat. Field Crop Abst. 25: 389-396.

Eggum, B.O., Kreft I. and Javornik, B., 1981. Chemical composition and protein quality of buckwheat (*Fagopyrum esculentum*). Qual. Plant Foods Human Nutr. 30: 175-179.

Farrell, D.J., 1976. The nutritive value of buckwheat (*Fagopyrum esculentum*). Proc. Aust. Soc. Anim. Prod. 11: 413-416.

Farrell, D.J., 1978. A nutritional evaluation of buckwheat (*Fagopyrum esculentum*). Anim. Feed Sci. Technol. 3: 95-108.

Ikeda, K. and Kusano, T., 1983. Purification and properties of the trypsin inhibitors from buckwheat seed. Agric. Biol. Chem. 47: 1481-1486.

Ikeda, K. Ohminami, H. and Kusano, T., 1983. Purification and properties of an aminopeptidase from buckwheat seed. Agric. Biol. Chem. 47: 1799-1805.

Lyman, C.M. Kuiken, K.A. and Hall, F., 1956. Essential amino acid content of farm feeds. J. Agric. Food Chem. 4: 1008-1013.

Morrison, F.B., 1956. Feeds and Feeding. 22nd ed. Morrison Publishing Company, Ithaca, New York, 1165 pp.

National Research Council, 1988. Nutrient Requirments of Domestic Animals, No. 2. Nutrient Requirements of Swine. 9th ed. National Academy of Sciences, Washington, D.C.

Pomeranz, G. and Robbins, G.S., 1972. Amino acid composition of buckwheat. J. Agric. Food Chem. 20: 270-274.

Sure, B., 1955. Nutritive value of proteins in buckwheat and their role as supplements to proteins in cereal grains. J. Agric. Food Chem. 3: 793-795.

Thacker, P.A., Anderson, D.M. and Bowland, J.P., 1983a. Nutritive value of common buckwheat as a supplement to cereal grains when fed to laboratory rats. Can. J. Anim. Sci. 63: 213-219.

Thacker, P.A., Anderson, D.M. and Bowland, J.P., 1983b. Chemical composition and nutritive value of buckwheat cultivars for laboratory rats. Can. J. Anim. Sci. 63: 949-956.

van Wyk, H.P., Verbeck, W.A. and Oosthuizen, S.A., 1952. Buckwheat in rations for growing pigs. Farming in South Africa 27: 399-402.

Wender, S.H., 1946. The action of photosensitizing agents isolated from buckwheat. Amer. J. Vet Res. 7: 486-489.

CHAPTER 8

Canola Meal

P. A. Thacker

INTRODUCTION

The development of low erucic acid, low glucosinolate cultivars of canola seed has led to the availability of a feed ingredient with considerable potential to replace soybean meal in diets for all classes of swine. Canola meal is a high quality product and when properly utilized, can be used to advantage in reducing feed costs for swine producers.

In order to be called canola, the oil must contain less than 2% erucic acid while the meal must contain less than 30 micromoles per gram of glucosinolates. Two types are currently grown. Westar is the most commonly grown variety of Argentine canola (*Brassica napus*), while Candle and Tobin are the most commonly grown varieties of Polish canola (*Brassica campestris*). The Argentine cultivars of canola are generally later maturing but higher yielding than the Polish varieties.

NUTRIENT CONTENT OF CANOLA MEAL

A chemical analysis of soybean meal and canola meal is presented in Table 8.1. The crude protein content of canola meal varies depending on the cultivar from which the meal is produced (Clandinin et al., 1981). Meal from cultivars of *B. campestris* contains approximately 35% crude protein while meal from cultivars of *B. napus* contains from 38 to 40% crude protein. Canola meal produced from a mixture of these types can be expected to contain 37 to 38% crude protein (Clandinin et al., 1981).

Table 8.1. Chemical Analysis of Soybean Meal and Canola Meal (as fed)

	Soybean Meal	Canola Meal
Dry Matter (%)	90.0	94.0
Crude Protein (%)	48.5	38.0
Ether Extract (%)	0.9	3.8
Crude Fiber (%)	3.4	13.1
D.E. (kcal/kg)	3680	2900
Lysine (%)	3.1	2.3
Threonine (%)	1.9	1.7
Methionine + Cystine (%)	1.4	1.2

National Research Council, 1988.

Since the protein content of canola meal is lower than soybean meal, higher levels of canola must be included in the diet to provide the same level of dietary protein as that which would be supplied by soybean meal. Using average values for the protein in barley (11.5%), canola meal (38.0%) and soybean meal (48.5%), approximately 25% more canola meal must be used when formulating a ration with canola meal as opposed to formulating with soybean meal.

One of the main factors that tends to limit the nutritional value of canola meal is its relatively low digestible energy content (Saben et al., 1971). The low level of digestible energy in canola meal is a reflection of its high crude fiber content (Kennelly et al., 1978). This results from the high proportion of hull in canola relative to the size of the seed (Bayley and Hill, 1975). The yellow seeded varieties of canola (Candle) have thinner hulls than the brown seeded varieties and as a result, they contain a lower crude fiber level (Bell and Shires, 1982).

Since canola meal contains about 15 to 25% less digestible energy than soybean meal, it is advisable to increase the energy content of rations containing canola meal. This can be done by mixing higher energy cereal grains such as corn or wheat into the ration. Inclusion of animal or vegetable fat in the diet would also increase its energy content.

The ether extract content of canola meal is higher than soybean meal (Clandinin et al., 1981). This occurs because gums are added during processing. These gums are obtained during the processing of canola oil and consist of glycolipids, phospholipids and variable amounts of triglycerides, sterols and fatty acids (Clandinin et al., 1981). Their addition provides a market outlet for the gums and improves the handling and pelleting characteristics of the meal.

The nutritive value of a protein supplement is determined to a large extent by its amino acid content. Of particular importance are the levels of lysine, threonine and the sulfur containing amino acids because these have been shown to be the most limiting amino acids in swine diets composed predominately of cereal grains (Sauer et al., 1977). Soybean meal contains more lysine than canola meal while the levels of the sulfur containing amino acids (methionine and cystine) and threonine are similar in soybean meal and canola meal (National Research Council, 1988).

Although the amino acid profile of canola meal compares favorably with soybean meal, the availability of the amino acids is lower. The availability of lysine is approximately ten percentage units lower in canola meal than soybean meal (Sauer et al., 1982). Since lysine is the first limiting amino acid in cereal grains, this reduction in availability means that higher levels of canola meal must be used to supplement a swine diet than the difference in lysine content between soybean meal and canola meal would indicate.

Canola meal contains higher levels of calcium, iron, magnesium, manganese and zinc than soybean meal (Table 8.2). Canola meal also contains almost twice as much phosphorus as soybean meal. Phosphorus is an expensive ingredient in swine nutrition, giving canola meal a distinct advantage over soybean meal in this regard.

Table 8.2. Vitamin and Mineral Content of Canola Meal and Soybean Meal (as fed)

	Canola Meal	Soybean Meal
Mineral		
Calcium (%)	0.7	0.3
Magnesium (%)	0.6	0.3
Manganese (%)	53.9	29.3
Phosphorus (%)	1.2	0.7
Potassium (%)	1.3	2.0
Copper (mg/kg)	10.4	21.5
Iron (mg/kg)	159.0	120.0
Selenium (mg/kg)	1.0	0.1
Zinc (mg/kg)	71.4	27.0
Vitamin		
Biotin (mg/kg)	0.7	0.3
Choline (%)	0.9	0.3
Folic Acid (mg/kg)	2.3	1.3
Niacin (mg/kg)	159.5	29.0
Pantothenic Acid (mg/kg)	9.5	16.0
Riboflavin (mg/kg)	3.7	2.9
Thiamine (mg/kg)	5.2	4.5

Clandinin et al., 1981.

The high phytic acid and fiber content of canola meal reduces the availability of phosphorus, calcium, magnesium, copper, manganese and zinc in canola meal (Nwokolo and Bragg, 1977; Keith and Bell, 1987). However, in spite of these lower mineral availabilities, canola meal is still a better source of available calcium, iron, manganese, phosphorus and magnesium than soybean meal. Selenium is another element that is becoming increasingly important in ration formulation. Both the content and availability of selenium are higher in canola meal than soybean meal (Bragg and Seier, 1974).

Although canola meal is generally not looked upon as a major source of vitamins in swine diets, it contains more choline, biotin, folic acid, niacin, riboflavin and thiamine than soybean meal (Clandinin et al., 1981). However, canola meal contains a lower level of pantothenic acid.

UNDESIRABLE CONSTITUENTS IN CANOLA MEAL

Prior to the general adoption of the new cultivars of canola, the presence of glucosinolates was the major factor limiting the use of rapeseed meal in swine rations (Bell, 1984). Rapeseed contains an enzyme called myrosinase which is capable of breaking down these glucosinolates into a variety of toxic compounds including isothiocyanates, oxazolidinethiones, nitriles and inorganic thiocyanate ion (Paik et al., 1980). These compounds cause the enlargement of the thyroid gland and inhibit the synthesis and secretion of the thyroid hormones (Christison and Laarveld, 1981; McKinnon and Bowland, 1979). These hormones play an essential role in the control of the body's metabolism and if deficient, may reduce the utilization of dietary nutrients causing poor growth and reproductive performance.

As a result of genetic selection, the glucosinolate content of canola meal has been reduced to about 15% of the level contained in the old rapeseed meal (Bell, 1984). In addition, during the processing of canola oil, the seed is cooked which inactivates the myrosinase enzyme and prevents the hydrolysis of the glucosinolates (Srivastava and Hill, 1977). The intact glucosinolates are relatively harmless (Bell, 1984). Therefore, provided the meal is properly processed, the presence of glucosinolates is no longer of major consequence when formulating rations with canola meal.

Two other groups of compounds found in canola meal which influence its feeding value are tannins and sinapine. Tannins are found at a level of about 3% in canola meal (Leung et al., 1979; Clandinin and Heard, 1961) and may adversely affect the digestibility of the protein and energy in the diet. Canola meal contains approximately 1.5% sinapine (Mueller et al., 1978), which is a bitter tasting compound that may reduce the palatability of rations containing high levels of canola meal.

FEEDING CANOLA MEAL

Starter Pigs

A considerable amount of effort has been devoted to the task of determining the optimum level of inclusion of canola meal in the diet of the starter pig (McKinnon and Bowland, 1977; Ochetim et al., 1980; McIntosh et al., 1986). Despite this effort, there is still a distinct lack of consensus as to the optimum level of inclusion of canola meal in the diet. A recent experiment conducted at the University of Alberta compared the performance of starter pigs in which 0, 25, 50, 75 and 100% of the dietary protein supplied by soybean meal was replaced by canola meal (Table 8.3). Performance was significantly reduced if the level of canola meal in the diet exceeded 9%.

Table 8.3. Performance of Starter Pigs (7-15 kg) Fed Diets Supplemented With Soybean Meal (SBM) and Canola Meal (CM)[1, 2]

Level of SBM (%)	25.2	18.6	12.0	5.7	0.0
Level of CM (%)	0.0	9.0	18.0	27.0	36.0
Daily Gain (g)	295	301	269	238	223
Daily Intake (g)	570	537	492	447	433
Feed Efficiency	1.94	1.79	1.85	1.89	1.98

[1]McIntosh et al., 1986.
[2]All diets formulated to contain 18% crude protein.

The reduction in performance when starter pigs are fed high levels of canola meal would appear to be the result of a reduction in feed intake. Taste preference trials conducted at the University of Alberta have shown that pigs given a choice between soybean meal and canola meal will consume significantly greater quantities of a diet containing soybean meal than they will of a diet containing canola meal (Baidoo et al., 1986). This experiment also showed that pigs can detect the presence of canola meal at dietary inclusion levels as low as 5%. The results of this and other recent experiments (Baidoo et al., 1987) indicate that canola meal should be limited to a level of 5% in starter pig diets.

Growing Pigs

The results of an experiment in which canola meal was used to replace 50 or 100% of the protein supplied by soybean meal in rations fed to growing pigs are presented in Table 8.4. Complete replacement of soybean meal with canola meal resulted in a significant reduction in daily gain and feed efficiency. However, pigs fed diets containing both soybean meal and canola meal had similar daily gains and feed efficiency in comparison with pigs fed soybean meal as the sole source of supplementary protein.

Table 8.4. Performance of Growing Pigs (20-41 kg) Fed Diets Supplemented with Soybean Meal (SBM) or Canola Meal (CM)[1]

Level of CM (%)	0	19.8	9.3
Level of SBM (%)	13.7	----	7.4
Daily Gain (kg)[3]	0.64[a]	0.55[b]	0.63[a]
Daily Intake (kg)	1.67	1.63	1.68
Feed Efficiency	2.63[a]	2.96[b]	2.67[a]
Cost/kg Gain ($)[2]	0.51	0.53	0.50

[1]McKinnon and Bowland, 1977.
[2]Cost per tonne of diet was $195.65 for the SBM diet, $179.80 for the CM diet and $189.87 for the blend.
[3]Means followed by same or no letter do not differ ($P<0.05$).

It is important to consider the cost per kilogram gain rather than the cost per tonne of ration when selecting ingredients for use in swine rations. Although the ration in which canola meal supplied the sole source of supplementary protein was the least expensive ration, the cost per kilogram gain for pigs fed this diet was the greatest. The lowest cost per kilogram gain was obtained when the diet contained both soybean meal and canola meal. Based on this and other recent experiments (Baidoo et al., 1987), it would appear that best results can be obtained if canola meal supplies only one-half of the supplementary protein in diets fed to growing pigs.

Finishing Pigs

For finishing swine (50-100 kg), the vast majority of the published information indicates that canola meal can be used to completely replace all of the supplementary protein supplied by soybean meal without adversely affecting feed intake, growth rate, feed conversion efficiency or carcass quality (Bell et al., 1981; Aherne and Lewis, 1978; Narendran et al., 1981). The results of an experiment in which canola meal was used to replace 50 or 100% of the soybean meal in rations fed to finishing swine are presented in Table 8.5. Complete replacement of soybean meal with canola meal did not affect pig performance. In addition, ration cost and cost per kilogram gain were the lowest when pigs were fed the diet in which canola meal completely replaced the supplementary protein supplied by soybean meal.

Table 8.5. Performance of Finishing Pigs (41-85 kg) Fed Diets Supplemented with Soybean Meal (SBM) or Canola Meal (CM)[1]

Level of CM (%)	0	12.6	6.1
Level of SBM (%)	9.2	----	4.8
Daily Gain (kg)	0.62	0.58	0.58
Daily Intake (kg)	2.36	2.24	2.25
Feed Efficiency	3.80	3.87	3.87
Cost/kg Gain[2]	0.69	0.65	0.68

[1]McKinnon and Bowland, 1977.
[2]Cost per tonne of diet was $181.55 for the SBM diet, $169.69 for the CM diet and $176.15 for the blend.

Breeding Stock

Recent experiments conducted at various research institutions have shown that canola meal can be used as part of all of the supplementary protein fed to breeding stock (Filipot and Dufour, 1977; Lewis et al., 1978). Research trials conducted at the University of Alberta showed no reduction in litter size, birth weight or weaning weight when canola meal was fed to sows for two successive parities (Table 8.6).

Table 8.6. Reproductive Performance of Sows Fed Soybean Meal (SBM) or Canola Meal (CM)

	SBM	SBM/CM	CM
Gestation Weight Gain (kg)	40.6	40.8	39.0
Pigs Born/Litter	9.7	9.9	9.7
Birth Weight (kg)	1.1	1.1	1.1
Pigs Weaned/Litter	7.7	8.9	8.5
Weaning Weight (kg)	4.8	4.8	4.5
Sow Lactation Weight Loss (kg)	7.0	10.0	6.3

Lewis et al., 1978.

Previous recommendations in which rapeseed meal was limited to 3% of the ration for pregnant and lactating sows were based on the high glucosinolate rapeseed available at the time (Bowland and Bell, 1972). Feeding this high glucosinolate rapeseed meal will cause reductions in litter size and conception rates if fed to breeding sows. As a consequence of the poor reproductive performance resulting from feeding high gulcosinolate rapeseed meal, many producers have been reluctant to include the new low glucosinolate varieties of canola meal in their breeding stock rations. These fears would appear to be unfounded.

SUMMARY

The development of low erucic acid, low glucosinolate cultivars of canola has led to the availability of a feed ingredient with considerable potential to replace soybean meal in diets for all classes of swine. Recent research has shown that canola meal can be used to replace approximately 25% of the supplementary protein in rations fed to starter pigs, 50% in rations fed to growing pigs and completely replace all of the soybean meal in diets fed to finishing pigs and breeding stock. Under farm conditions, it may be safer to restrict the inclusion of canola meal to 5% in starter diets and 10% in the diet of growing-finishing pigs and breeding stock. Canola meal is a high quality product and when properly utilized can reduce feed costs. Using typical feed grain and protein supplement prices, canola meal is competitive with soybean meal if it can be purchased at 65 to 70% of the cost of soybean meal on a unit weight basis.

REFERENCES

Aherne, F.X. and Lewis, A.J., 1978. The nutritive value of Tower rapeseed meal for swine. Anim. Feed Sci. Technol. 3: 235-242.

Baidoo, S.K., McIntosh, M.K. and Aherne, F.X., 1986. Selection preferences of starter pigs fed canola meal and soybean meal supplemented diets. Can. J. Anim. Sci. 66: 1039-1049.

Baidoo, S.K., Mitaru, B.N., Aherne, F.X. and Blair, R., 1987. The nutritive value of canola meal for early weaned pigs. Anim. Feed Sci. Technol. 18: 45-53.

Baidoo, S.K., Aherne, F.X., Mitaru, B.N. and Blair, R., 1987. Canola meal as a protein supplement for growing-finishing pigs. Anim. Feed Sci. Technol. 18: 37-44.

Bayley, H.S. and Hill, D.C., 1975. Nutritional evaluation of low and high fiber fractions of rapeseed meal using chickens and pigs. Can. J. Anim. Sci. 55: 223-232.

Bell, J.M., Anderson, D.M. and Shires, A., 1981. Evaluation of Candle rapeseed meal as a protein supplement for swine. Can. J. Anim. Sci. 61: 453-461.

Bell, J.M. and Shires, A., 1982. Composition and digestibility for pigs of hull fractions from rapeseed cultivars with yellow or brown seed coats. Can. J. Anim. Sci. 62: 557-565.

Bell, J.M., 1984. Toxic factors in rapeseed meal and progress toward overcoming their effects. J. Anim. Sci. 58: 996-1010.

Bowland, J.P. and Bell, J.M., 1972. Rapeseed meal for pigs. Canadian Rapeseed Meal in Poultry and Animal Feeding. Rapeseed Association of Canada, Winnipeg, Manitoba.

Bragg, D.B. and Seier, L., 1974. Mineral content and biological availability of selenium in rapeseed meal. Poult. Sci. 53: 22-27.

Christison, G.I. and Laarveld, B., 1981. Thyroid hormone response to thyrotropin-releasing hormone by pigs fed canola, rapeseed or soybean meals. Can. J. Anim. Sci. 61: 1023-1029.

Clandinin, D.R. and Heard, J., 1961. Effect of sinapine, the bitter substance in rapeseed meal on the growth of chickens. Poult. Sci. 40: 484-487.

Clandinin, D.R., Robblee, A.R., Singer, S.J. and Bell, J.M., 1981. Composition of canola meal. In: D.R Clandinin, ed. Canola Meal For Livestock and Poultry. Canola Council of Canada Publ. No. 59, pp. 8-11.

Filipot, P. and Dufour, J.J., 1977. Reproductive performance of gilts fed rapeseed meal (cv. Tower) during gestation and lactation. Can. J. Anim. Sci. 57: 567-571.

Keith, M.O. and Bell, J.M., 1987. Effect of canola meal on tissue trace mineral concentrations in growing pigs. Can. J. Anim. Sci. 67: 133-140.

Kennelly, J.J., Aherne, F.X. and Lewis, A.J., 1978. The effects of isolation, or varietal differences in, high fiber hull fraction or low glucosinolate rapeseed meals on rat or pig performance. Can. J. Anim. Sci. 58: 743-752.

Leung, J., Fenton, T.W., Mueller, M.M. and Clandinin, D.R., 1979. Condensed tannins of rapeseed meals. J. Food Sci. 44: 1313-1316.

Lewis, A.J., Aherne, F.X. and Hardin, R.T., 1978. Reproductive performance of sows fed low glucosinolate (cv. Tower) rapeseed meals. Can. J. Anim. Sci. 58: 203-208.

McIntosh, M.K., Baidoo, S.K., Aherne, F.X. and Bowland, J.P., 1986. Canola meal as a protein supplement for 6-20 kg pigs. Can. J. Anim. Sci. 66: 1051-1056.

McKinnon, P.J. and Bowland, J.P., 1977. Comparison of low glucosinolate-low erucic acid rapeseed meals and soybean meal as sources of protein for starting, growing and finishing pigs and young rats. Can. J. Anim. Sci. 57: 663-678.

McKinnon, P.J. and Bowland, J.P., 1979. Effects of feeding low and high glucosinolate rapeseed meal and soybean meal on thyroid function of young pigs. Can. J. Anim. Sci. 59: 589-596.

Mueller, M.M., Ryl, E.B., Fenton, T. and Clandinin, D.R., 1978. Cultivar and growing location differences on the sinapine content of rapeseed. Can. J. Anim. Sci. 58: 579-583.

Narendran, R., Bowman, G.H., Leeson, S. and Pfeiffer, W., 1981. Effect of different levels of Tower rapeseed meal in corn-soybean meal based diets on growing-finishing pig performance. Can. J. Anim. Sci. 61: 213-216.

National Research Council, 1988. Nutrient Requirements of Domestic Animals, No. 2. Nutrient Requirements of Swine. 9th ed. National Academy of Sciences, Washington, D.C.

Nwokolo, E.N. and Bragg, D.B., 1977. Influence of phytic acid and crude fiber on the availability of minerals from four protein supplements in growing chicks. Can. J. Anim. Sci. 57: 475-477.

Ochetim, S., Bell, J.M., Doige, C.E. and Young, C.G., 1980. The feeding value of Tower rapeseed meal for early weaned pigs. I. Effects of methods of processing and of dietary levels. Can. J. Anim. Sci. 60: 407-421.

Paik, I.K., Robblee, A.R. and Clandinin, D.R., 1980. Products of the hydrolysis of rapeseed glucosinolates. Can. J. Anim. Sci. 60: 481-493.

Saben, H.S., Bowland, J.P. and Hardin, R.T., 1971. Digestible and metabolizable energy values for rapeseed meals and soybean meals fed to growing pigs. Can. J. Anim. Sci. 51: 419-425.

Sauer, W.C., Cichon, R. and Misir, R., 1982. Amino acid availability and protein quality of canola and rapeseed meal for pigs and rats. J. Anim. Sci. 54: 292-297.

Sauer, W.C., Stothers, S.C. and Phillips, G.D., 1977. Apparent availabilities of amino acids in corn, wheat and barley for growing pigs. Can. J. Anim. Sci. 57: 585-597.

Srivastava, V.K. and Hill, D.C., 1977. Effect of mild heat treatment on the nutritive value of low glucosinolate, low erucic acid rapeseed meal. In: G. Sarwar and J.M. Bell, eds. 5th Progress Report. Research on Canola Seed Oil, Meal and Meal Fractions. Canola Council of Canada, Winnipeg, pp. 117-122.

CHAPTER 9

Canola Seed: Full-Fat

F.X. Aherne and J.M. Bell

INTRODUCTION

In recent years, there has been increasing interest in including fats and oils in swine diets. This renewed interest has been stimulated by a desire to reduce dust levels in swine barns and by an increase in the use of genetically lean, high producing animals whose appetites may not have kept pace with their genetic potential for growth or milk production. Also, the results of many experiments have demonstrated that the addition of fat or oil to swine diets will improve the growth rate and feed efficiency of starter, grower and finisher pigs and will reduce weight loss of lactating sows and increase the energy content of their milk.

It is therefore not surprising that there has been considerable interest in the use of full-fat canola seed (low glucosinolate, low erucic acid rapeseed) as a source of energy in swine diets. Canola seed is a rich source of energy, is often available locally and has the advantage of being a solid material that can be handled easily in on-farm feed mills.

GROWING RAPESEED

World rapeseed production currently ranks third among oilseeds with nearly 20 million tonnes being produced annually (Meilke, 1987). The major producing countries, ranked in descending order, are China, Canada, India, Poland, France and West Germany. Canada's production was converted in the middle 1970s to cultivars having low erucic acid levels in the oil and low glucosinolate levels in the meal. These nutritionally superior cultivars were subsequently trade-named Canola types, a term now being adopted in other countries to apply to cultivars having similar oil and meal characteristics.

All rapes belong to the Brassicae tribe of plants but two species now represent the main cultivated oilseed types: *Brassica napus* and *B. campestris*. The *B. campestris* types appear to have evolved separately in two regions, Europe-Mediterranean and Asia, and consequently have some differences in nutritional characteristics. Differences among the types may be important in relation to the feeding of whole or entire seeds to animals.

Rapeseed is a cool season crop and requires more available moisture than wheat. It prefers loam soils but will produce good crops on light soils if rainfall and fertility are adequate. Rapeseed is also moderately tolerant to soil acidity. Winter (fall sown) and spring cultivars are grown depending on the climatic conditions, the latter tending to be grown in the drier and cooler regions.

NUTRIENT CONTENT OF CANOLA SEED

Canola seed contains approximately 40% ether extract, 22% crude protein and a gross energy of 5855 kcal/kg (Robblee et al., 1983; Bayley and Summers, 1975). However, these values may vary due to cultivar and environmental effects. The crude fiber level in canola seed is approximately two percentage units higher than in soybeans. The D.E. value of canola seed is about 4300 kcal/kg compared to 4000 kcal/kg in soybeans. The lysine level, as a percentage of the seed, is lower in canola seed than in soybeans but the methionine plus cystine level in canola seed is considerably higher. A comparison of the nutrient content of canola seed, canola meal, soybean meal and soybeans is presented in Table 9.1.

Table 9.1. Composition of Seed and Meal from Soybean and Canola (90% DM)

Components	Soybean[1] Seed	Soybean[1] Meal	Canola[2] Seed	Canola[2] Meal
Crude Protein (%)	36.7	44.0	21.7	37.7
Crude Fiber (%)	5.2	7.3	7.4	11.8
Ether Extract (%)	18.8	1.1	39.74	3.53
Lysine (%)	2.25	2.90	1.20	1.94
Methionine + Cystine (%)	1.01	1.18	0.95	1.14
D.E. (kcal/kg)	4035	3490	4330	2900

[1]National Research Council, 1988.
[2]Bell and Keith (unpublished).

The mineral content of canola seed and meal as well as soybean seed and meal are presented in Table 9.2. Canola seed is a richer source of calcium, magnesium, sulfur, selenium and zinc than are soybeans (Bell, 1980). However, the content of potassium and copper is lower.

Table 9.2. Mineral Content of Seed and Meal from Soybean and Canola (90% DM)

	Soybean[1] Seed	Soybean[1] Meal	Canola[2] Seed	Canola[2] Meal
Calcium (%)	0.26	0.30	0.39	0.63
Magnesium (%)	0.22	0.29	0.32	0.51
Phosphorus (%)	0.61	0.65	0.64	1.02
Potassium (%)	1.75	2.11	0.76	1.22
Sulfur (%)	0.22	0.42	0.53	0.85
Copper (mg/kg)	16.0	23.0	3.56	5.70
Iron (mg/kg)	80.0	140.0	88.0	141.0
Manganese (mg/kg)	30.0	30.6	30.7	49.2
Selenium (mg/kg)	0.11	0.10	0.69	1.10
Zinc (mg/kg)	16	52	42	68

[1]National Research Council, 1988.
[2]Bell and Keith (unpublished).

Canola seed is a rich source of biotin, choline, niacin and riboflavin (Table 9.3.). However, the content of folic acid and pantothenic acid is low.

Table 9.3. Vitamin Content of Seed and Meal from Soybean and Canola (90% DM)

	Soybean[1] Seed	Soybean[1] Meal	Canola[2] Seed	Canola[2] Meal
Vitamin E (mg/kg)	----	2.4	9.06	14.5
Biotin (mg/kg)	0.30	0.32	0.67	1.07
Choline (mg/kg)	2495	2609	4185	6700
Folacin (mg/kg)	3.6	0.6	1.4	2.3
Niacin (mg/kg)	23	28	100	160
Pantothenic Acid (mg/kg)	16.1	16.3	5.9	9.5
Riboflavin (mg/kg)	2.7	2.9	3.6	5.8
Thiamine (mg/kg)	----	6.0	3.2	5.2
Pyridoxine (mg/kg)	----	6.0	4.5	7.2

[1]National Research Council, 1988.
[2]Bell and Keith (unpublished).

Canola seed contains twice as much oil as do soybeans. About 80% of canola oil is made up of oleic and linoleic acids and linolenic acid accounts for about 9% (Table 9.4). Therefore, canola oil may be considered to be superior to other vegetable oils as a human food oil.

Table 9.4. Fatty Acid Composition of Canola Oil (%)

C16:0	Palmitic	3.7
C16:1	Palmitoleic	0.4
C18:0	Stearic	1.5
C18:1	Oleic	58.0
C18:2	Linoleic	22.9
C18:3	Linolenic	9.3
C20:0	Arachidic	1.1
C20:1	Gadoleic	1.7
C22:0	Behenic	0.3
C22:1	Erucic	0.1
C24:0	Lignoceric	0.1
C24:1	Nervonic	0.3

Bell and Keith, 1982.

UNDESIRABLE CONSTITUENTS IN CANOLA SEED

Prior to the general adoption of the new cultivars of canola seed, the presence of glucosinolates was the major factor limiting the use of rapeseed in swine rations. The glucosinolates usually measured in rapeseed or canola include 3-butenyl-, 4-pentenyl-, and 2-hydroxy-3-butenyl-glucosinolates (Appelqvist, 1972; Bell, 1984). These glucosinolates are hydrolyzed by the enzyme myrosinase, present in canola and rapeseed, to produce glucose, sulphate, isothiocyanates, nitriles, amines and oxazolidinethiones (Bell and Belzile, 1965; Bjorkman, 1976; Kjaer, 1976). Intestinal microorganisms may also hydrolyze glucosinolates to a limited extent. The nature and amount of these various products depends on the nature of the glucosinolates present, the source of the myrosinase and the reaction conditions (Belzile et al., 1963; Paik et al., 1980; Bell, 1980). These substances have been shown to depress feed intake and growth rate in pigs and to reduce plasma levels of T_3 and T_4 (thyroxine compounds), which eventually results in enlargement of the thyroid gland.

Remarkable progress has been made since the discovery in 1967 that the variety Bronowski from Poland had a low glucosinolate content. These low glucosinolate, low erucic acid cultivars of rapeseed (Canola) must contain less than 30 micromoles of glucosinolates and the oil must contain less than 2% erucic acid according to official government standards.

Two other groups of compounds may have an influence on the feeding value of canola seed. The first of these are tannins (polyflavoid compounds) which tend to accumulate in the seeds of the rape plant (Leung et al., 1979). The other is sinapine, a choline ester of 4-hydroxy-3,5-dimethoxycinnamic acid (Larson et al., 1983). Both of these compounds in isolation may reduce palatability but there is little data on what effect their presence has on the performance of swine.

FEEDING RAPESEED AND CANOLA SEED TO SWINE

A. RAPESEED

Because of the general similarity between rapeseed and canola seed and because of the importance of glucosinolate level in feeding swine, it is appropriate to examine the results of feeding trials conducted using rapeseed before discussing those in which canola was used. Apparently, the first investigation of the nutritional value of unprocessed whole rapeseed for pigs was conducted by Bowland (1971). Pigs were fed diets containing 0, 5 or 10% rapeseed (*B. campestris*) from 8 to 86 kg liveweight. The addition of 5 or 10% rapeseed to the diet depressed daily feed intake by 10.4% and 20.4%, respectively. Bowland (1971) estimated that feed intake was depressed approximately 2% for every 1% of rapeseed added to the diet. Growth rate was reduced ($P<0.05$) by 7.6 and 10.6% when diets contained 5 or 10% rapeseed. Feed efficiency or carcass quality were not significantly affected by the addition of rapeseed to the diet.

In a subsequent experiment, Bowland (1972) reported that the addition of 5 or 10% unprocessed ground rapeseed (*B. campestris*) to the diets of pigs from 7 to 80 kg liveweight also significantly reduced growth rate. Feed intake was decreased by 8 and 22% when rapeseed was added to the diet at the 5 or 10% level, respectively. Feed efficiency and carcass quality were not significantly affected by the addition of rapeseed to the diet. The backfat of pigs fed the rapeseed supplemented diets contained more C18:2, C18:3 and C22:1 and less C16:0, C18:0 and C18:1 fatty acids than did the control pigs.

Bowland and Newell (1974) found that the addition of 10% rapeseed of *B. campestris* or *B. napus* (both low erucic acid varieties) to the diets of pigs from 17 to 89 kg liveweight, resulted in about a 6.5% reduction in daily feed intake and an 11% reduction in daily gain compared with a soybean meal control. Rapeseed did not significantly affect the carcass parameters measured but there was an increase in the polyunsaturated fatty acid content of the backfat.

Castell and Mallard (1974) reported that feed intake and growth rate of pigs fed diets supplemented with 0 to 12% rapeseed (*B. campestris* cv. Span) were lower than those for pigs fed a control diet (Table 9.5). Addition of rapeseed at levels exceeding 4% of the diet increased the degree of unsaturation of the backfat.

Bayley and Summers (1975) fed diets containing 40% extruded rapeseed (*B. campestris*) or 40% extruded soybeans to pigs from 25 to 75 kg liveweight. Digestibility coefficients were determined when the pigs weighed 30 and 70 kg liveweight. Apparent digestibility of nitrogen was similar for extruded soybeans and rapeseed. Apparent fat digestibility was 74.6 and 71.4% for the 30 kg and 70 kg pigs, respectively, suggesting that the oil in rapeseed was well digested. The digestible energy (D.E.) values for the extruded soybeans and extruded rapeseed were 4310 kcal/kg and 4830 kcal/kg for the 30 kg pigs and 4560 kcal/kg and 4510 kcal/kg for the 70 kg pigs. These values for the D.E. value of rapeseed are considerably lower than the value of 5150 kcal/kg reported by Bowland (1972). It is possible that the extrusion of the rapeseed at 120°C reduced the D.E. value.

Table 9.5. Effects of Dietary Rapeseed on Pig Performance (29-91 kg)

	% Rapeseed				
	0	4	8	12	SE
Daily Gain (kg/day)	0.80	0.77	0.74	0.74	0.03
Feed Intake (kg/day)	2.90	2.70	2.57	2.60	----
Feed Efficiency	3.63	3.51	3.47	3.51	0.06

Castell and Mallard, 1974.

Nasi et al. (1985) reported that heat treatment reduced the digestibility of organic matter and crude protein. Extrusion of rapeseed did not improve the biological value of the protein or the anti-thyroid activity in experiments with rats (Smithard and Eyre, 1986). Fenwick et al. (1986) reported that extrusion at 150°C inactivated myrosinase, but had no effect on total glucosinolate content unless 5% alkali and 1% ferrous sulphate were added before extrusion. These authors concluded that extrusion is unlikely to be useful in rapeseed detoxification.

Rapeseed of high glucosinolate type, (mixed *B. napus* and *B. campestris*) was "micronized" at 196°C for 90 seconds, ground and fed to pigs in digestibility trials (Lawrence, 1978). Micronization reduced the OZT and ITC contents but the digestibility of the micronized rapeseed was lower than for corresponding amounts of rapeseed meal and oil. Ochetim et al. (1980) reported that increasing the level of autoclaved rapeseed in the diets from 0 to 20% of the diet had no significant effect on growth rate, feed intake or efficiency of utilization of feed. Digestibility coefficients for energy or protein were not significantly reduced until the level of rapeseed in diet reached 20%. It was concluded that rapeseed with total glucosinolate content of 4.7 to 8.5 μmol/g of fat-free seed requires appropriate processing to destroy myrosinase activity, if ground whole seed is to be used as a major dietary energy and protein source for early weaned pigs without adverse effects on the thyroid gland or on pig performance.

Organic acids have also been used in attempts to improve the nutritive value of rapeseed. However, treatment of rapeseed with 0.5 to 4.0% organic acids (acetic, propionic) before grinding did not consistently improve the nutritive value of rapeseed for pigs (Bowland, 1972; Bowland and Newell, 1974).

In summary, rapeseed varies in glucosinolate content because of genetic and environmental factors. Most experimental results show unfavorable results with market pigs when ground rapeseed is fed, with reduced daily feed intakes, reduced daily gain and poorer feed efficiency. Research on the use of rapeseed meal indicates that it would also be unwise to use rapeseed in the diets of breeding pigs.

B. CANOLA SEED

Starter Pigs

Increasing the dietary levels of canola seed from zero to 15% of the diet of four-week old pigs had no significant effect on growth rate, feed intake or feed efficiency (Table 9.6). At a dietary level of 30%, a marked reduction ($P<0.05$) in gain and feed intake occurred. Dry heating the canola seed in a "jet sploder" (116°C, 60 sec.) did not improve its nutritive value for young pigs.

Table 9.6. Effects of Canola Seed on Weanling Pig Performance

	Daily Gain (g)	Daily Feed (g)	Feed Efficiency
Experiment 1			
Control	436	665	1.53
Canola Seed (7.5%)	462	704	1.52
Canola Seed (7.5%, jet-sploded)	423	668	1.58
Experiment 2			
Control	498	638	1.28
Canola Seed (15%)	512	649	1.26
Canola Seed (30%)	415	507	1.22

Shaw, 1987.

Growing-Finishing Pigs

The first investigation of the feeding value of low glucosinolate canola seed (Tower) for growing pigs was conducted by Castell (1977). Pigs fed diets containing 10% Tower canola seed (having about 13 μmol glucosinolates/g oil-free seed) had significantly superior growth rates compared with pigs fed 10% low erucic acid rapeseed containing higher levels of total glucosinolates (31 to 46 μmol oil-free seed). In another experiment, Castell (1977) observed that levels of 15% Tower canola seed in the diet of pigs from 23 to 89 kg liveweight did not significantly reduce feed intake, growth rate or feed efficiency relative to pigs fed a control diet. However, Castell (1977) calculated that each increase of 0.4 μmol total glucosinolate per kg of diet reduced daily gain by approximately 3.5%. Although not significant, there was a trend toward reduced feed intake as the level of canola seed in the diet increased.

Candle (*B. campestris*) canola seed at dietary levels of 0, 3, 6, 9, 12 and 15% was fed ground in pelleted diets to pigs from 25 to 87 kg (Castell and Falk, 1980). Level of canola seed in the diet did not affect rate of gain or feed intake. Feed efficiency tended to improve ($P<0.05$) with the addition of canola seed to the diet (Table 9.7). There were no significant treatment effects on most of the carcass traits measured but the level of unsaturation in the backfat of pigs fed the canola seed supplemented diets was greater than that of pigs fed the control diet. Levels of ground canola seed in barley diets up to 20% permitted swine performance equal to that obtained from the soybean meal control, with growing-finishing pigs (Froseth and Honeyfield, 1982).

Table 9.7. Performance of Growing Pigs (25-90 kg) Fed Diets Supplemented with Canola Seed

	Level of Canola Seed (%)					
	0	3	6	9	12	15
Daily Gain (kg/day)	0.67	0.72	0.73	0.70	0.70	0.70
Feed Intake (kg/day)	2.09	2.17	2.10	2.19	2.03	2.04
Feed Efficiency	3.11	3.01	2.87	3.12	2.90	2.91

Castell and Falk, 1980.

Salo (1980) also reported no difference in the growth rate or efficiency of feed utilization of pigs fed 10% canola seed of either *B. campestris* (Candle) or *B. napus* (Regent) varieties and pigs fed the control diet. The D.E. values of the Candle and the Regent canola seed were 5177 kcal/kg and 5605 kcal/kg on a dry basis, respectively. The large differences in D.E. between the Candle and Regent varieties of canola seed could have been due to the coarse grinding of the seed which allowed more of the typically smaller Candle seeds to escape grinding. Bell et al. (1985) reported that grinding significantly improved the digestibility of both energy and protein in canola seed.

Froseth and Peters (1981) have also reported that pigs fed a diet containing 20% extruded canola seed, during the growing period from 39 to 95 kg, grew significantly faster than pigs fed a 20% raw canola seed, but there were no other significant diet responses.

Breeding Stock

Levels of 0, 5, 10, 15, 20 and 25% *B. napus* (Tower) canola seed were incorporated into sow diets, commencing at day 109 of gestation and continuing until 21 days postpartum (Spratt and Leeson, 1985). Sow performance was not affected by 5 or 10% canola seed levels but at 15% canola seed, daily feed intake decreased and 7 to 21 day post-partum sow weight loss increased from 5.7 kg at 10% canola seed to 27.2 kg. There were no significant effects of treatment on piglet performance or fatty acid content of milk fat (Table 9.8).

Table 9.8. Reproductive Performance of Sows Fed Canola Seed

	Canola Seed in Diet (%)			
	0	5	10	15
Lactation Feed Intake (kg/day)	4.86	4.60	4.94	2.06
Pigs Born Alive	11.00	9.33	8.50	8.83
Sow Weight Loss, Day 7-21 (kg)	8.80	5.80	5.70	27.20
Milk Fat, Day 21 (%)	6.60	7.43	7.35	10.04

Spratt and Leeson, 1985.

C. FROST DAMAGED CANOLA SEED

Castell (1977) concluded that there was little economic benefit to be obtained from using canola seed in growing-finishing pig diets in contrast to its value as a source of edible oil. However, dry growing conditions, early frosts and other weather conditions may result in the production of lower grade canola seed which is undesirable for crushing or export. Oil from frost damaged immature canola seed is high in chlorophyll which imparts a green color to the oil and is more costly to refine.

Bell et al. (1985) conducted a digestibility trial using 57 kg pigs fed diets containing 0, 15, or 30% canola seed blended to contain 20, 45 and 65% frost damaged seed. The D.E. values determined for the frost-damaged canola seed were 3728 to 4230 kcal/kg. The digestibility coefficient for energy was increased from 32 to 65% by grinding the seed, but degree of frost damage had little effect on digestibility. Bell et al. (1985) reported that ammoniation of the seed made only a slight improvement in digestibility. They estimated that ground frost-damaged canola seed was worth 20 to 50% more monetarially than good barley.

In a subsequent experiment (Bell and Keith, 1986), canola seed having 10, 45 and 65% frost damage was fed at levels of 10, 20 or 30% to pigs from 23 to 100 kg liveweight. It appeared that frost-damaged canola seed had improved palatability and less glucosinolate. There were no significant differences in growth rate, feed intake or feed gain ratio of pigs fed 10 or 20% canola seed in the diet. At the 30% level of inclusion of canola seed growth rate and feed intake were significantly lower than that of pigs fed the 10 and 20% canola seed diets. Bell and Keith (1986) estimated the D.E. value of the frost-damaged seed as 3905 to 3996 kcal/kg.

An important fact to consider regarding the use of raw canola seed is the quality of the seed used. Bell et al. (1985) reported that the more frost damaged the canola seed, the higher the content of weed seeds and other inert matter and thus palatability and performance problems are more likely when such seed is included in pig diets.

SUMMARY

From the results of experiments with pigs fed canola seed, it is evident that pig performance has been significantly better than that observed with pigs fed rapeseed. It can be assumed that much of this improved performance is due to the reduction in the glucosinolate level in the canola seed. There is ample evidence that the canola seed should be ground before incorporation into pig diets. The majority of studies suggest that canola seed does not need to be heated before being fed to pigs. Of the heat treatments used with canola seed, autoclaving appears to be the most effective way of destroying the myrosinase present in canola seed but extrusion or use of dry heat at high temperatures were also effective. It may be concluded that the diets of starter and growing-finishing pigs may contain up to 15% ground canola seed in the diet with no significant reductions in growth rate, feed intake or efficiency of feed utilization. However, sows should not receive more than 10% canola seed in their diet.

REFERENCES

Appelqvist, L.A., 1972. Chemical constituents of rapeseed. In: Rapeseed. Elsevier Publishing Company, Amsterdam, pp. 123-173.

Bayley, H.S. and Summers, J.D., 1975. Nutritional evaluation of extruded full fat soybeans and rapeseed using pigs and chickens. Can. J. Anim. Sci. 55: 441-450.

Bell, J.M., 1980. Characteristics of Canola Rapeseed. Proceedings 15th Annual Pacific Northwest Animal Nutrition Conference, Spokane, pp. 63-80.

Bell, J.M., 1984. Nutrients and toxicants in rapeseed meal: A review. J. Anim. Sci. 58: 996-1010.

Bell, J.M. and Belzile, R.J., 1965. Goitrogenic properties. In: J.P. Bowland, D. Clandinin and L.R. Welter, eds. Rapeseed Meal For Livestock and Poultry: A review. Can. Dept. Agric., Ottawa, Publ. No. 1257: 45-60.

Bell, J.M. and Keith, M.O., 1982. Gross energy values of oils from high and low erucic acid rapeseed oils and of rapeseed gums. Can. Inst. Food Sci. Technol. J. 15: 221-224.

Bell, J.M. and Keith, M.O., 1986. Growth, feed utlization and carcass quality responses of pigs fed frost-damaged canola seed (low glucosinolate rapeseed) as affected by grinding, pelleting and ammoniation. Can. J. Anim. Sci. 66: 181-190.

Bell, J.M., Keith, M.O. and Kowalenko, W.S., 1985. Digestibility and feeding value of frost-damaged canola seed (low glucosinolate rapeseed) for growing pigs. Can. J. Anim. Sci. 65: 735-743.

Belzile, R., Bell, J.M. and Welter, L.R., 1963. Growth depressing factors in rapeseed oil meal. V. The effects of myrosinase activity on the toxicity of the meal. Can. J. Anim. Sci. 43: 169-173.

Bjorkman, R., 1976. Properties and functions of plant myrosinases. The Biology and Chemistry of the Cruciferae. Academic Press. New York, 191 pp.

Bowland, J.P., 1971. Rapeseed as an energy and protein source in diets for growing pigs. Can. J. Anim. Sci. 51: 503-510.

Bowland, J.P., 1972. Unprocessed rapeseed treated with propionic acid in diets of growing pigs: Performance, energy and protein digestibility, and nitrogen retention, carcass measurement, and fatty acid composition of backfat. Can. J. Anim. Sci. 52: 553-562.

Bowland, J.P. and Newell, J.A., 1974. Ground rapeseed from low erucic acid (Lear) cultivars Span and Zephyr with or without organic acid treatment as a dietary ingredient for growing-finishing pigs. Can. J. Anim. Sci. 54: 455-464.

Castell, A.G., 1977. Effects of cultivar on the utilization of ground rapeseed in diets for growing-finishing pigs. Can. J. Anim. Sci. 57: 111-120.

Castell, A.G. and Falk, L., 1980. Effects of dietary canola seed on pig performance and backfat composition. Can. J. Anim. Sci. 60: 795-797.

Castell, A.G. and Mallard, T.M., 1974. Utilization of ground seed or meal from low erucic acid rape (*Brassica campestris* cv. Span) in diets for growing-finishing pigs. Can. J. Anim. Sci. 54: 443-454.

Fenwick, G.R., Spingsk, E.A., Wilkinson, A.P., Heany, R.K. and Legoy, M.A., 1986. Effect of processing on the antinutrient content of rapeseed. J. Sci. Food Agric. 37: 735-741.

Froseth, J.A. and Honeyfield, D.C., 1982. Extruded full-fat canola seed for growing-finishing pigs. Swine Day Proceedings, Report No. 5, Washington Agricultural Experimental Station, Pullman, Washington.

Froseth, J.A. and Peters, D.M., 1981. Feeding rapeseed meal and raw or extruded whole rapeseed to growing-finishing pigs. Washington State University, Pullman, Swine Day Producer Report No. 11.

Kjaer, A., 1976. Glucosinolates in the cruciferae. In: The Biology and Chemistry of the Cruciferae. Academic Press, New York, 207 pp.

Larson, L.M., Olsen, O., Ploger, A. and Sorensen, H., 1983. Phenolic choline esters in rapeseed: Possible factors affecting nutritive value and quality of rapeseed meal. Proc. 6th International Rapeseed Conference, Paris, pp. 1577-1582.

Lawrence, T.L.J., 1978. Effects of micronization on the digestibility of whole soya beans and rapeseeds for the growing pig. Anim. Feed Sci. Technol. 3: 179-189.

Leung, J., Fenton, T.W., Mueller, M.M. and Clandinin, D.R., 1979. Condensed tannins of rapeseed meal. J. Food Sci. 44: 1313-1316.

Meilke, S., 1987. Oil World: Statistics Up-date. Hamburg: ISTA. Mielke Gmbh, West Germany.

Nasi, M., Alaviuhkola, T. and Suomi, K., 1985. Rapeseed meal of low- and high-glucosinolate type fed to growing-finishing pigs. J. Agric. Sci. Finland 57: 263-269.

National Research Council, 1988. Nutrient Requirements of Domestic Animals, No. 2. Nutrient requirements of swine. 9th ed. National Academy of Sciences, Washington, D.C.

Ochetim, S., Bell, J.M., Doige, C.E. and Youngs, C.G., 1980. The feeding value of Tower rapeseed for early-weaned pigs. I. Effect of methods of processing and dietary levels. Can. J. Anim. Sci. 60: 407-421.

Paik, I.K., Robblee, A.R. and Clandinin, D.R., 1980. Products of the hydrolysis of rapeseed glucosinolates. Can. J. Anim. Sci. 60: 481-493.

Robblee, A.R., Rebolledo, M., Shires, A. and Clandinin, D.R., 1983. The use of full fat canola seed in poultry rations. Publ. No. 61, Canola Council, Winnipeg, 7th Progress Report, pp. 153-158.

Salo, M.L., 1980. Nutritive value of full fat rapeseeds for growing pigs. J. Sci. Agric. Soc. Finland 52: 1-6.

Shaw, J., 1987. Full fat canola seed for young pigs. M.Sc. Thesis. The University of Alberta. Edmonton, Alberta.

Smithard, R.R. and Eyre, M.D., 1986. The effects of dry extrusion of rapeseed with other feedstuffs upon its nutritional value and anti-thyroid activity. J. Sci. Food Agric. 37: 136-140.

Spratt, R.S. and Leeson, S., 1985. The effect of raw ground full-fat canola on sow milk composition and piglet growth. Nutr. Rep. Int. 31: 825-831.

CHAPTER 10

Cassava Leaf Meal

V. Ravindran

INTRODUCTION

Cassava (*Manihot esculenta*) ranks among the top ten food crops in the world. It is an all-season crop of the humid tropics where its starchy root provides the staple food for over 500 million people (Lancaster et al., 1982). Although the cassava root is essentially grown for human food, there is increasing interest in its use as an animal feed. This aspect has been considered elsewhere in this book (Oke, 1990).

The cassava plant also produces a lush crop of high protein foliage. The potential yields of cassava leaves may amount to as much as 34 tonnes of dry matter per hectare. Most of the leaves are currently returned to the soil as a green manure and are therefore under utilized as a feed resource. The leaves could easily be processed into a sun-cured hay or dehydrated leaf meal with excellent storage qualities.

HARVESTING AND PROCESSING OF CASSAVA LEAVES

Cassava leaves are harvested by stripping the leaves (with petioles) from the stems. The green foliage is chopped through a forage chopper and wilted for two to three days by spreading it on the floor of a room equipped with cross ventilation. The chopped material must be turned at least twice a day to avoid mold formation. The wilted material is then dried to less than 12% moisture level (8-12 hours under ideal conditions) and ground into a semi-powdery form. Both sun-drying and artificial drying are equally effective, but sun-drying would be the method of choice under farm situations in developing countries. The rate and effectiveness of sun-drying in specific locations depends upon the ambient temperature, wind and relative humidity.

The potential yield of cassava leaves varies considerably depending on cultivar, plant density, harvesting frequency, soil fertility and climate (Moore, 1976; Montaldo and Montilla, 1976; Gomez and Valdivieso, 1984). Leaf production could be enhanced by harvesting cassava leaves during the growing season, but this would adversely affect root yield (Normanha, 1962; Ahmad, 1973). Since cassava is cultivated primarily for its roots, it is imperative that leaf harvesting does not reduce root yields.

Several studies have demonstrated that it is now possible to harvest cassava leaves while maintaining an acceptable yields of roots. Ravindran (1985) defoliated once during a seven-month growing season and obtained a DM harvest of 6.75 tonnes/ha while maintaining 86% of the normal yields of roots. Dahniya et al. (1981) recommended a harvesting frequency of two to three months, starting from four months, for best all-round yields. However, the variation that appears to exist among cultivars in their tolerance to defoliation needs to be taken into consideration before making any recommendation of harvesting frequency.

When the cultivation of cassava is aimed exclusively towards leaf production, the plant density could be increased and the harvesting frequency could be shorter. Under such conditions, annual leaf dry matter yields of over 34 tonnes/ha can be obtained (Montilla, 1976). This corresponds to a possible production of more than 6 tonnes of protein/ha per year. Whether the aim of cassava cultivation in an area should be roots, leaves, or an all-round production of both, depends on the relative prices of cassava roots, cassava leaf meal and traditional feedstuffs.

NUTRIENT COMPOSITION OF CASSAVA LEAF MEAL

The proximate composition of cassava leaf meal compares favourably with that of dehydrated alfalfa meal, the most widely used legume meal for swine feeding, and of coconut meal, the major protein supplement in swine rations in many Asian and Pacific countries (Table 10.1). However, due to its higher fiber and ash content, the metabolizable energy value of cassava leaf meal is lower than that of coconut meal.

The protein content of cassava leaves is extremely high for a non-leguminous plant. Cassava leaf meal contains an average of 21% protein. However, the reported values for protein content range from 16.7 to 39.9%, depending on variety, stage of maturity, soil fertility and climate. The stage of maturity at harvest is the major factor contributing to this variability (Ravindran and Ravindran, 1988). Almost 85% of the crude protein fraction is reported to be true protein (Eggum, 1970).

Table 10.1. Proximate Composition of Cassava Leaf Meal, Coconut Meal and Alfalfa Meal

	Cassava Leaf Meal		Coconut Meal[1]	Alfalfa Meal[1]
Dry Matter (%)	93.0		93.0	93.1
Crude Protein (%)	21.0	(16.7 - 39.9)[2]	22.0	20.0
Ether Extract (%)	5.5	(3.8 - 10.5)	6.0	3.5
Crude Fiber (%)	10.0	(4.8 - 29.0)	12.0	20.0
Ash (%)	8.5	(5.7 - 12.5)	7.0	10.5
M.E. (kcal/kg)	2.2[3]		2.5	2.0

[1]Allen, 1984.
[2]Ranges reported by Rogers 1959; Rogers and Milner, 1963; Ross and Enriquez, 1969; Eggum, 1970; Ramos-Ledon and Popenoe, 1970; Yeoh and Chew, 1976; Ravindran and Ravindran, 1988.
[3]Ravindran (unpublished data).

Cassava leaf protein is deficient in the sulphur-containing amino acids, possibly marginal in tryptophan but rich in lysine (Rogers and Milner, 1963; Eggum, 1970; Yeoh and Chew, 1976). It is noteworthy that the essential amino acid profile of cassava leaf meal is either similar or superior to those of coconut meal and alfalfa meal (Table 10.2). In particular, the high lysine content of cassava leaf meal may become useful under tropical situations where typical swine rations are deficient in lysine. Eggum (1970), using rat bioassays, studied the nutritional availability of individual amino acids in cassava leaves. Only 59% of the methionine was available, resulting in a low biological value of 49 to 57%. Supplementation with methionine improved the biological value to 80%.

Cassava leaf meal is a good source of minerals, especially of calcium and microminerals (Table 10.3). Cassava leaves are excellent sources of carotene (FAO, 1972) and contain significant amounts of riboflavin (Caldwell and Enoch, 1972). However, considerable losses of these vitamins may occur during processing and storage (Lancaster and Brooks, 1983). Cassava leaf meal has also been reported to contain an unidentified growth factor for poultry (Ravindran et al., 1983).

Table 10.2. Essential Amino Acid Contents of Cassava Leaf Meal, Coconut Meal and Alfalfa Meal (% of Protein)

	Cassava Leaf Meal[1]	Coconut Meal[2]	Alfalfa Meal[2]
Arginine	5.3	10.5	4.9
Histidine	2.3	1.3	2.1
Isoleucine	4.5	4.6	4.9
Leucine	8.2	6.8	7.5
Lysine	5.9	2.5	4.4
Methionine	1.9	1.6	1.7
Cystine	1.4	0.9	1.2
Phenylalanine	5.4	3.6	5.2
Threonine	4.4	2.7	4.4
Tryptophan	2.0	0.9	2.3
Valine	5.6	4.6	6.0

[1]Eggum, 1970.
[2]Allen, 1984.

Table 10.3. Mineral Composition of Cassava Leaf Meal, Coconut Meal and Alfalfa Meal

	Cassava Leaf Meal[1]	Coconut Meal[2]	Alfalfa Meal[2]
Macrominerals (%)			
Calcium	1.35	0.07	1.50
Potassium	1.28	1.83	2.50
Magnesium	0.42	0.30	0.32
Phosphorus	0.45	0.63	0.27
Sodium	0.02	0.09	0.08
Microminerals (mg/kg)			
Zinc	149	49	19
Manganese	52	18	34
Iron	259	68	281
Copper	12	46	9

[1]Ravindran et al., 1982.
[2]Allen, 1984.

UNDESIRABLE CONSTITUENTS IN CASSAVA LEAF MEAL

The lack of tradition in using cassava leaves for animal feeding is related to the presence of high levels of two cyanogenic glucosides (linamarin and lotoaustralin) in the fresh material. The reported values range from 20 to 80 mg hydrocyanic acid per 100 g of fresh leaf weight (Lancaster and Brooks, 1983). These levels are substantially higher than the normal range of hydrocyanic acid reported for fresh cassava roots (Coursey, 1973; Yeoh and Oh, 1979). The existence of cyanogenic glucosides has made some form of processing a prerequisitive for the use of cassava leaves in animal feeding. Recent studies have shown that it is possible to produce cassava leaf meal with low cyanide levels using simple processing techniques (Ravindran et al., 1987a). Simple sun-drying alone eliminated almost 90% of the initial cyanide content. When combined with chopping and wilting, cyanide in the final meal was reduced to levels which are safe for monogastric animals (Table 10.4).

Table 10.4. Hydrocyanic Acid Content (mg/kg DM) of Cassava Leaf Meal As Influenced by Processing Methods[1]

	Full Leaves		Chopped Leaves	
Fresh Material	1436	(0.0)[2]	1045	(0.0)
Sun Dried for 2-3 Days	173	(88.0)	109	(92.4)
Sun Dried + 1 Day Wilting	141	(90.2)	88	(93.9)
Sun Dried + 2 Days Wilting	114	(92.1)	72	(95.0)
Sun Dried + 3 Days Wilting	93	(93.5)	53	(96.3)

[1]Ravindran et al., 1987a.
[2]Values in parentheses represent the reduction in HCN as a percentage of initial level in freshly harvested leaves.

The presence of condensed tannins in cassava leaves also represents grounds for some concern (Reed et al., 1982). Tannins, particularly the condensed types, are known to lower protein digestibility by forming indigestible tannin-protein complexes and/or by inhibiting enzyme activities (Price and Butler, 1980). However, the reported values for tannins (0.25%) in cassava leaf meal are low and within safety limits for monogastrics when cassava leaf meal is fed at low levels (Ravindran et al., 1987a).

FEEDING CASSAVA LEAF MEAL

Recently, there has been an upsurge of interest in evaluating cassava leaf meal as a poultry feed (Hutagalung et al., 1974; Wyllie and Chamanga, 1979; Ravindran et al., 1983; Ravindran et al., 1986). However, corresponding information regarding its use in swine feeding is limited.

Starter Pigs

Cassava leaf meal is not recommended for use in diets fed to starter pigs. Its high fiber content and low energy level are likely to limit growth and reduce the efficiency of feed utilization when fed to pigs of this weight range. Other alternatives are available and producers would be wise to choose a higher energy feedstuff as the foundation for their starter diets. There is also some concern that the hydrocyanic acid levels in cassava leaf meal may reduce palatability. Since starter pigs are often reluctant to eat solid feed anyway, feedstuffs of questionable palatability should be avoided.

Growing-Finishing Pigs

Early studies of feeding fresh cassava leaves showed that palatability was depressed and growth performance was lowered with increasing proportions of cassava leaves in swine rations (Mahendranathan, 1971). The adverse effects were evidently due to the high hydrocyanic acid levels in fresh leaves (Lee and Hutagalung, 1972). However, more recent studies have shown that with proper processing techniques, it is now possible to produce cassava leaf meal with much lower cyanide levels (Ravindran et al., 1987a).

Rajaguru et al. (1979) and Ravindran et al. (1987b) evaluated cassava leaf meal as a substitute for coconut meal in rations fed to growing swine. The results showed that properly processed cassava leaf meal can replace up to 66% of coconut meal (26% of the total diet) in swine rations without adverse effects on performance (Table 10.5). Most efficient gains were obtained at 33% replacement (13% of the total diet), suggesting that the use of low levels of cassava leaf meal in feed formulations will permit greater savings in feed cost compared to higher levels. The findings also indicated that cassava leaf protein is utilized efficiently, although other nutrients in cassava leaf meal are not as digestible as those in coconut meal.

Attempts to utilize cassava leaf meal as a replacement for other common protein sources in growing-finishing rations have been less encouraging. Alhassan and Odoi (1982) reported depressions in gains and feed efficiency when sun-dried cassava leaf meal was included at 20 and 30% levels in rations for growing-finishing swine. Cassava leaf meal was used to replace parts of peanut meal, fish meal and corn in the basal ration. Ravindran (1986) substituted 10, 20 and 30% cassava leaf meal for a corn-soybean meal basal ration and reported that the gains and feed efficiency of growing pigs were lowered linearly with increasing levels of leaf meal.

The utilization of cassava leaf meal-based rations can be considerably improved by supplementation with methionine and energy (Ross and Enriquez, 1969; Ravindran et al., 1986). Supplemental methionine serves both to overcome the inherent deficiency of sulphur-containing amino acids per se and provides a readily available source of labile sulphur for cyanide detoxification (Maner and Gomez, 1973). Ravindran (1986) studied the effects of supplementing cassava leaf meal based diets with methionine and coconut oil.

The performance of pigs on rations containing 10% cassava leaf meal was significantly improved by methionine and energy supplementation (Table 10.6).

Table 10.5. Effects of Replacing Coconut Meal by Cassava Leaf Meal in Rations for Growing Swine

	Level of Cassava Leaf Meal (%)			
	0	13	26	40
Performance				
Daily Gain (kg)	0.38	0.44	0.39	0.32
Feed Efficiency	3.30	2.80	3.20	3.90
Apparent Digestibility (%)				
Dry Matter	78	75	71	66
Cell Wall	69	64	59	53
Energy	83	80	76	73
Nitrogen	49	49	47	44
Apparent Protein Utilization (%)				
Net Protein Utilization	49	50	48	44
Biological Value	62	65	67	66

Ravindran et al., 1987b.

Table 10.6. Effects of Methionine and Energy Supplementation on the Utilization of Cassava Leaf Meal (CLM) by Growing Swine

	Daily Gain (kg)	Feed Efficiency
Control (Corn + Soybean Meal)	0.54	2.6
10% CLM	0.46	3.0
10% CLM + 0.2% Methionine	0.48	3.0
10% CLM + 0.2% Methionine + 1.5% Oil	0.53	2.6

Ravindran, 1986.

Breeding Stock

There is unexplored potential for the use of cassava leaf meal as a source of protein in rations fed to breeding stock. Evidence suggests that the energy in fibrous feedstuffs, such as alfalfa meal, is well utilized by sows (Danielson and Noonan, 1975; Pollman et al., 1979; Danielson, 1982). Higher levels of up to 40% cassava leaf meal may be safely fed to sows in gestation.

SUMMARY

The available data indicate that the potential of cassava leaf meal as a swine feed in the tropics is too great to be ignored. Low level usage of cassava leaf meal in swine rations can be successfully accomplished, but awareness of proper processing techniques is essential. The use of cassava leaf meal is not recommended for starter pigs due to its high fiber content. However, properly processed cassava leaf meal could be used to replace up to 66% of coconut meal (or 26% of the total ration) in balanced rations fed to growing-finishing swine. In corn-soybean meal rations, cassava leaf meal may be included up to 10% level, provided that care is taken to balance for energy and methionine. Higher levels of up to 40% cassava leaf meal may be safely fed to sows in gestation.

Cassava leaf meal represents an untapped resource that could be developed into a feed with all the impact of alfalfa meal in temperate countries. The economics of collection and processing make cassava leaf meal less desirable at present, but with escalating prices of traditional protein supplements, its eventual use in tropical swine rations seems inevitable.

REFERENCES

Ahmad, M.I., 1973. Potential fodder and tuber yield of two varieties of tapioca. Malaysian Agric. J. 49: 166-174.

Alhassan, W.S. and Odoi, F., 1982. Use of cassava leaf meal in diets for pigs in the humid tropics. Trop. Anim. Health Prod. 14: 216-218.

Allen, R.D., 1984. Feedstuffs Ingredient Analysis Table. Feedstuffs 56(30): 25-30.

Caldwell, M.J. and Enoch, I.C., 1972. Riboflavin content of Malaysian leaf vegetables. Ecol. Food Nutr. 1: 309-312.

Coursey, D.G., 1973. Cassava as a food: Toxicity and technology. In: B.L. Nestel and R. MacIntyre, eds. Chronic Cassava Toxicity. International Development Research Centre, Ottawa, Monograph No. 010e pp. 27-36.

Dahniya, M.T., Oputa, C.O. and Hahn, S.K., 1981. Effects of harvesting frequency on leaf and root yields of cassava. Exp. Agric. 17: 91-95.

Danielson, M., 1982. Utilization of alfalfa in swine diets. Feed Management 33: 30-33.

Danielson, M. and Noonan, J.J., 1975. Roughages in swine gestation diets. J. Anim. Sci. 41: 94-99.

Eggum, O.L., 1970. The protein quality of cassava leaves. Brit. J. Nutr. 24: 761-769.

FAO, 1972. Food Composition Table for Use in East Asia. Food and Agriculture Organization, Rome and U.S. Dept. of Health, Education and Welfare, Washington, D.C.

Gomez, G. and Valdivieso, M., 1984. Cassava for animal feeding: Effect of variety and plant age on production of leaves and roots. Anim. Feed Sci. Technol. 11: 49-55.

Hutagalung, R.I., Jalaludin, S. and Chang, C.C., 1974. Evaluation of agricultural products and by-products as animal feeds. II. Effects of levels of dietary cassava leaf and root meals on performance, digestibility and body composition of broilers. Malaysian Agric. Res. 3: 49-59.

Lancaster, P.A. and Brooks, J.E., 1983. Cassava leaves as human food. Econ. Bot. 37: 331-348.

Lancaster, P.A., Ingram, J.S., Lim, M.Y. and Coursey, D.G., 1982. Traditional cassava-based foods: Survey of processing techniques. Econ. Bot. 36: 12-45.

Lee, K.C. and Hutagalung, R.I., 1972. Nutritional value of tapioca leaf for swine. Malaysian Agric. Res. 2: 38-47.

Mahendranathan, T., 1971. The effect of feeding tapioca (*Manihot utilissima* Pohl) leaves to pigs. Malaysian Agric. J. 48: 60-68.

Maner, J.H. and Gomez, G., 1973. Implications of cyanide toxicity in animal feeding studies using high cassava rations. In: B.L. Nestel and R. MacIntyre, eds. Chronic Cassava Toxicity. International Development Research Centre, Ottawa, Monograph No. 010e, pp. 113-120.

Montaldo, A. and Montilla, J.J., 1976. Production of cassava foliage. Proc. 4th Symp. Intern. Symp. Tropical Root Crops, Cali, Columbia, pp. 142-143.

Montilla, J.J., 1976. Utilization of whole cassava plant in animal feed. Proc. 1st Intern. Symp. Feed Composition, Animal Nutrient Requirements and Computerization of Diets. Logan, Utah, pp. 98-104.

Moore, C.P., 1976. The utilization of cassava forage in ruminant feeding. Proc. Int. Symp. Tropical Livest. Prod. Acapulco, Mexico, 21 pp.

Normanha, E.S., 1962. Meal of stalks and leaves of cassava. Agronomica (Venezuela) 14: 16-19.

Oke, O.L., 1990. Cassava meal. In: P.A. Thacker and R.N. Kirkwood, eds. Nontraditional Feed Sources for Use in Swine Production, Butterworths, Boston, pp. 103-112.

Pollman, D.S., Danielson, M.D. and Peo, J.R., 1979. Value of high fibre diets for gravid swine. J. Anim. Sci. 48: 1385-1393.

Price, M.L. and Butler, L.G., 1980. Tannins and Nutrition. Purdue University Agriculture Experimental Station Bulletin No. 272, West Lafayette, Indiana.

Rajaguru, A.S.B., Ravindran, V. and Ranaweera Banda, R.M., 1979. Manioc leaf meal as a source of protein for fattening swine. J. Natl. Sci. Coun. Sri Lanka 7: 105-110.

Ramos-Ledon, L.J. and Popenoe, J., 1970. Comparative chemical composition of cultivars of Manihot esculenta Crantz and some related species. Proc. Trop. Reg. Amer. Soc. Hort. Sci. 14: 232-234.

Ravindran, V., 1985. Development of Cassava Leaf Meal as an Animal Feed. Ph.D. Thesis, Virginia Tech. University, Blacksburg, U.S.A. 129 pp.

Ravindran, V., 1986. Perspectives on the use of non-conventional feed resources. Seminar on Utilization of Fibrous Feed Resources in Animal Feeding. Digana, Sri Lanka 10 pp.

Ravindran, G. and Ravindran, V., 1988. Changes in the nutritional composition of cassava (*Manihot esculenta* Crantz) leaves during maturity. Food Chem. 27: 299-309.

Ravindran, V., Kornegay, E.T., Webb, K.E., Jr. and Rajaguru, A.S.B., 1982. Nutrient characterization of some feedstuffs of Sri Lanka. J. Natl. Agric. Soc. Ceylon 19: 19-32.

Ravindran, V., Kornegay, E.T. and Cherry, J.A., 1983. Feeding values of cassava tuber and leaf meals. Nutr. Rep. Int. 28: 189-196.

Ravindran, V., Kornegay, E.T., Rajaguru, A.S.B., Potter, L.M. and Cherry, J.A., 1986. Cassava leaf meal as a replacement for coconut oil meal in broiler diets. Poult. Sci. 65: 1720-1727.

Ravindran, V., Kornegay, E.T. and Rajaguru, A.S.B., 1987a. Influence of processing methods and storage time on the cyanide potential of cassava leaf meal. Anim. Feed Sci. Technol. 17: 227-234.

Ravindran, V., Kornegay, E.T., Rajaguru, A.S.B. and Notter, D.R., 1987b. Cassava leaf meal as a replacement for coconut oil meal in pig diets. J. Sci. Food Agric. 41: 45-53.

Reed, J.S., McDowell, R.E., Van Soest, P.J. and Horvath, P.J., 1982. Condensed tannins: A factor limiting the use of cassava forage. J. Sci. Food Agric. 33: 213-220.

Rogers, D.J., 1959. Cassava leaf protein. Econ. Bot. 13: 261-263.

Rogers, D.J. and Milner, M., 1963. Amino acid profile of manioc leaf protein in relation to nutritive value. Econ. Bot. 17: 211-216.

Ross, E. and Enriquez, F.Q., 1969. The nutritive value of cassava leaf meal. Poult. Sci. 48: 846-853.

Wyllie, D. and Chamanga, P.J., 1979. Cassava leaf meal in broiler diets. Trop. Anim. Prod. 4: 232-240.

Yeoh, H.H. and Chew, M.Y., 1976. Protein content and amino acid composition of cassava leaf. Phytochemistry 15: 1597-1599.

Yeoh, H.H. and Oh, H.Y., 1979. Cyanide content of cassava. Malaysian Agric. J. 52: 24-28.

CHAPTER 11

Cassava Meal

O.L. OKE

INTRODUCTION

Cereal grains constitute the most important class of feedstuff used to supply energy to pigs. Therefore, it is not surprising that in developed countries such as the United States, about 80% of the corn produced is used as animal feed. However, in the developing countries, most of the corn goes for human food and little is used as animal feed (Oke, 1975). As a consequence, alternative sources of energy must be developed for use in livestock production. One feedstuff that has attracted a considerable amount of attention as an energy source is cassava meal (also called manioc, tapioca and yucca).

GROWING CASSAVA

Cassava (*Manihot esculentis* Crantz) is a perennial woody shrub which is grown almost entirely in the tropics. It is highly adaptable to the infertile tropical soil and can produce a yield of about six tonne/ha without much input. Under intensive cultivation, yields of 20 to 30 tonne/ha are possible. Once established, cassava becomes resistant to drought and insect pests, especially locusts which devastate farm lands in the tropics.

Traditionally, cassava has been regarded as a low-grade subsistence crop or famine relief crop. However, cassava is one of the world's most productive farm crops, yielding edible nutrients equivalent to an average of over 13 million kcal/acre, compared to 9 million for yams and 1 million each for guinea corn and maize. World production is about 120 million tonne a year. With a moisture content of 65%, this is equivalent to about 42 million tonne of dry matter or 40-50 million tonne of grain per annum.

NUTRIENT CONTENT OF CASSAVA

About 65% of the cassava root is made up of water, but the actual level depends on variety, age at harvest, soil type, climatic conditions and the health of the plant (Oke, 1967). Of the dry matter, about 90% is made up of starch and sugar, giving cassava a digestible energy content of about 4,000 kcal/kg dry matter, which is very similar to that of maize (4055 kcal/kg). Because of the fibrous peel, the crude fiber and ash content are much higher than those of maize and this may lead to lower digestibility.

Unlike maize, cassava contains only small amounts of protein (up to 2.6% of the dry matter; Table 11.1). Only about 50 to 60% of this is true protein, the rest being made up of nitrates, nitrites, cyanide and free amino acids (Oyenuga, 1955). In addition, cassava protein is deficient in the sulphur containing amino acids (Table 11.2). Therefore, since the amount of protein is so low and the quality is poor, it is likely wise to ignore the potential contribution of cassava protein as a source of amino acids in practical diet formulation.

Table 11.1. Chemical Analysis of Cassava Root Products (% as fed)

	Cassava Chips	Cassava Meal	Cassava Refuse	Cassava Flour
Moisture	11.7	11.2	80.0	14.9
Crude Protein	1.9	2.6	0.4	0.3
Crude Fiber	3.0	5.6	1.6	0.1
Ether Extract	0.7	0.6	0.1	0.1
Ash	2.1	6.1	0.3	0.2

Lim, 1978.

The mineral content of cassava is very low (Table 11.3). In addition, the presence of factors in cassava may directly or indirectly affect the mineral balance of the ration. Maust et al. (1972) fed pigs maize-based diets or diets in which 36% of the maize was replaced by cassava. Pigs on the cassava-based diets developed parakeratosis during the fourth week but this was eliminated by addition of $ZnCO_3$ (even when the zinc content met the requirement).

Table 11.2. Amino Acid Content of Cassava Meal and Corn (% as fed)

	Corn	Cassava Meal
Arginine	0.40	0.18
Cystine	0.15	0.02
Histidine	0.26	0.03
Isoleucine	0.35	0.05
Leucine	1.19	0.06
Lysine	0.25	0.07
Methionine	0.18	0.02
Phenylalanine	0.46	0.04
Threonine	0.34	0.04
Tryptophan	0.07	0.02
Valine	0.46	0.05

Oke, 1984.

Table 11.3. Mineral Content of Cassava Meal and Corn (% as fed)

	Corn	Cassava Meal
Calcium	0.10	0.13
Magnesium	0.12	0.04
Phosphorus	0.41	0.15
Potassium	0.38	1.38

Oke, 1967.

Based on the foregoing discussion, it should be evident that cassava meal should be regarded only as a source of dietary energy. Therefore, successful ration formulation using cassava meal is dependent on supplementation with a very good source of protein, adequate minerals and quality vitamins.

UNDESIRABLE CONSTITUENTS IN CASSAVA

A major factor limiting the use of cassava meal as an animal feed is the high content of cyanide in the tuber. Cassava contains the cyanogenic glucoside linamarin (and smaller quantities of its closely related compound lotanstralin), which on ingestion are hydrolysed to hydrocyanic acid (cyanide). The normal range of cyanide in cassava is about 15 to 500 mg/kg fresh weight. Cyanide is one of the most toxic substances known and could result in death if taken in acute doses. However, at chronic dose levels, the body has an efficient mechanism for the detoxification of the cyanide to the much less toxic thiocyanate (Oke, 1969). Since the sulphur needed for the detoxification of cyanide is obtained from methionine in the ration, the presence of cyanogenic glucosides could lead to a deficiency of this essential amino acid if poor or marginal sources of methionine like oilseeds are used as the main source of supplemental protein. This deficiency will reduce animal performance (Oke, 1978).

Thiocyanate impedes the thyroid uptake of iodine, leading to increased excretion of iodine by the kidney, lowering of thyroxine and increasing thyroid-stimulating hormone. Affected female pigs seem to have lower thyroxine than males, demonstrating the female's greater sensitivity to cyanide (Tewe et al., 1984). This may create problems during pregnancy if adequate iodine is not provided in the ration. In addition, thiocyanate has been found to cross the placental barrier in pigs and so it can affect the fetus, leading to lower performance of the breeding herd as well as reducing overall productivity (Tewe and Maner, 1981).

PROCESSING CASSAVA MEAL

Since cassava contains cyanogenic glucosides which may cause toxicity as well as retard performance, the glucosides need to be removed prior to feeding. Detoxification can be effected by several methods. According to Okeke et al. (1985), the minimum processing times needed to achieve safe cyanide levels for monogastric animals are 6.95 minutes, 20.16 minutes, 15.38 hours, 3.24 days and 7.61 days for boiling, roasting, soaking, ensiling and sun-drying, respectively. The residual effect of the remaining cyanide can be overcome by methionine supplementation.

Cassava meal tends to produce a powdery diet which Henry (1971) reported to be ulcerogenic to the gastric mucosa. This can be overcome by pelleting. Pelleting has the added advantages of eliminating irritation to the respiratory organs, reducing lesions in the cardiac region as well as reducing eye infections which can be caused by powdery feed. Furthermore, the heat generated by the steam and high pressure during pelleting releases the cellulose from the lignin-cellulose bond, thereby increasing the digestibility of the starch and fiber. This also causes destruction of growth inhibitors of microbial origin (molds, fungi and other contamination during sun drying or storage). Pelleting of cassava-based diets is therefore highly recommended.

Problems with dust can also be overcome by the addition of up to 8% palm oil to form a paste (Hew, 1975). The effect of palm oil may be due to improved palatability and better absorption or retention of other nutrients.

FEEDING CASSAVA MEAL TO SWINE

Starter Pigs

There has been very little work conducted on the use of cassava meal as an energy source for starter pigs. Aumaitre (1969) compared corn, wheat, barley, dehulled oats and cassava meal in 20% protein diets fed to 5 to 9 week old pigs. Pigs fed the cassava meal diet gained more rapidly (416 g/day) than pigs fed barley (386 g/day), oats (380 g/day), corn (354 g/day) or wheat (360 g/day). However, due to potential problems with palatability, it might be wise to limit the use of cassava meal in starter diets. A maximum of 45% cassava meal in a pig starter ration has been suggested (Muller et al., 1972).

Growing-Finishing Pigs

Sonaiya and Omole (1983) conducted a study to determine the effects of using cassava meal in diets fed to growing pigs. Their control diet was based on corn and was supplemented with groundnut cake, fish meal and brewer's dried grains. The experimental diets, containing 15 to 60% cassava meal, were formulated to be isocaloric and isonitrogenous by altering the amount of corn, groundnut cake, fish meal and brewer's dried grain in the diet. The overall results of this experiment indicate that cassava meal can replace as much as 60% of the corn in a growing ration without any adverse effects on performance (Table 11.4).

Table 11.4. Performance of Growing Pigs (9-40 kg) Fed Cassava Meal

	Level of Cassava Meal (%)				
	0	15	30	45	60
Daily Gain (kg)	0.40	0.48	0.50	0.54	0.48
Daily Intake (kg)	1.40	1.40	1.19	1.40	1.39
Feed Efficiency	3.47	2.93	2.40	2.60	2.92

Sonaiya and Omole, 1983.

Gomez et al. (1984) reported that when cassava was substituted for corn in a swine ration, the performance progressively improved as the level of cassava in the feed was increased. Thus, it required 339 kg of feed to produce 100 kg weight with corn alone whereas it required 337 or 331 kg, respectively, with 20 and 30% cassava substitution (Table 11.5).

Some workers have reported better digestibility of cassava-based diets for swine than of cereal-based diets. Partridge (1985) reported ileal and fecal digestibility of dry matter, protein and energy in diets containing 15 and 30% cassava meal compared with corn. There was little difference between the control and 15% cassava diets for either ileal or fecal digestibility. However, the 30% diet generally had lower ileal digestibilities and higher fecal digestibilities than did the other two diets (Table 11.6). This indicates that a greater proportion of the digestion of dry matter, protein and energy occurs in the large intestine and would be of questionable value to the pig.

Table 11.5. Performance of Growing-Finishing Pigs (20-90 kg) Fed Least-Cost Diets with Varying Levels of Cassava Meal

	Level of Cassava Meal (%)		
	0	20	30
Daily Gain (kg)	0.77	0.82	0.78
Daily Intake (kg)	2.55	2.77	2.54
Feed Efficiency	3.39	3.37	3.31

Gomez et al., 1984.

Table 11.6. Ileal and Fecal Digestibility Coefficients (%) for Dry Matter, Crude Protein and Energy for Pigs Fed Cassava or Corn

	Level of Cassava (%)		
	0	15	30
Dry Matter (Ileum)	0.77	0.78	0.73
Dry Matter (Overall)	0.82	0.84	0.84
Energy (Ileum)	0.78	0.80	0.75
Energy (Overall)	0.81	0.83	0.84
Crude Protein (Ileum)	0.80	0.80	0.75
Crude Protein (Overall)	0.81	0.81	0.83

Partridge, 1985.

Because of the low levels of methionine in cassava, methionine is the first limiting amino acid in a cassava-based ration. As a consequence, pig performance will be significantly better if a methionine-rich protein supplement such as fish meal is used in the diet in comparison with methionine-poor soybean meal (Hew and Hutagalung, 1972).

Apart from its function as an essential amino acid, methionine also acts as a sulphur donor for the detoxification of cyanide. As a consequence, the requirement for methionine in the diet is increased. Portela and Maner (1972) reported that the performance of pigs fed cassava-based diets could be improved by the addition of methionine and fat to the diet (Table 11.7).

The reported effects of feeding cassava meal on carcass quality are conflicting. Kok and Ribeiro (1943) and Modebe (1963) observed an increase in fat deposition and a reduction in loin eye area as a result of inclusion of cassava in the diet of finishing pigs. However, Oyenuga and Opeke (1957) and Oyenuga (1961) did not observe any deleterious effects. More recently, Obioha (1988) substituted 100% of the maize in a pig ration with cassava peel without any decline in performance. Indeed, the whole-cassava based feed was superior in dressing percentage and trimness to the whole-maize ration, with feed cost savings at higher levels of cassava supplementation. Similar results have been reported by Sonaiya and Omole (1977).

Table 11.7. Effect of Methionine and Fat Supplementation on the Utilization of Cassava Meal by Growing Pigs

	Daily Gain (kg)	Feed Efficiency
Control Corn + Soybean Meal	0.74	2.81
Cassava Meal (55%)	0.76	2.65
Cassava Meal + 0.1% DL-Methionine	0.83	2.46
Cassava Meal + 0.2% DL-Methionine	0.82	2.49
Cassava Meal + 10% Beef Tallow	0.85	2.24
Cassava Meal + Tallow + 0.1% DL-Methionine	0.80	2.35
Cassava Meal + Tallow + 0.2% DL-Methionine	0.74	2.21

Portela and Maner, 1972.

Muller et al. (1972) suggested that the maximum amount of cassava in a diet should depend on the ash content; if the ash content is less than 2.2% (and crude fiber 2.8%) then the maximum amount of cassava would be 60% for growers and 75% for finishers. The corresponding figures for an ash content greater than 5% (crude fiber 5%) should be 20 and 40%, respectively.

Breeding Stock

There is limited information on the use of cassava meal in diets fed to breeding stock. Gomez et al. (1984) conducted a trial in which gilts were fed either a corn-soybean meal control diet, a cassava meal-soybean diet or a similar cassava based diet supplemented with 0.3% methionine. During gestation, the gilts received 1.8 kg of diet per day while they were fed ad libitum in lactation. The results of this experiment indicate that cassava meal can be fed to sows in gestation and lactation without any adverse effects on performance (Table 11.8).

Table 11.8. Performance of Gilts Fed Cassava Meal in Gestation and Lactation

	Control	Cassava	Cassava + Methionine
Pigs Born Alive	8.5	9.1	9.4
Birth Weight (kg)	1.1	1.1	1.1
Pigs Weaned	7.1	8.2	8.0
Weaning Weight (kg at 56 days)	16.7	16.2	16.5
Preweaning Mortality (%)	16.4	9.9	14.9
Sow Lactation Feed Intake (kg)	267	321	312

Gomez et al., 1984.

SUMMARY

As pointed out, since cassava has a negligible amount of protein, the economics of its replacing maize in a swine feed will depend not only on the price of cassava with respect to maize, but also on the protein supplement that will be needed to balance the cassava-based diet. The ultimate goal will therefore vary from one country to another. Most of the present studies indicate that a 40% incorporation into the ration is economically feasible and so this level is recommended. This coincides with the level that is presently used by countries in the European Economic Community that import cassava from Thailand for use as animal feed.

REFERENCES

Aumaitre, A., 1969. Valeur alimentaire du manioc et de differentis cereales dans les regime las croissance des animaux. Annals Zootech. 18: 385-398.

Gomez, G., Santos, J. and Valdivieso, M., 1984. Evalualtion of methionine supplementation to diets containing cassava meal for swine. J. Anim. Sci. 58: 812-820.

Henry, V., 1971. Effects nutritionnels de l'incorporation de cellulose purifice dans le regime du porcen croissance. Annals Zootech. 19: 117-118.

Hew, V.F., 1975. The effect of some local carbohydrate sources such as cassava and sugar on performance and carcass characteristics of growing-finishing pigs. M.Sc. Thesis, University of Malaya.

Hew, V.F. and Hutagalung, R.I., 1972. The utilization of tapioca root meal in swine feeding. Malaya Agric. Res. 1: 124-130.

Kok, E.A. and Ribeiro, G.A., 1943. Cassava meal compared with maize meal for fattening pigs. Nutr. Abst. Rev. 18: 5129.

Lim, H.K., 1978. Composition data for feeds and concentrates. Malaysian Agric. J. 46: 63-79.

Maust, L.E., Pond, W.G. and Scott, W.L., 1972. Energy value of a rice bran-cassava meal diet with and without supplemental zinc for pigs. J. Anim. Sci. 35: 953-957.

Modebe, A.N.A., 1963. Preliminary trial on the value of dried cassava (*Manihot utilissima*) for pig feeding. J. West African Sci. Assoc. 7: 127-133

Muller, Z., Chou, K.C., Nah, K.C. and Tan, T.K., 1972. Study of nutritive value of cassava in economic ration for growing finishing pigs in the tropics. In: UNDP/SF Project SIN 67/105, Pig and Poultry Research and Training Institute, Singapore. 672: 1-35.

Obioha, P.C., 1988. Alternative Sources of Feeds. Workshop held in Ilorin, Nigeria. Federal Government Task Force.

Oke, O.L., 1967. The present state of nutrition in Nigeria. World Rev. Nutr. Diet. 8: 25-61.

Oke, O.L., 1969. The role of hydrocyanic acid in nutrition. World Rev. Nutr. Diet. 11: 170-198.

Oke, O.L., 1975. A case for vegetable protein in developing countries. World Rev. Nutr. Diet. 23: 259-295.

Oke, O.L., 1978. Problems in the use of cassava as animal feed. Anim. Feed Sci. Technol. 3: 345-380.

Oke, O.L., 1984. The use of cassava as pig feed. Nutr. Abst. Rev. 54: 301-314.

Okeke, G.C., Obioha, P.C. and Udeagu, A.E., 1985. Processing of cassava. Nutr. Rep. Int. 32: 139-145.

Oyenuga, V.A., 1955. Nigerian feedstuffs. Ibadan University, Ibadan, Nigeria.

Oyenuga, V.A., 1961. Nutritive value of cereal and cassava diets for growing and fattening pigs in Nigeria. Brit. J. Nutr. 15: 327-338.

Oyenuga, V.A. and Opeke, L.K., 1957. The value of cassava rations for pork and bacon production. West African J. Biol. Chem. 1: 3-7.

Partridge, I.G., 1985. The digestion of diets containing manioc by young growing pigs. Anim. Feed Sci. Technol. 12: 119-123.

Portela, J. and Maner, J.H., 1972. In: B.L. Nestel and R. MacIntyre, eds. Chronic Cassava Toxicity. International Development Research Centre, London, Monograph No. 010e, pp. 113-120.

Sonaiya, E.B. and Omole, T.A., 1977. Cassava peel for finishing pigs. Nutr. Rep. Int. 16: 479-486.

Sonaiya, E.B. and Omole, T.A., 1983. Cassava meal and cassava peel meal in diets for growing pigs. Anim. Feed Sci. Technol. 8: 211-220.

Tewe, O.O., Afolabi, O.A., Grissom, F.E., Littleton, G.K. and Oke, O.L., 1984. Effect of varying dietary cyanide levels on serum thyroxine and protein metabolites in pigs. Nutr. Rep. Int. 30: 1249-1253.

Tewe, O.O. and Maner, J.K., 1981. Performance and pathophysiological changes in pregnant pigs fed cassava diets containing different levels of cyanide. Res. Vet. Sci. 30: 142-151.

CHAPTER 12

Cocoyams

E. Nwokolo

INTRODUCTION

The cocoyam (also called taro or tannia), is well known in most tropical and subtropical countries and is found throughout Africa, Asia, the Pacific region, the Carribean and South America. In these areas, the cocoyam constitutes a major source of calories in the human diet. However, in the last decade, there has been a large increase in the utilization of cocoyams in swine nutrition. The cocoyams most commonly used are poor quality corms not suitable for human consumption or the leaves which are fed in the form of silage.

COCOYAM PRODUCTION

Cocoyam Corms

The cocoyam is an herbaceous perennial, reaching 1.5 to 2.0 meters in height, possessing an above ground stem with a short internode (O'Hair, 1984). The leaves are very large and may be sagittate or hastate. The plant is borne on an enlarged, starchy corm beneath the ground. The weight of the corm usually varies from 0.2 to 5 kg, although a mother corm, with several lateral corms (cormels), may weigh up to 10 kg.

The cocoyam is a tropical crop that grows very well at daily average temperatures of 21° to 27°C (de la Pena, 1983) but may withstand temperatures approaching but not below 0°C (O'Hair, 1984). Cocoyams grow well in diverse soil conditions ranging from upland, well drained soils to waterlogged, soggy soil types in high rainfall areas (Onwueme, 1978). About 62% of world production of cocoyam is in Africa, 32% in Asia, 6% in the Pacific regions and less than 1% in the Carribean and South America. Nigeria is the largest producer of cocoyam in Africa, followed by Ghana (FAO, 1985).

Cocoyams are usually planted in small farm units but may be established on a large scale in commercial plantations. Large scale production of cocoyams may require irrigation and considerable application of fertilizer. Plant spacing affects total yield per plot (de la Pena, 1983). A planting density of 100,000 plants per hectare yields about 52 metric tonnes of corms per hectre while a planting density of 10,000 plants per hectare yields 34 metric tonnes per hectare. Average yields of cocoyam from farmer's plots range from a low of 5.5 metric tonnes per hectare in unfertilized and unirrigated plots in Africa and Asia (FAO, 1985), to a high of 31 metric tonnes per hectare in fertilized and irrigated plots in Hawaii (de la Pena, 1983). Potential yield from cocoyam approximates the expected yield from cassava in irrigated, fertilized plots (30 MT/ha) and establishes cocoyam as a tuber crop with a great potential for commercial exploitation in most tropical countries.

Cocoyam Silage

The cocoyam is a highly vegetative plant and by increasing the ratio of nitrogen in the fertilizer, tremendous growth of leaves, petioles and stem can be achieved at the expense of corm development. Carpenter and Steinke (1983) assumed a moderate yield of 2 kg of cocoyam tops (stem, leaves and petiole) per plant, harvested every 31 weeks from fertilized plots and calculated a realistic yield of 52.5 metric tonnes of fresh herbage per hectare per year.

NUTRIENT CONTENT OF COCOYAM

The nutrient composition of cocoyam corms, leaves and silage is presented in Table 12.1. Cocoyam corms are essentially carbohydrate sources in human and animal diets. Their carbohydrate content varies from 6.5% in immature fresh corms (Rashid and Daunicht, 1979) to 32.5% in mature fresh corms (Watt and Merrill, 1963). Dried corms have a carbohydrate content of between 82 and 83.5% (Chowdhury and Hussain, 1979). The starch granules are small and appreciably more waxy than other small-grain starches observed in cereal sources (Goering, 1971; Goering and De Haas, 1972). Payne et al. (1941) noted that starch granules from cocoyam are about one-tenth the size of those from potato, a possible explanation for the extremely high digestibility (98.2%) reported by Langworthy and Duel (1922) for cocoyam starch. Szylit and co-workers (1977) demonstrated that starch grains from cocoyam are less than 1 micron in diameter while those from cassava are 12 microns. Cocoyam silage is a good source of dietary nutrients, although it is unfortunately very low in dry matter content. Dry matter is generally very low (8.5-9.5%) and protein content moderate (14-18%).

Table 12.1. Nutrient Composition of Fresh Corms, Mature Dried Corms, Leaves and Silage of Cocoyam (as fed)

	Fresh Corms	Dried Corms	Leaves	Silage
Moisture (%)	64.4	10.5	89.0	91.0
Protein (%)	2.2	3.1	2.5	1.7
Fat (%)	0.2	0.7	1.0	0.5
Carbohydrate (%)	32.0	82.3	5.3	---
Fiber (%)	1.0	2.4	2.1	---
Ash (%)	1.2	2.2	1.3	---
Calcium (mg/100g)	16	480	95	---
Phosphorus (mg/100g)	47	338	130	---
Iron (mg/100g)	0.9	---	2.0	---
Ascorbic acid (mg/100g)	8.0	---	37	---

O'Hair, 1984; Fetuga and Oluyemi, 1976; Carpenter and Steinke, 1983.

The amino acid composition of both cocoyam corm and cocoyam silage are shown in Table 12.2. The concentration of the essential amino acids are several times higher in cocoyam silage than in cocoyam corms. Cystine and methionine are very low in cocoyam corm and silage, but lysine is three times

Table 12.2. Amino Acid Content (% of DM) of Cocoyam Corm and Silage

	Cocoyam Corm[1]	Cocoyam Silage[2]
Arginine	0.35	1.02
Histidine	0.07	0.24
Isoleucine	0.13	0.65
Leucine	0.29	1.07
Lysine	0.15	0.51
Methionine + Cystine	0.15	0.15
Phenylalanine	0.20	0.62
Threonine	0.16	0.46
Tryptophan	0.05	----
Tyrosine	0.14	0.11
Valine	0.24	0.65

[1]Standal, 1983.
[2]Wang, 1983.

higher in the silage than in the corm. The sulfur-amino acids constitute 4% of all amino acids while the essential amino acids constitute 41% of all amino acids. In an excellent protein source like soybean meal, the sulfur amino acids also constitute about 3 to 4% of all amino acids but the essential amino acids constitute about 53% of total amino acids. For effective utilization of cocoyam corms and silage in swine nutrition, a fairly high level of supplementation of cocoyam-based diets with lysine and with sulfur-amino acids is necessary. Supplementation with lysine may be achieved from synthetic sources or from the addition of low levels of fish meal to cocoyam-based diets.

The mineral and vitamin content of cocoyam corms and cocoyam silage, shown in Table 12.3, indicate that silage is a significantly better source of many minerals than cocoyam corm. Similarly, cocoyam leaves are a better source of vitamins than cocoyam corms. However, there is a high content of oxalic acid or calcium oxalate in the leaves, petioles and corms of cocoyam and oxalic acid is known to interfere with mineral absorption.

Table 12.3. Mineral and Vitamin Content of Cocoyams (DM Basis)

	Corm[1]	Leaf[1]	Silage[2]
Minerals			
Calcium (%)	0.15	1.31	3.15
Phosphorus (%)	0.52	0.47	0.50
Iron (%)	0.00	0.06	0.00
Sodium (%)	0.03	----	----
Potassium (%)	2.04	----	3.36
Magnesium (%)	----	----	0.88
Copper (mg/kg)	----	----	17.90
Zinc (mg/kg)	----	----	166
Vitamins			
Carotene (IU/kg)	67	28,550	----
Thiamine (mg/kg)	0.09	0.22	----
Riboflavin (mg/kg)	0.03	0.26	----
Niacin (mg/kg)	0.40	1.10	----
Ascorbic acid (mg/kg)	0.00	12	----

[1]Gopalan et al., 1977.
[2]Wang, 1983.

UNDESIRABLE CONSTITUENTS IN THE COCOYAM

A major hinderance to the utilization of cocoyam as an animal feed in the raw state is the presence of an irritating substance in the corms, leaves and petiole of the plant (Tang and Sakai, 1983). When the raw corm, leaf or petiole comes in contact with the skin, a burning sensation follows immediately. The sensation may persist for up to 15 minutes and is primarily responsible for refusal of raw cocoyam by swine and other monogastric animals but not by sheep or goats. The chemical compound that causes acridity in cocoyam is still unknown in spite of almost 50 years of attempts to discover its composition. O'Hair (1984) reported that in highly colored cultivars of cocoyam, some water-soluble saponins or glucosaponins may be found which are toxic to animals consuming these cocoyam cultivars.

Whatever the nature of the toxic substance causing acridity in cocoyam, cooking, heating and fermentation of cocoyam seem effective in improving the nutritional quality of this feed ingredient. Prolonged cooking is necessary prior to human consumption of cocoyam corms, the combination of high temperature (100°C) and long duration (about 1 hour) being necessary to inactivate the toxic components in cocoyam. Fermentation also inactivates the active principle. Murai and co-workers (1958) observed that once acridity is destroyed or removed, cocoyam is an excellent carbohydrate food and a source of minerals and vitamins.

The high level of calcium oxalate crystals found in cocoyam has also been implicated in causing the acridity of cocoyam. A physical explanation for acridity is thought to be the presence of calcium oxalate monohydrate crystals existing in needle-like monoclinic forms and calcium oxalate polyhydrate in tetragonal form. It has been speculated that raphides, which are needle-like crystals of calcium oxalate, or a chemical toxin within the raphide idioblast, may be responsible for acridity in cocoyam and other edible aroids (Tang and Sakai, 1983).

PROCESSING OF COCOYAMS FOR SWINE FEEDING

Cocoyam Corms

Corms destined for utilization as swine feed need to be cooked prior to feeding. Lateral corms are in demand as human food and therefore it is the central mother corms which are usually processed into swine feed. For processing, the mother corms are cleaned of all adhering soil and the roots clipped. They are washed, sliced into 1 cm thick pieces that are roughly 5 cm by 5 cm. These are packed in woven baskets and immersed in drums or vats of boiling water. Cooking lasts from 45 minutes to 1 hour, after which the basket is removed, dipped in a drum of fresh or running water and the contents spread out to dry. Drying may be achieved in the sun, by the use of a solar dryer or in a forced-drought oven. Drying is not necessary if the corms are mashed and fed as a slop. Cooked dried cocoyam slices may subsequently be milled. Cooked, dried cocoyam slices have a storage life of many months when kept in jute bags under warehouse conditions, and are similar to cassava chips.

In a field test conducted between 1984 and 1985, we experimented with feeding first raw and subsequently cooked cocoyams to growing hogs on a swine farm in Port Harcourt, Nigeria. Raw cocoyams were fed in replacement of 50% of grains, to four pens of growing-fattening pigs (15 per pen). Growth rate was reduced and the incidence of scours increased when the pigs were fed raw cocoyam. When the cocoyams were cooked prior to feeding, normal growth was achieved and scouring stopped, indicating a relationship between feeding raw cocoyam and the poor growth accompanied by scouring.

Cocoyam Silage

Cocoyam tops are a rich source of soluble carbohydrates and protein, which are ingredients necessary for the production of good quality silage. The leaves, stem and petiole of cocoyam may be ensiled alone but usually, their dry matter content is too low. This problem can be solved by ensiling cocoyam tops with other drier-leaf materials, or dry concentrates such as wheat bran, palm kernel cake or spent brewer's grains. When the vegetative parts of cocoyam are ensiled in combination with various dry products, a silage of moderate dry matter content results. Wang (1983) reported that a combination of 50% cocoyam vegetation and 50% sugarcane trash yielded silage of about 25% dry matter and 8 to 9% protein, while 50% cocoyam vegetation and 50% pangola grass resulted in a silage of 24% dry matter and 7% protein. Cocoyam silage may also be produced with guinea grass or rice bran, resulting in a product with 22 to 23% dry matter and a 9 to 12% protein content. Cocoyam silage produced in this way is an excellent feed for swine, supplying a considerable portion of the pig's requirements for nutrients. However, cocoyam silage is not recommended for pigs in the early growing stages, but rather should be fed to pigs in the late growing or fattening stage.

FEEDING COCOYAMS TO SWINE

Cocoyam Corms

There are relatively few experiments providing details of results of feeding cocoyam corms to swine. Thus, while the potential of the crop has been recognized for decades, few actual experiments have been conducted, most of the use being on-farm by small-scale producers determined to cut down feed costs. Oyenuga and Fetuga (1975) reported experiments in which metabolizable energy of cooked cocoyams was determined with young pigs. Metabolizable energy was 3.53 kcal/g, which is similar to the metabolizable energy values reported by the same workers for sweet potato (3.60 kcal/kg).

In a growth study with pigs, Soldevila and Vincente-Chandler (1978) showed that cocoyams could not maintain growth in pigs without a great deal of protein supplementation. That cocoyam used alone cannot sustain growth in pigs is expected, since most tubers and roots are extremely poor sources of protein and need considerable protein supplementation in diets of all domestic animals. Indeed, neither wheat, corn, barley nor oats can sustain reasonable growth in pigs when used as a sole feed.

Cocoyam Silage

Workers from the University of Hawaii have provided most of the data available on the utilization of cocoyam silage as a swine feed. Wang (1983) reported growth data from 31 and 35 day-feeding trials as well as from 42 and 56 day-feeding trials involving young pigs fed silage as 0, 20, 30 or 40% of the dry matter intake. The data from the 31-day and 35-day trials shown in Table 12.4 indicated that cocoyam silage could provide up to 20% of the dry matter intake of growing pigs without any significant depression in growth parameters.

Table 12.4. Performance of Pigs Fed Cocoyam Silage

	Commercial Pig Grower	Taro Silage 20% of DM	Taro Silage 30% of DM	Taro Silage 40% of DM
31-Day Trial				
Number of Pigs	4	4	4	4
Initial Weight (kg)	46.4	47.4	48.2	48.5
Final Weight (kg)	70.0	70.0	67.3	63.2
Daily Gain (kg)	0.76	0.73	0.62	0.47
35-Day Trial				
Number of Pigs	9	9	9	9
Initial Weight (kg)	33.3	32.8	33.0	32.4
Final Weight (kg)	59.4	51.3	52.0	48.7
Daily Gain (kg)	0.62	0.54	0.54	0.47

Wang, 1983.

The effects of inclusion of cocoyam silage in diets of pregnant pigs are demonstrated in Tables 12.5 and 12.6. Cocoyam silage had no influence on duration of gestation but rather diminished weight gain during gestation, preventing overcondition, a potential problem with pregnant sows. The various indices of reproductive performance (litter size, litter weight at birth, average number weaned and average daily gain of litter) were excellent in gilts fed up to 40% cocoyam silage on a dry matter basis. However, when cocoyam silage constituted 70% of the dry matter intake, there was considerable depression in reproductive performance. This could be due to the inability of a high cocoyam silage diet to meet nutrient requirements during gestation.

Table 12.5. Growth Data of Gilts Fed Cocoyams During Gestation

	Commercial Pig Grower	Cocoyam Silage (20% of DM)	Cocoyam Silage (40% of DM)
Number of Days	115.3	113.8	115.8
Weight at Breeding (kg)	90.2	91.5	88.7
Weight at Farrowing (kg)	152.7	144.0	139.2
Total Weight Gain (kg)	62.5	52.5	50.5
Daily Gain (kg)	0.54	0.46	0.44

Wang, 1983.

Table 12.6. Reproductive Performance of Gilts Fed Cocoyam Silage

		Cocoyam Silage Diets (% of DM)			
	Commercial Diet	20	35	40	70
Number of Gilts	4	4	4	4	4
Pigs Born Alive	11.3	12.3	9.5	11.0	8.5
Birth Weight (kg)	0.97	0.93	0.98	0.91	1.03
Pigs Weaned	9.25	9.25	9.00	10.25	8.00
Litter Growth (kg/day)	0.23	0.25	0.14	0.20	0.16

Wang, 1983.

SUMMARY

Cocoyams intended for feeding of swine should be cooked prior to drying to ensure destruction of anti-nutritional factors. Raw cocoyams should never be fed to pigs. It is recommended that cooked, dried cocoyams not be incorporated into the diets of starter pigs or those in the early grower phase. They may however be fed to sows in gestation and late lactation.

Cocoyam silage may be fed to pigs in the growing and fattening stages, as well as to gilts in gestation and lactation. At levels of 20 to 40% of dry matter, cocoyam silage supports adequate growth rate in young pigs, and can be used for gilts in gestation and for brood sows in lactation.

REFERENCES

Carpenter, J.R. and Steinke, W.E., 1983. Utilization of taro as animal feed. In: J.K. Wang, ed. Taro: A Review of *Colocasia esculenta* and its Potentials. The University of Hawaii Press, Honolulu, pp. 269-300.

Chowdhury, B. and Hussain, M., 1979. Chemical composition of the edible parts of aroids grown in Bangladesh. Indian J. Agric. Sci. 49: 110-115.

de la Pena, R.S., 1983. Agronomy of Taro. In: J.K. Wang, ed. Taro: A Review of *Colocasia esculenta* and its Potentials. The University of Hawaii Press, Honolulu, pp. 167-179.

FAO, 1985. Production Yearbook. Food and Agriculture Organization of the United Nations, Rome.

Fetuga, B.L. and Oluyemi, J.A., 1976. The metabolizable energy of some tropical tuber meals for chicks. Poult. Sci. 55: 868-873.

Goering, K.J., 1971. Properties of the small granule starch from *Colocasia esculenta*; a comparison of the small granule starch from dasheen tubers with that obtained from cereal type sources. Cereal Sci. Today 16: 306 (Abstract).

Goering, K.J. and De Haas, B., 1972. New starches. VIII. Properties of the small granule starch from Colocasia esculenta. Cereal Chem. 49: 712-719.

Gopalan, C., Ramasastri, B.V. and Balasubramanian, S.C., 1977. Nutritive Value of Indian Foods. The National Institute of Nutrition, Indian Council of Medical Research, Hyderabad, India.

Langworthy, C.F. and Deuel, H.J., 1922. Digestibility of raw rice, arrow root, canna, cassava, taro, tree fern and potato starches. J. Biol. Chem. 52: 251-261.

Murai, M., Pen, F. and Miller, C.D., 1958. Some Tropical South Pacific Island Foods. University of Hawaii Press, Honolulu.

O'Hair, S.K., 1984. Farinaceous Crops. In: F.W. Martin, ed. Handbook of Tropical Food Crops. CRC Press. Boca Raton, Florida, pp. 109-137.

Onwueme, I.C., 1978. The Tropical Tuber Crops: Yam, Cassava, Sweet Potato and Cocoyam. John Wiley and Sons, Chichester, New York.

Oyenuga, V.A. and Fetuga, B.L., 1975. Chemical composition, digestibility and energy values of some varieties of yam, cassava, sweet potatoes and cocoyams for pigs. Nigerian J. Sci. 9: 63-110.

Payne, J.H., Ley, G.J. and Akau, G., 1941. Processing and chemical investigation of taro. Hawaii Agricultural Experimental Station Bulletin, Honolulu, No. 86, pp. 1-41.

Rashid, M.M. and Daunicht, H.J., 1979. Chemical composition of nine edible aroid cultivars of Bangladesh. Scientia Horticulturae 10: 127-134.

Soldevilla, M. and Vincente-Chandler, J., 1978. Tanier corms used as feed for growing pigs. J. Agric. Univ. Puerto Rico 62: 283-289.

Standahl, B.R., 1983. Nutritive value of taro. In: J.K. Wang, ed. Taro: A Review of *Colocasia esculenta* and its Potentials. The University of Hawaii Press, Honolulu, pp. 141-147.

Tang, C. and Sakai, W.W., 1983. Acridity of taro and related plants in *Araceae*. In: J.K. Wang, ed. Taro: A Review of *Colocasia esculenta* and its Potentials. The University of Hawaii Press, Honolulu, pp. 148-163.

Wang, J.K., 1983. Improvements of Taro For Food and Feed Uses in the Tropics. Report for the Period July 1, 1982-December 31, 1983. USDA, Washington, Grant No. 58-9AHZ-9-445.

Watt, B.K. and Merrill, A.L., 1963. Composition of Food. US Department of Agriculture Handbook No. 8, Washington, D.C.

CHAPTER 13

Copra Meal

P.J. Thorne, D.J.A. Cole and J. Wiseman

INTRODUCTION

The coconut palm (*Cocos nucifera*) is widely distributed throughout the tropics. Parts of the coconut plant fulfill a multitude of functions including use as fuel and as a building material. Under village conditions, parts of the plant are used as a feed for humans. However, the primary coconut product to enter world trade is copra, its dried kernel, which is used as the raw material for coconut oil production.

Around 30 to 40% by weight of the copra which is used in oil extraction remains as a residue, this being copra meal. Countries with well developed coconut oil-producing industries (e.g. Philippines, Indonesia, India) have long appreciated the desirability of using copra meal as a component of livestock diets. However, this product is still underutilized and there would appear to be considerable potential to increase its use in swine diets, particularly when fed in combination with other feed ingredients.

PRODUCTION OF COPRA MEAL

Copra meal is produced by expeller extracting dried coconut kernels to remove the coconut oil. Copra meal has a residual oil content of about 8%. However, it is not economically viable to remove the remaining oil by solvent extraction due to the relatively low value of coconut oil. The residual material produced by the expeller is in the form of dried chunks of cake and should be ground to a meal before inclusion in swine diets.

NUTRIENT CONTENT OF COPRA MEAL

Copra meal is a rather variable commodity (Table 13.1). The variation in the nutrient content of copra meals is fundamentally a function of differences in residual oil content. This is largely a consequence of differences in efficiency of oil extraction depending on the efficiency of the process used.

Table 13.1. Nutrient Content of Copra Meals (% of Meal)[1]

Nutrient	Expeller Meal (Philippines)	Expeller Meal (Solomon Islands)	Expeller Meal (England)[2]
Dry Matter	90.2	89.4	95.4
Crude Protein	20.8	19.2	19.0
Ether Extract	8.1	11.7	16.1
Crude Fiber	11.6	8.8	8.8
Ash	6.2	5.1	4.1
Calcium	0.10	0.05	0.05
Phosphorous	0.59	0.49	0.45
Lysine	0.42	0.70	0.58
Methionine + Cystine	0.36	0.77	0.49

[1]Produced using a small scale non-industrial expeller.
[2]Thorne, 1990 (unpublished).

An experiment, conducted by Thorne (1986), illustrates the relationship between the residual oil content of copra meals and their energy value to the pig. The results are summarized in Table 13.2. Most of the copra meals encountered in practice fall between 9 and 16% residual oil content. However, some meals produced by small scale expeller extraction processes or using ill-maintained or ill-managed equipment may have residual oil contents above 20%. As a result, copra meal is potentially a valuable source of dietary energy for pigs.

Copra meal can also make a significant contribution to the protein requirements of swine although, with its crude protein content of around 20%, it cannot be considered as a high protein feed. Some previous studies have regarded it in this light with disappointing results (Grieve et al., 1966; Malynicz, 1973). Unfortunately, scrutiny of the amino acid content of copra meal protein shows its balance to be far from ideal (Table 13.3). Lysine is particularly deficient. There is also a considerable excess of arginine which has been shown to depress the growth rate of pigs (Southern and Baker, 1982; Hagemeier et al., 1983).

Table 13.2. Digestible Energy (D.E.) Value of Copra Meals in Relation to Their Residual Oil Contents

Copra Meal	Residual Oil Content (% Ether Extract)	D.E. (kcal/kg)
Solvent Extracted	0.4	3020
Expeller Extracted	9.1	3700
Expeller Extracted	16.1	4340
Expeller Extracted	22.0	4540
Expeller Extracted	33.3	4690
Unpressed Copra	66.2	7060

Thorne, 1986.

Table 13.3. The Balance Relative to Lysine of Amino Acids in Copra Meal

	Ideal Protein[1]	Expeller Copra Meal (Solomon Islands)
Lysine	100	100
Methionine + Cystine	50	114
Threonine	60	75
Tryptophan	15	19
Isoleucine	55	89
Leucine	100	162
Histidine	33	73
Phenylalanine + Tyrosine	96	186
Valine	70	136
Arginine	100	371

[1]ARC, 1981.

The protein quality of copra meals is affected by processing efficiency (Butterworth and Fox, 1963). In general, more efficient oil extraction processes perform more work to extract a greater proportion of the oil. This results in increased generation of heat causing more severe damage to the amino acids in the copra meal. As a consequence, poor digestibility and imbalance of essential amino acids in copra meal protein must be taken into account when this feed is considered as an ingredient in pig diets. Their combined effects probably result in copra meal's contributing little more to the animals' protein requirement than do dietary cereals.

The fatty acid composition of coconut oil is worthy of brief comment. It is unusual in being composed predominantly of short chain saturated fatty acids (50% C12:0; 15% C14:0). Such fats are relatively easily absorbed (Freeman, 1984) which is probably instrumental in maintaining the energy value of copra meal to the pig. They are also readily deposited in pig backfat (Christensen, 1963; Thorne, 1986) resulting in firm, white fat and pork of enhanced sensory appeal. However, increasing the proportions of saturated fatty acid in meat products may be undesirable in terms of consumer acceptability on health grounds.

UNDESIRABLE CONSTITUENTS IN COPRA MEAL

Copra meal inoculated under laboratory conditions supports the growth of *Aspergillus spp.*, some members of the genus being responsible for aflatoxin production. Levels of aflatoxin contamination of copra meal above those normally permitted for pig feeds have been reported in the field. However, there have been no reports of aflatoxicosis in pigs fed high levels of copra meal and it therefore seems unlikely that the problem would ever become widespread. Aflatoxin contamination of copra is most likely to result from improper procedures during drying. Such poor quality copra would normally be rejected for oil production during grading and would therefore not be involved in copra meal production.

The high crude fiber level of copra meal must also be regarded as a negative factor for swine feeding. Studies have demonstrated the relatively poor digestibility (50-70%) of copra meal fed to pigs (Creswell and Brooks, 1971b; Mee and Brooks, 1973; Kuan et al., 1982; Thorne, 1986). It seems likely that this may be linked in part with the fiber content of copra meal although the effects of heat damaged protein and poorly digested carbohydrates are probably also implicated.

FEEDING COPRA MEAL TO SWINE

The most successful approaches to feeding high levels of copra meal to growing/finishing swine make use of complimentary feeds to balance the copra meal's inherent deficiencies. A large scale study was conducted in the Philippines using commercial copra cake to supplement a corn-soy-fish meal diet (Thorne et al., 1988). Diets tested progressive increments of copra meal inclusion from 0 to 50%. These diets were iso-nitrogenous and iso-energetic, but no attempts were made to sustain levels of essential amino acids as the inclusion level of copra meal increased. However, an additional treatment was used in which lysine, methionine and tryptophan were added to a 50% copra meal diet. Performance data are summarized in Table 13.4.

There was a linear reduction in daily gain as the level of copra meal in the diet increased. This reduction in gain was associated with a reduction in both feed intake and efficiency. The poor protein quality of copra meal would result in a progressive deficiency in the levels of the early limiting dietary amino acids with increased inclusion level. However, attention to dietary amino acid balance improved performance of pigs fed 50% copra meal to about 80% of that achieved with the copra meal-free control diet.

Table 13.4. Influence of Copra Meal on Growing Pig Performance[1]

	Level of Copra Meal (%)				
	0	10	30	50	50 + AA[2]
Daily Gain (kg)	0.76	0.70	0.60	0.46	0.62
Daily Intake (kg)	2.01	1.94	1.83	1.65	1.76
Feed Efficiency	2.62	2.79	3.09	3.61	2.85

[1]Thorne et al., 1988.
[2]Diet supplemented with lysine, methionine and tryptophan.

Creswell and Brooks (1971a) found that pigs receiving diets containing 40% copra meal with added synthetic lysine did not perform as well as those fed a control diet formulated without copra meal. Unfortunately, there was no comparison in this study with an un-supplemented copra meal diet, but the possibility of amino acid supplementation as a counter-measure to poor protein quality in diets incorporating high levels of copra meal was indicated.

The overall results of these studies demonstrate an important general point about diet formulation using relatively low quality feeds - these feeds are unlikely to result in optimum performance if diet formulation is based on crude protein alone. "Book values" for crude protein requirements are based on experimentation with high quality protein sources and are thus likely to overestimate the nutritive value of amino acids supplied from poorer quality protein sources. If low quality protein is to form a significant proportion of the dietary protein supply, lysine and methionine, at least, should be included in the formulating parameters.

SUMMARY

Experiments, based on substitution of copra meal for other dietary ingredients, generally suggest an inclusion level of approximately 30% as being optimal in the diets of growing swine (Grieve et al., 1966; Malynicz, 1973). This approach to feed evaluation is not ideal as it fails to consider the incorporation of the feed into balanced diets - that is, into diets formulated with complimentary feeds. Despite this, it is probably inappropriate to make hard and fast recommendations on maximum inclusions of copra meal in pig diets. These will be dependent on the other feed raw materials that are available to compliment it and should become apparent when diet formulation is practiced using appropriate parameters. The feed can be used to the greatest extent when high quality protein sources are available to supplement it. When this is not possible, reducing the levels of individual raw materials in a formulation while increasing the number of different feeds used is more likely to result in a balanced diet. In conclusion, decisions about levels of inclusion of copra meal should be based on consideration of:

1. The quality of copra meal available, which is variable.

2. Availability of complimentary feeds, particularly sources of good quality protein.

Table 13.5 summarizes problems that may be encountered when copra meal is used in swine diets along with possible solutions to them.

Table 13.5. Problems Encountered in Diet Formulation with Copra Meal and Strategies for Avoiding Them

Amino Acid Imbalance	i.	Supplement with high quality protein source (e.g., fish meal, fish silage, soybean meal, offal hydrolysates or silages).
	ii.	Supplement with synthetic amino acids.
Essential Fatty Acid Deficiency	i.	Include maize in formulations.
	ii.	Add small quantities of oils high in essential fatty acids.
High Fiber Content	i.	Use complimentary feeds low in fiber (e.g., maize meal).
	ii.	Ensure adequate water supply and ad libitum access to feed.
	iii.	Add small quantities of fats or oils to increase energy content.
Aflatoxin	i.	Ensure correct procedures during production and storage (e.g., minimize water content).
Variable Composition	i.	Raw material analysis.
	ii.	Visual assessment (e.g., a dark expeller meal is likely to have been pressed hard and therefore have a low residual oil content and heat damaged protein).

REFERENCES

Agricultural Research Council, 1981. The Nutrient Requirements of Pigs, ARC, London.

Butterworth, M.H. and Fox, H.C., 1963. The effects of heat treatment on the nutritive value of coconut meal and the prediction of nutritive value by chemical methods. Brit. J. Nutr. 17: 445-452.

Christensen, K.D., 1963. Various fatty acids in the fat tissues of pigs of the Danish Landrace breed fed with coconut fat or soybean oil. Acta Agric. Scand. 13: 245-249.

Creswell, D.C. and Brooks, C.C., 1971a. Effect of coconut meal on coturnix quail and of coconut meal and coconut oil on performance, carcass measurements and fat composition in swine. J. Anim. Sci. 33: 370-375.

Creswell, D.C. and Brooks, C.C., 1971b. Composition, apparent digestibility and energy evaluation of coconut oil and coconut meal. J. Anim. Sci. 33: 366-369.

Freeman, C.P., 1984. The digestion, absorbtion and transport of fats: non-ruminants. In: J. Wiseman, ed. Fats in Animal Nutrition, Butterworths, London.

Grieve, G.C., Osbourn, D.F. and Gonzales, F.O., 1966. Coconut oil meal in growing and finishing rations for swine. Trop. Agric. (Trinidad) 43: 257-261.

Hagemeier, D.L., Libal, G.W. and Wahlstrom, R.C., 1983. Effects of excess arginine on swine growth and plasma amino acid levels. J. Anim. Sci. 57: 99-105.

Kuan, K.K., Mak, T.F., Razak Alimon and Farrell, D.J., 1982. Chemical composition and digestible energy of some feedstuffs determined with pigs in Malaysia. Trop. Anim. Prod. 7: 315-321.

Malynicz, G.L., 1973. Coconut meal for growing pigs. Papua New Guinea Agric. J. 24: 142-144.

Mee, J.M.L. and Brooks, C.C., 1973. Amino acid availability of coconut meal protein in swine. Nutr. Rep. Int. 8: 261-269.

Southern, L.L. and Baker, D.H., 1982. Performance and concentration of amino acids in plasma and urine of young pigs fed diets with excesses of either lysine or arginine. J. Anim. Sci. 55: 857-866.

Thorne, P.J., 1986. The Use of Copra Meal in Pig Diets. PhD Thesis, University of Nottingham.

Thorne, P.J., Wiseman, J., Cole, D.J.A. and Machin, D.H., 1988. Use of diets containing high levels of copra meal for growing/finishing pigs and their supplementation to improve animal performance. Trop. Agric. (Trinidad) 65: 197-201.

CHAPTER 14

Corn Gluten Feed

P.J. Holden

INTRODUCTION

Swine have major dietary requirements for protein and energy. These needs are commonly met by the use of cereal grains and protein supplements, with cereal grains providing most of the energy required by the pig while soybean meal usually provides the major portion of the dietary protein. However, other alternatives, such as processing by-products, can often substitute nutritionally into a swine diet. One of these alternatives is corn gluten feed.

The main parts of a corn kernel are the hull, the germ, the gluten and the starch. Most of the oil is contained in the germ. The gluten is the yellowish portion found on either side of the kernel and contains most of the protein while the remaining whole powdery part is the starch.

Wet milling of yellow dent corn involves its separation into the four major components indicated in Table 14.1. Corn gluten feed is defined as "that part of the commercial shelled corn that remains after extraction of the larger part of the starch, gluten and germ by the processes employed in the wet milling manufacture of corn starch or syrup." It is a blend of the hulls, evaporated steepwater and germ meal. The steepwater is the concentrated solubles resulting from the initial soaking stage of the wet milling process while the germ meal is the remainder of the germ after the oil has been extracted.

Table 14.1. Products of Corn Wet Milling (DM)[1]

Corn Starch	67.2%
Corn Gluten Feed	19.6%
Corn Gluten Meal, 60% protein	5.7%
Corn Germ[2]	7.5%

[1]Long, 1985.
[2]50% corn oil.

NUTRIENT CONTENT OF CORN GLUTEN FEED

The major protein consideration for swine feeding is amino acid balance. Since both cereals and protein sources are contributors of amino acids, the levels of certain amino acids in both are critical. Corn gluten feed lies between a cereal grain and a protein source in this regard (Table 14.2). Although the protein content of corn gluten feed is relatively high (23.3%), the balance of amino acids is poor. Therefore, it cannot be used as a protein source unless the diets are balanced for the amino acids lysine and tryptophan, both of which are limiting.

Table 14.2. Chemical Analysis of Corn Gluten Feed (as fed)

	Corn Gluten Feed	Corn	Soybean Meal
Dry Matter (%)	90.00	88.00	90.00
Crude Protein (%)	23.30	8.50	48.50
Lysine (%)	0.64	0.25	3.12
Threonine (%)	0.79	0.36	1.90
Tryptophan (%)	0.15	0.09	0.69
Calcium (%)	0.18	0.03	0.26
Phosphorus (%)	0.99	0.28	0.64
Fat (%)	2.70	3.60	0.90
Fiber (%)	6.80	2.30	3.40
D.E. (kcal/kg)	3155	3530	3680

National Research Council, 1988.

Corn gluten feed is relatively low in energy due in part to its relatively low fat content and high fiber level. Ewan (unpublished data), estimated the metabolizable and net energy content of corn gluten feed to be 2695 and 1694 kcal/kg, respectively, on an as-fed basis. This net energy value is 74% of the value of corn.

The calcium content of corn gluten feed (0.18%) is much higher than that of corn (0.03%). Although the phosphorus content of corn gluten feed is high (0.99%), it has an availability estimate of only 59% (Burnell et al., 1989). This compares to a value for dicalcium phosphate of 98% availability.

As a consequence, corn gluten feed should not be considered a major source of dietary phosphorus in a swine feed and an inorganic source should also be fed.

UNDESIRABLE CONSTITUENTS IN CORN GLUTEN FEED

Corn gluten feed contains no known anti-nutritional factors. However, when used as a large portion of the diet, the severe deficiency of available lysine and tryptophan will reduce feed intake. Adding synthetic lysine will alleviate the lysine deficiency. However, synthetic tryptophan is not currently economically available to meet the tryptophan deficiency.

FEEDING CORN GLUTEN FEED TO SWINE

Starter Pigs

Although high in protein, the poor availability of the lysine and tryptophan in corn gluten feed, in combination with its high fiber and low digestible energy, limits its usefulness in swine diets. The deficiency of amino acids and low energy content are of particular concern during the starter phase and therefore corn gluten feed is not recommended for use with pigs weighing less than 20 kg body weight.

Growing-Finishing Pigs

Corn gluten feed also has limited potential for use during the growing phase. Yen et al. (1971) fed graded levels of corn gluten feed to 23 kg pigs for a period of 42 days and observed a reduction in gain and feed efficiency as the level of corn gluten feed in the diet increased (Table 14.3).

Table 14.3. Performance of Growing Pigs (23 kg) Fed Corn Gluten Meal

	Level of Corn Gluten Meal (%)			
	0	10	20	30
Daily Gain (kg)	0.58	0.55	0.56	0.54
Daily Intake (kg)	1.52	1.58	1.58	1.59
Feed Efficiency	2.62	2.87	2.82	2.94

Yen et al., 1971.

Poor feed efficiency has also been observed during the finishing phase when diets containing corn gluten feed are fed. Yen et al. (1971) substituted corn gluten feed for corn and soybean meal in diets fed from 47 kg until the pigs reached market weight. Although growth rate was not appreciably altered as a result of including corn gluten feed in the diet, feed efficiency declined rather dramatically (Table 14.4).

Table 14.4. Performance of Finishing Pigs (47 kg) Fed Corn Gluten Meal

	Level of Corn Gluten Meal (%)			
	0	10	20	30
Daily Gain (kg)	0.58	0.58	0.61	0.57
Daily Intake (kg)	2.14	2.17	2.31	2.19
Feed Efficiency	3.68	3.74	3.78	3.84

Yen et al., 1971.

Zimmerman and Honeyman (1986) fed diets containing 25 or 50% corn gluten feed to growing-finishing pigs (Table 14.5). The diets contained 0.70% lysine for the first four weeks and 0.61% for the remainder of the experiment. As corn gluten feed was added, the energy content of the diet decreased. Increasing corn gluten feed levels had no effect on feed intake but daily gain was slightly depressed and feed per unit of gain increased significantly. The depressed feed efficiency would be expected based on Ewan's estimate that corn gluten feed supplies 74% of the net energy of corn.

Table 14.5. Performance of Market (36-100 kg) Pigs Fed Corn Gluten Feed

	Level of Corn Gluten Feed (%)		
	0	25	50
Daily Gain (kg)	0.76	0.69	0.65
Daily Intake (kg)	2.55	2.48	2.56
Feed Efficiency	3.42	3.59	3.95
Dressing Percentage	72.7	71.2	70.0
Backfat (cm)	0.36	0.33	0.30

Zimmerman and Honeyman, 1986.

The performance of pigs fed corn gluten feed has been shown to be significantly improved by supplemental tryptophan or tryptophan plus lysine (Yen et al., 1971). It has been suggested that the tryptophan in corn gluten feed is largely unavailable to the pig (Yen et al., 1971). LaRue et al. (1987), estimated the ileal apparent digestibility of tryptophan in corn gluten feed to be only 32% compared to 81% for tryptophan in soybean meal.

Honeyman and Zimmerman (1985) observed that market hogs fed from 91 to 113 kg bodyweight utilized a diet containing 75% corn gluten feed as well as one with 50% corn gluten feed. Daily gains were 0.90 and 0.91 kg and feed:gain ratios were 4.50 and 4.81, respectively. Pigs fed a diet containing 92.5% corn gluten feed gained 0.62 kg/day with an efficiency of 6.17. When marketed, corn gluten feed-fed pigs have been shown to yield significantly lower dressing percentages but have significantly leaner carcasses (Zimmerman and Honeyman, 1986; Cromwell et al., 1987). The reduction in dressing percentage has been suggested to be the result of a greater gut fill due to the higher crude fiber level in corn gluten feed (Cromwell et al., 1987).

Honeyman and Zimmerman (1985) indicated that swine producers should use dry pelleted corn gluten feed if possible, as its use reduces many of the feed intake problems associated with dry and wet ground corn gluten feed and improves handling and storage of the feed. Pellets should be reground before feeding to assure a uniform diet. In addition, Yen et al. (1971), observed that pelleting corn gluten feed improved nitrogen retention from 42.6 to 53.0%. They concluded that the major factor for improved retention was an increased tryptophan availability.

Breeding Stock

Honeyman and Zimmerman (1987) fed gestating sows different levels of corn gluten feed for three reproductive cycles. The three treatments were: 1) corn-soybean meal control fed at 1.8 kg/day; 2) corn gluten feed-soybean meal fed at 2.27 kg/day; and 3) corn gluten feed-soybean meal fed at 2.6 kg/day. Diet 2 assumed corn gluten feed to have 80% of the net energy of corn and diet 3 assumed the energy value to be 70%. Sows were fed a corn-soybean meal diet ad libitum during lactation. Dietary protein was increased from 12% to over 20% when corn gluten feed replaced the corn but the available lysine and tryptophan levels were decreased.

The performance of the sows is shown in Table 14.6. Sow weight differences suggest that sows fed corn gluten feed at 2.27 kg/day weighed consistently less than those fed the corn-soybean meal diet. Sows receiving 2.6 kg daily of the corn gluten feed diet had gains similar to the control sows. This suggests that corn gluten feed has approximately 70% of the net energy value of corn for gestating sows. No other statistical differences were observed although there was a trend for gestating sows fed corn gluten feed to farrow and wean more pigs. The authors concluded that corn gluten feed can provide most of the energy and some of the amino acids for gestating sows and that the trend for increased litter size should be further investigated.

Table 14.6. Performance of Sows Fed Corn Gluten Feed (CGF) Diets

	Control	CGF (2.27 kg/day)	CGF (2.6 kg/day)
Total Litters	66	62	65
Sow Weight at Breeding (kg)	172	159	168
Sow Weight at Farrowing (kg)	200	186	200
Sow Weight at Weaning (kg)	190	179	190
Lactation Feed Intake (kg)	121	128	122
Pigs Born Alive	10.8	11.8	11.3
Pigs Weaned	7.8	8.3	8.2

Honeyman and Zimmerman, 1987.

SUMMARY

Table 14.7 indicates maximum levels of corn gluten feed that can be included in swine diets and still produce satisfactory performance. These percentages assume that corn gluten feed is primarily replacing corn and that adequate levels of available lysine and phosphorus are maintained.

Table 14.7. Maximum Recommended Levels of Corn Gluten Feed in Swine Diets

Stage	Maximum Level
Gestation	90
Lactation	5
Starter	5
Grower-Finisher	25

Several factors are used in devising relative feeding values of ingredients for swine feeds. These factors include energy, protein (or lysine) and phosphorus. Table 14.8 indicates the relative value of corn gluten feed with varying prices of corn, soybean meal (44% protein) and dicalcium phosphate as sources of metabolizable energy, lysine and phosphorus, respectively. Note that the value is more affected by changing prices of corn rather than soybean meal. Changing the price of dicalcium phosphate has only a minimal effect (Table 14.8, footnote 1). This indicates that the major factor in evaluating the use of corn gluten feed is its energy value and not the lysine or phosphorus content.

Table 14.8. Relative Value of Corn Gluten Feed

	Corn ($/100 kg)			
	$8.00	$10.00	$12.00	$14.00
Soybean Meal 44%, $/100 kg	Corn Gluten Feed Value, $/100 kg			
$20.00	$7.10	$ 8.47	$ 9.99	$11.44
$25.00	7.30	8.67	10.19	11.64
$30.00	7.50	8.87	10.38	11.83
$35.00	7.70	9.09	10.60	12.06

[1]Dicalcium phosphate cost = $33/100 kg. Increasing price of dicalcium phosphate to $38/100 kg raises each value $0.06.
[2]Assumes 50% availability of lysine and phosphorus in Corn Gluten Feed.

REFERENCES

Burnell, T.W., Cromwell, G.L. and Stahly, T.S., 1989. Bioavailability of phosphorus in triticale, hominy feed and corn gluten feed for pigs. J. Anim. Sci. 67 (Suppl. 1): 262.

Cromwell, G.L., Stahly, T.S. and Randolph, J.H., 1987. Corn gluten feed for growing-finishing swine. University of Kentucky, Lexington, Swine Research Report, pp. 12-14.

Ewan, R.C., 1989. Energy values of feed ingredients. Unpublished data, Iowa State University, Ames, Iowa.

Honeyman, M. and Zimmerman, D., 1985. Research on Line: Corn Gluten Feed. Pork '85, p. 9.

Honeyman, M. and Zimmerman, D., 1987. Corn gluten feed in gestating sow diets. Iowa State University, Ames, Swine Research Reports. AS Leaflet R487.

LaRue, D.C., Knabe, D.A. and Tanksley, T.D., 1987. Ileal digestibility of tryptophan in protein and grain-by-product feedstuffs for swine. J. Anim. Sci. 65 (Suppl. 1): 125.

Long, J. E., 1985. The wet milling process: Products and co-products. Corn Gluten Conference for Livestock. Ames, Iowa.

National Research Council, 1988. Nutrient Requirements of Domestic Animals, No. 2. Nutrient Requirements of Swine. 9th ed. National Academy of Sciences, Washington, D.C.

Yen, J.T., Baker, D.H., Harmon, B.G. and Jensen, A.H., 1971. Corn gluten feed in swine diets and effect of pelleting on tryptophan availability to pigs and rats. J. Anim. Sci. 33: 987-991.

Zimmerman, D. and Honeyman, M., 1986. Corn gluten feed as an ingredient in diets of growing-finishing pigs. Iowa State University Swine Research Reports. AS580-C, Ames, Iowa.

CHAPTER 15

Cottonseed Meal

T.D. Tanksley, Jr.

INTRODUCTION

Cotton has been grown for several thousands of years as a source of fiber that is used as a textile material. However, for every 100 kg of cotton fiber produced, the cotton plant also yields approximately 160 kg of cottonseed. With processing, typical yields from cottonseed are 50% meal, 22% hulls, 16% oil, 7% linters, with a 5% loss. Although much of the meal is utilized in ruminant diets, the price relationship between cottonseed meal and other high-protein feedstuffs often provides an excellent opportunity for pork producers to use cottonseed meal in order to reduce feed costs.

Unfortunately, some producers and feed manufacturers have developed a psychological barrier toward including cottonseed meal in swine diets. Too many times, the popular conception of cottonseed meal is based on a fear of gossypol toxicity or results of feeding trials conducted many years ago in which cottonseed meal, often of poor quality, was the only source of supplemental protein. However, excellent pig performance has been obtained when limited amounts of cottonseed meal have been fed in combination with other sources of high-quality protein. Therefore, although the limitations of cottonseed meal must be recognized, as a result of improved processing methods and extensive nutritional research, there is much more potential to utilize cottonseed meal in practical growing-finishing, gestation and lactation diets than is currently being realized.

The development of glandless cotton varieties (devoid of gossypol-containing pigment glands) yield cottonseed that is essentially free of gossypol. Glandless meals provide exciting possibilities for non-ruminant nutrition. Current acreage is limited but further genetic improvement of glandless cotton varieties that have the equivalent production potential of glanded varieties may mean more cottonseed will be available for use in swine diets.

PRODUCTION OF COTTONSEED MEAL

The three most common types of processing methods used to produce cottonseed meal are screwpress (expeller), direct solvent and prepress solvent. Typically, screwpress meals are relatively high in residual lipids and low in free gossypol but are also somewhat lower in protein quality. A screwpress can be regulated to produce a higher quality meal, but such a procedure generally results in a lower oil yield. In addition, if excessive processing temperatures are used, reactions between the epsilon amino group of lysine and various carbohydrates and pigments in the cottonseed meal take place, thereby making lysine less available to the animal in the already lysine deficient meal.

Prepress solvent meals are generally low in residual lipids and free gossypol and high in protein quality but precise manufacturing conditions must be maintained for this to be accomplished. Because of differences in processing conditions among plants, there is actually a wide range of protein quality among prepress solvent meals. Therefore, feed manufacturers, who purchase in large quantities, often buy prepress solvent meals on specifications to ensure high protein quality and low free gossypol. Nitrogen solubility in 0.02 N NaOH and the determination of protein lysine with free epsilon amino groups have been two of the more common laboratory methods employed to measure protein quality in cottonseed meal.

The use of direct solvent extraction has increased steadily in the past three decades and this gives a more uniform product. Direct solvent meals are generally high in protein quality but are also high in free gossypol, because low temperatures are used during processing. However, since research has demonstrated an effective method of blocking the harmful effect of free gossypol, direct solvent meals can be used with complete safety.

Traditionally, solvent processed meals have been preferred to screwpress meals because of their higher protein quality. Even though recent research suggests that some over-processed solvent meals can be as low or lower in protein quality (based on ileal digestible lysine) than screwpress meals (Knabe et al., 1989), solvent meals are still preferred over screwpress meals.

NUTRIENT CONTENT OF COTTONSEED MEAL

Ideally, cottonseed meal fed to swine should have a high protein quality and a low free gossypol content. Research and field experience have demonstrated that cottonseed meals differ in nutritional value for swine due to the type of process used in their production and the specific processing conditions employed (temperature, time and moisture) at each plant.

Proximate analyses for three types of meal are given in Table 15.1. These are average values and the standard deviations shown indicate that rather wide variations in nutrient content may be found among meals made by the same process.

Table 15.1. Proximate Analyses of Three Types of Cottonseed Meals[1]

	Screwpress	Direct Solvent	Prepress Solvent
Dry Matter (%)	92.3 ± 1.8	91.1 ± 1.5	90.8 ± 1.8
Crude Protein (%)	41.4 ± 0.8	42.1 ± 1.5	41.7 ± 0.7
Ether Extract (%)	3.9 ± 0.4	2.1 ± 1.2	0.8 ± 0.5
Crude Fiber (%)	12.6 ± 1.8	11.3 ± 2.1	12.7 ± 1.8
Ash (%)	6.0 ± 0.5	6.2 ± 0.3	6.4 ± 0.4
Total Gossypol (%)	0.96 ± 0.1	0.98 ± 0.1	1.04 ± 0.2
Free Gossypol (%)	0.03 ± 0.01	0.24 ± 0.13	0.04 ± 0.02
N-Solubility[2]	42.6 ± 11.7	71.4 ± 4.0	57.0 ± 5.2
EAF Lysine[3]	0.97 ± 0.15	1.43 ± 0.04	1.34 ± 0.14
M.E. (kcal/kg)	2685	2630	2592

[1]Smith, 1970.
[2]Solubility in 0.02 N NaOH.
[3]Epsilon Amino Free Lysine; Rao et al., 1963.

The essential amino acid content of cottonseed meals are given in Table 15.2. Amino acid data indicate that the most important nutritional difference among the three mill processes is in total lysine content. Processing conditions used in the production of screwpress meal appears to destroy some lysine.

A comparison between the ileal digestibilities for selected amino acids in soybean meal and some cottonseed meals is shown in Table 15.3. Since rather wide differences have been found among individual samples of both screwpress and direct solvent cottonseed meals, data for them is pooled in Table 15.3. Digestibilities of amino acids in the screwpress and direct solvent meals were consistently lower than those for soybean meal with the widest difference found for lysine (56 vs. 86%). Also, the range in digestibilities for all amino acids in cottonseed meal was much higher than for the soybean meals. Among the cottonseed meal amino acids, lysine digestibility was most variable (42-69%), which further complicates diet formulation.

Although there is essentially no glandless cottonseed meal available for livestock feeding at the present time, digestibility values are shown for a laboratory-processed and a commercially-processed glandless meal. Lysine digestibility was higher for the laboratory-processed than the commercially-processed meal (87 vs. 77%) but values were similar for other amino acids. Digestibilities were consistently higher for the glandless than the glanded cottonseed meal. This is expected since no gossypol would be present in the glandless meals to form the indigestible gossypol-protein complexes during processing.

Table 15.2. Amino Acid Content of Three Types of Cottonseed Meal (% as fed)

	Screwpress	Direct Solvent	Prepress Solvent
Arginine	4.48	4.87	4.57
Histidine	1.09	1.15	1.13
Isoleucine	1.31	1.37	1.33
Leucine	2.29	2.47	2.42
Lysine	1.63	1.81	1.74
Methionine	0.54	0.56	0.56
Methionine + Cystine	1.27	1.24	1.28
Phenylalanine	2.22	2.28	2.24
Phenylalanine + Tyrosine	3.36	3.48	3.39
Threonine	1.32	1.37	1.31
Tryptophan	0.49	0.51	0.48
Valine	1.88	1.91	1.86

Smith, 1970; National Research Council, 1988; Texas A & M Nutrition Lab.

Table 15.3. Apparent Ileal Digestibilities of Selected Amino Acids in Soybean (SBM) and Cottonseed (CSM) Meals (%)

			Glandless CSM	
	SBM (44%)[1]	Screwpress/ Direct Solvent CSM (41%)[2]	Laboratory Processed (62%)[3]	Commercially Processed (43%)[4]
Isoleucine	83 (78-85)	66 (58-76)	84	85
Lysine	86 (81-89)	56 (42-69)	87	77
Methionine	--------	69 (65-80)	83	85
Threonine	75 (70-79)	65 (56-76)	79	76
Tryptophan	79 (71-84)	75 (71-84)	83	79

[1]Knabe et al., 1989.
[2]Tanksley and Knabe, 1984.
[3]Tanksley et al., 1981.
[4]LaRue et al., 1985.

The mineral and vitamin content of cottonseed meal are shown in Tables 15.4 and 15.5.

Table 15.4. Mineral Content of Three Types of Cottonseed Meal

	Screwpress	Direct Solvent	Prepress Solvent
Calcium (%)	0.18	0.17	0.17
Phosphorus (%)	0.97	1.07	1.10
Magnesium (%)	0.43	0.41	0.41
Potassium (%)	1.24	1.20	1.26
Sodium (%)	0.04	0.04	0.04
Copper (mg/kg)	21	19	22
Iron (mg/kg)	139	110	120
Manganese (mg/kg)	22	21	19
Zinc (mg/kg)	60	61	64

Smith, 1970; National Research Council, 1988; Texas A & M Nutrition Lab.

Table 15.5. Vitamin Content (mg/kg) of Three Types of Cottonseed Meal (as fed)

	Screwpress	Direct Solvent	Prepress Solvent
Vitamin E	16	16	16
Biotin	0.53	0.55	0.55
Folic Acid	1.3	1.4	1.4
Niacin	40	41	43
Pantothenic Acid	11.8	12.7	12.2
Pyridoxine	5.2	5.4	4.9
Riboflavin	4.2	4.8	4.5
Choline	2790	2764	2910

Smith, 1970; National Research Council, 1988; Texas A & M Nutrition Lab.

LIMITATIONS TO THE USE OF COTTONSEED MEAL

Protein Quality

In a comprehensive study, Knabe et al. (1979) determined that the poorer performance of growing pigs fed sorghum-cottonseed meal diets was due primarily to low dietary lysine. Not only is cottonseed meal (41% CP) relatively low in lysine (1.70%) compared to 44% soybean meal (2.80%), but its protein quality is further lowered if the cottonseed meal is processed at high temperatures, which results in some of the lysine becoming indigestible. High processing temperatures promote a reaction between free gossypol and free amino groups in the protein to form an indigestible complex (Baliga and Lyman, 1957; Smith, 1972). Supplementation of cereal grain-cottonseed meal based growing-finishing diets with synthetic lysine has consistently improved pig performance (Aguirre et al., 1960; Hale and Lyman, 1962; Bell and Larsen, 1963; Noland et al., 1968). Supplementation of similar diets with methionine (Stephenson and Neuman 1951; Wallace et al., 1955; Chamberlain and Lidvall, 1962; Hintz and Heitman, 1967) or tryptophan (Chamberlain and Lidvall, 1962) failed to improve pig performance.

Free Gossypol

Gossypol is a highly reactive polyphenolic dinaphthaldehyde compound which is present in the oil glands of cottonseed. Gossypol is divided into bound gossypol (Carruth, 1947; Boatner et al., 1947) which is nontoxic to monogastrics (Lyman et al., 1944) and free gossypol which is toxic. Although several factors affect the free gossypol tolerance of pigs (age of pig, percent dietary protein and lysine), feed consumption and daily gains are decreased and toxicity symptoms usually occur when free gossypol levels approach 100 mg/kg (Holley et al., 1955; Hale and Lyman, 1957; Clawson et al., 1961; Kornegay et al., 1961). Therefore, diets containing up to 100 mg/kg of free gossypol can be fed to swine without depressed performance or fear of gossypol toxicity.

Free gossypol is not usually a concern when screwpress and prepress solvent cottonseed meals are included in swine diets. Based on the average free gossypol content of the three types of cottonseed meals (Table 15.1), a diet could include up to 17% screwpress or a low-gossypol prepress solvent meal and still contain less than 100 mg/kg free gossypol. Unfortunately, the free gossypol content of direct solvent meals varies widely and realistically may range from 0.04 to 0.40%. Therefore, when direct solvent meal is used, including only 3.3% in the diet would give 100 mg/kg free gossypol (based on 0.3% free gossypol). As a consequence, when such meals are used, an effort should be made to purchase from a supplier that provides the free gossypol content of the meal. When the free gossypol content is not known, most nutritionists use 0.3% free gossypol as a safety factor in formulating diets. Recent processing innovations have enabled some plants to produce a low gossypol direct solvent meal.

Diets with over 100 mg/kg free gossypol need not be a problem since iron salts are highly effective in blocking the effects of gossypol (Withers and Carruth, 1917; Robison, 1934; Clawson et al., 1967; Knabe et al., 1979). Knabe et al. (1979) found no significant difference in performance between animals fed diets containing 90, 300 and 500 mg/kg free gossypol, when 500 mg/kg iron salt was added. Therefore, in diets containing over 100 mg/kg of free gossypol, it has been suggested that the toxic effects of gossypol can be completely eliminated by including iron on a 1:1 weight ratio of iron to free gossypol (NCPA, 1966).

The most effective iron salt in blocking gossypol is ferrous sulfate. Two types of ferrous sulfate have been used successfully. Exsiccated (dried) ferrous sulfate is more expensive than copperas sulfate but contains a higher percentage of iron (32 vs. 20%) and is less of an oxidant. Each 100 grams of dried ferrous sulfate included per tonne of complete feed provides 32 mg/kg of iron in the complete diet. Each 100 grams of copperas sulfate included per ton provides 20 mg/kg of iron in the complete feed. On this basis, if more than 33 kg of direct solvent meal is added per ton of diet, dried ferrous sulfate should be added to maintain a 1:1 iron to free gossypol ratio.

Energy Content

Commercially-processed cottonseed meals contain 11 to 13% crude fiber which lowers their digestible energy content compared with soybean meal. Husby and Kroening (1971) found that two prepress solvent cottonseed meals containing 8.4 and 10.7% crude fiber had higher digestible and metabolizable energy values than a prepress solvent meal containing 13% crude fiber. In another experiment, Knabe et al. (1979) fed cottonseed meal diets containing approximately 3 or 4.5% crude fiber. Crude fiber level had minimal effect on feed intake and gain, but pigs fed the low fiber diet had a 2.3 to 3.8% improvement in feed efficiency than those fed the high fiber diet.

FEEDING COTTONSEED MEAL TO SWINE

Starter Pigs

There is very little information on the use of cottonseed meal in diets fed to starter pigs. Until more research is conducted, it would seem wise to limit the use of cottonseed meal in diets fed to pigs of this weight range.

Growing-Finishing Pigs

Attempts to use cottonseed meal as the only supplemental protein in pig diets, when synthetic lysine was added to equal the digestible lysine content of soybean meal-supplemented diets, have resulted in inferior pig performance compared with soybean meal-supplemented diets. This suggests that other amino acids in addition to lysine are limiting due to total amount or digestibility. However, excellent performance has been obtained when cottonseed meal was fed in combination with soybean meal or fish meal or a combination of both soybean and fish meals. Many early workers have clearly demonstrated that good performance can be obtained when cottonseed meal is used in combination with other high-quality, high-protein feedstuffs provided its nutrient content and the presence of free gossypol are considered in diet formulation. (Sewell et al., 1955; Haines et al., 1957; Hale and Lyman, 1957; Hale and Lyman, 1961; Tanksley and Lyman, 1966).

Table 15.6 shows grower and finisher diets which have been used successfully in commercial practice. All diets have been formulated on a digestible lysine basis. Therefore, these diets should support performance similar to diets supplemented with soybean meal alone. The amino acid profile is somewhat higher and more balanced in the diets containing fish meal.

Table 15.6. Grower and Finisher Swine Diets Using Solvent Processed Soybean Meal (SBM), Cottonseed Meal (CSM) and Fish Meal (FM)[1]

	Grower (16%)		Finisher (14%)	
	50% SBM 50% CSM	25% SBM 25% FM 50% CSM	50% SBM 50% CSM	25% SBM 25% FM 50% CSM
Corn or Sorghum	754.6	779.6	817.0	835.0
CSM (Solvent, 41%)	111.0	108.0	80.0	78.0
Lysine-HCl (98%)	1.7	1.4	0.6	0.4
SBM (Solvent, 44%)	103.0	50.5	75.0	37.0
FM (Menhaden, 60%)	---	37.0	---	26.5
Ground Limestone	10.0	7.7	9.4	7.9
Dicalcium Phosphate	12.7	8.8	11.8	9.0
Sodium Chloride	3.0	3.0	3.0	3.0
Mineral Premix[2]	1.5	1.5	1.5	1.5
Vitamin Premix[3]	1.5	1.5	1.0	1.0
Dried $FeSO_4.H_2O$	1.0	1.0	0.7	0.7
Calculated Analysis				
Protein (%)	16.0	16.0	14.0	14.0
Total Lysine (%)	0.80	0.79	0.60	0.60
Digestible Lysine (%)	0.64	0.64	0.47	0.47
Calcium (%)	0.70	0.70	0.65	0.65
Total Phosphorus (%)	0.60	0.60	0.55	0.55
M.E. (kcal/kg)	3200	3231	3243	3264
Free Gossypol (mg/kg)[4]	333	324	240	234
Added Iron (mg/kg)	320	320	224	224

[1]Ileal lysine digestibility used: corn, 80%; SBM, 86%; FM, 83%; CSM, 56%.
[2]Trace mineral premix containing 6.7% zinc, 6.7% iron, 1.35% manganese, 0.67% copper, 0.02% iodine and 0.02% selenium.
[3]Vitamin premix containing the following per kg: A, 3.3 million IU; D_3, 330,000 IU; E, 16,500 IU; K_3, 506 mg; riboflavin, 3.3 g; pantothenic acid, 9.9 g; niacin, 11 g; choline, 220 g; B_{12}, 20 mg.
[4]Based on 0.3% free gossypol in solvent CSM.

Breeding Stock

Haught et al. (1977) compared 12% protein gestation and 16% protein lactation diets containing either 50% cottonseed meal and 50% soybean meal or soybean meal as the supplemental protein source. Litter size and pig birth weights were similar between treatments suggesting that satisfactory performance can be obtained when gestation diets contain equal supplemental protein from cottonseed and soybean meal. Pigs weaned/litter was not affected by treatment, but pig weaning weights were somewhat lower (7.0 vs.

7.96 kg/pig). This suggests that lactation diets should contain less than 50% cottonseed meal. Their diets were formulated on a protein basis which caused the cottonseed meal-supplemented lactation diet to have about 0.1% less total lysine than the soybean meal diet which probably caused the lowered feed intake of sows fed the cottonseed meal lactation diet and the lighter weaning weights. Limiting cottonseed meal to 25% of the supplemental protein during lactation and formulating diets on a digestible lysine basis should give performance similar to an all soybean meal supplemented lactation diet. Table 15.7 shows diets which have been successful in commercial practice.

Table 15.7. Sow Diets Using Solvent Processed Soybean Meal (SBM) and Cottonseed Meal (CSM) and Fish Meal (FM)[1]

	Gestation (13%)		Lactation (15%)	
	50% SBM 50% CSM	25% SBM 25% FM 50% CSM	50% SBM 50% CSM	25% SBM 25% FM 50% CSM
Corn or Sorghum	831.1	847.8	783.7	803.4
CSM (Solvent, 41%)	66.0	65.0	48.0	47.0
Lysine-HCl (98%)	1.1	0.9	0.9	0.2
SBM (Solvent, 44%)	62.0	30.5	133.0	88.0
FM (Menhaden, 60%)	---	22.0	---	32.5
Ground Limestone	10.9	7.6	9.6	8.0
Dicalcium Phosphate	21.8	19.6	17.9	14.0
Sodium Chloride	3.0	3.0	3.0	3.0
Mineral Premix[2]	1.5	1.5	1.5	1.5
Vitamin Premix[3]	2.0	2.0	2.0	2.0
Dried $FeSO_4.H_2O$	0.6	0.6	0.7	0.4
Calculated Analysis				
Protein (%)	13.0	13.0	15.0	15.0
Total Lysine (%)	0.58	0.58	0.72	0.70
Digestible Lysine (%)	0.46	0.46	0.59	0.59
Calcium (%)	0.91	0.91	0.80	0.80
Total Phosphorus (%)	0.72	0.72	0.65	0.65
ME (kcal/kg)	3213	3236	3233	3259
Free Gossypol (mg/kg)[4]	198	195	144	141
Added Iron (mg/kg)	192	192	128	128

[1]Ileal lysine digestibility used: corn, 80%; SBM, 86%; FM, 83%; CSM, 56%.
[2]Trace mineral premix containing 6.7% zinc, 6.7% iron, 1.35% manganese, 0.67% copper, 0.02% iodine and 0.02% selenium.
[3]Vitamin premix containing the following per kg: A, 3.3 million IU; D_3, 330,000 IU; E, 16,500 IU; K_3, 506 mg; riboflavin, 3.3 g; pantothenic acid, 9.9 g; niacin, 11 g; choline, 220 g; B_{12}, 20 mg.
[4]Based on 0.3% free gossypol in solvent CSM.

SUMMARY

The major limitations of cottonseed meal for swine are its lower content and digestibility of lysine compared with soybean meal, its low digestible energy content and the presence of free gossypol. However, cottonseed meals can be fed to swine with excellent results when fed in combination with other high-quality, high-protein feedstuffs and when the diets are formulated to compensate for the limitations of cottonseed meal. Cottonseed meal can provide one-half of the supplemented protein in growing, finishing and gestation diets and one-fourth of the supplemental protein in lactation diets, but all diets should be formulated on a digestible lysine basis based on the amount of lysine present in an all soybean meal-supplemented diet. Free gossypol levels greater than 100 mg/kg in the complete diet should have iron (ferrous sulfate, dried) added on a 1:1 weight basis to inactivate the free gossypol. When this is done, gossypol has essentially no effect on pig performance. Solvent processed cottonseed meal is generally higher in protein quality than screwpress meals and is preferred in swine diets.

Based on present information, the following guidelines appear valid for the use of solvent-processed cottonseed meals in swine diets:

1. Use in combination with other high quality, high-protein feedstuffs in finishing, growing, gestation and lactation diets.

2. The combination of high-protein feedstuffs and amount of supplemental protein supplied by each feed that have proven most successful for growing, finishing and gestation diets include:

(a) one-half soybean meal and one-half cottonseed meal.
(b) one-fourth soybean meal, one-fourth fish meal and one-half cottonseed meal.

3. Combinations for lactation diets include:

(a) three-fourths soybean meal and one-fourth cottonseed meal.
(b) one-half soybean meal, one-fourth fish meal and one-fourth cottonseed meal.

4. Iron, in the form of ferrous sulfate, should be added (1:1 weight ratio to free gossypol) to inactivate free gossypol in excess of 100 mg/kg. Maximum level of supplemental iron added is 500 mg/kg. Attempt to determine the amount of free gossypol in direct solvent meals from the plant of manufacture. Many solvent meals have low free gossypol ($<$ 0.05%) but if free gossypol is unknown, use 0.3% in calculations as a safety factor.

REFERENCES

Aguirre, A., Wallace, H.D. and Combs, G.E., 1960. Effects of lysine supplementation of high-gossypol cottonseed oil meal rations for baby pigs. J. Anim. Sci. 19: 1246 (Abstract).

Baliga, B.P. and Lyman, C.M., 1957. Preliminary report on the nutritional significance of bound gossypol in cottonseed meal. J. Amer. Oil Chem. Soc. 34: 21-24.

Bell, J.T. and Larsen, L.M., 1963. Amino acid supplementation of cottonseed meal protein for swine. J. Anim. Sci. 22: 832 (Abstract).

Boatner, C.H., Hall, C.M., Rollins, M.L. and Castillon, E.E., 1947. The pigment glands of cottonseed. II. Nature and properties of the gland walls. Botan. Gaz. 108: 484-494.

Carruth, F.E., 1947. Stability of bound gossypol to digestion. J. Amer. Oil Chem. Soc. 24: 58.

Chamberlain, C.C. and Lidvall, E.R., 1962. Lysine, methionine and tryptophan supplementation of degossypolized cottonseed meal for growing-finishing swine. J. Anim. Sci. 21: 990 (Abstract).

Clawson, A.J., Smith, F.H. and Albrecht, J.E., 1967. Effects of calcium hydroxide and ferrous sulfate on the toxicity and tissue residues of dietary gossypol in pigs. J. Anim. Sci. 26: 214 (Abstract).

Clawson, A.J., Smith, F.H., Osbonne, J.C. and Barrick, E.R., 1961. Effect of protein source, autoclaving and lysine supplementation on gossypol toxicity. J. Anim. Sci. 20: 547-552.

Haines, C. E., Wallace, H.D. and Koger, M., 1957. The value of soybean oil meal, low gossypol (degossypolized) solvent processed cottonseed meal, low gossypol expeller processed cottonseed meal, and various blends thereof in the ration of growing-fattening swine. J. Anim. Sci. 16: 12-19.

Hale, F. and Lyman, C.M., 1957. Effect of protein level in the ration on gossypol tolerance in growing-fattening pigs. J. Anim. Sci. 16: 364-369.

Hale, F. and Lyman, C.M., 1961. Lysine supplementation of sorghum grain-cottonseed meal rations for growing-fattening pigs. J. Anim. Sci. 20: 734-736.

Hale, F. and Lyman, C.M., 1962. Effective utilization of cottonseed meal in swine rations. J. Anim. Sci. 21: 998 (Abstract).

Haught, D.G., Tanksley, T.D., Hesby, J.H. and Gregg, E.J., 1977. Effect of protein level, protein restriction and cottonseed meal in sorghum-based diets on swine reproductive performance and progeny development. J. Anim. Sci. 44: 249-256.

Hintz, H.F. and Heitman, H., 1967. Amino acid and vitamin supplementation to barley-cottonseed meal diets for growing-finishing swine. J. Anim. Sci. 26: 474-478.

Holley, K.T., Harms, W.S., Storherr, R.W. and Gray, S.W., 1955. Cottonseed meal in swine and rabbit rations. Georgia Agricultural Experimental Station, Tifton, Mimeo Series N.S. 12.

Husby, F.M. and Kroening, G.H., 1971. Energy value of cottonseed meal for swine. J. Anim. Sci. 33: 592-594.

Knabe, D. A., Tanksley, T.D. and Hesby, J.H., 1979. Effect of lysine, crude fiber and free gossypol in cottonseed meal on the performance of growing pigs. J. Anim. Sci. 49: 134-142.

Knabe, D.A., LaRue, D.C., Gregg, E.J., Martinez, G.M. and Tanksley, T.D., 1989. Apparent digestibility of nitrogen and amino acids in protein feedstuffs by growing swine. J. Anim. Sci. 67: 441-458.

Kornegay, E.T., Clawson, A.J., Smith, F.H. and Barrick, E.R., 1961. Influence of protein source on toxicity of gossypol in swine rations. J. Anim. Sci. 20: 597-602.

LaRue, D.C., Knabe, D.A. and Tanksley, T.D., 1985. Commercially processed glandless cottonseed meal for starter, grower and finisher swine. J. Anim. Sci. 60: 495-502.

Lyman, C. M., Holland, B.R. and Hale, F., 1944. Processing cottonseed meal: A manufacturing method for eliminating toxic qualities. Ind. Eng. Chem. 36: 188-190.

NCPA. 1966. Proceedings: Conference on inactivation of gossypol with mineral salts. National Cottonseed Producers Association, Inc.

National Research Council, 1988. Nutrient Requirements of Domestic Animals No. 2. Nutrient Requirements of Swine. 9th ed. National Academy of Sciences, Washington, D.C.

Noland, P.R., Funderburg, M., Atteberry, J. and Scott, K.W., 1968. Use of glandless cottonseed meal in diets for young pigs. J. Anim. Sci. 27: 1319-1321.

Rao, S.R., Carter, F.L. and Frampton, V.L., 1963. Determination of available lysine in oilseed meal proteins. Anal. Chem. 35: 1927-1930.

Robison, W.L., 1934. Cottonseed meal for pigs. Ohio Agricultural Experimental Station, Wooster, Bulletin No. 534.

Sewell, R.F., Keen, B.C. and Carmon, J.L., 1955. An evaluation of several blends of soybean oil meal, peanut oil meal, and degossypolized cottonseed meal as protein supplements in swine rations. J. Anim. Sci. 14: 1222 (Abstract).

Smith, F.H., 1972. Effect of gossypol bound to cottonseed protein on growth of weanling rats. J. Agric. Food Chem. 20: 803-804.

Smith, K.J., 1970. Special cottonseed products report. Nutrient composition of cottonseed meal. Feedstuffs 42 (16): 19-20.

Stephenson, E.L. and Neumann, A.L., 1951. The use of detoxified cottonseed meal as protein supplement for growing pigs. Assoc. Southern Agric. Workers Proc. 48: 73.

Tanksley, T.D. and Lyman, C.M., 1966. Recent studies on the use of cottonseed meal and guar meal in swine rations at Texas A & M. Proceedings 21st Annual Texas Nutrition Conference, College Station, October, p. 38.

Tanksley, T.D., Knabe, D.A., Purser, K., Zebrowska, T. and Corley, J.R., 1981. Apparent digestibility of amino acids and nitrogen in three cottonseed meals and one soybean meal. J. Anim. Sci. 52: 769-777.

Tanksley, T.D. and Knabe, D.A., 1984. Ileal digestibilities of amino acids in pig feeds and their use in formulating diets. In: D.J.A. Cole, ed. Recent Advances in Animal Nutrition, Butterworths, London, pp. 75-95.

Wallace, H.D., Cunha, T.J. and Combs, G.E., 1955. Low-gossypol cottonseed meal as a source of protein for swine. University of Florida Agricultural Experimental Station, Gainesville, Bulletin No. 566.

Withers, W.A. and Carruth, F.E., 1917. Iron as an antidote to cottonseed meal injury. J. Biol. Chem. 32: 245-257.

CHAPTER 16

Crab Meal

T.A. Van Lunen and D.M. Anderson

INTRODUCTION

In various areas of the world, by-products of the crab fishing industry are available for use as a livestock feed. These include crab meal, crab shells, whole dehydrated blue crabs (*Callinactes sapidus*), Chinese crabs (*Eriocheir sinensis*), common crabs (*Pagurus ehrenborgi*) or whole fresh rock crab (*Cancer irroratus*) (International Network of Feed Information Centers, 1980). Primary sources for these products include both the east and west coasts of North America and many coastal areas in the Orient. At present, supply far exceeds demand.

Crab meal is the most common by-product available and is a dry product composed of shells, viscera and unextracted meat from various types of crabs which have been processed, usually for canning. Crab meal offers some potential as a protein source for swine since its protein content approaches that of other, more commonly used protein sources, such as canola meal.

PRODUCTION OF CRAB MEAL

Crab meal is produced by drying the shells, viscera and unextracted meat from various types of crabs. The crab waste is usually sun dried, but sometimes gas or oil fired burners are used. After drying, the product is ground in a hammer mill through a 1 to 2 mm screen. Nutrient content can be improved if the crab waste is first put through a 40-mesh screen to remove the larger portions of shell.

NUTRIENT CONTENT OF CRAB MEAL

The nutrient content of crab meal can be rather inconsistent due to the variability of the raw material used to produce it. The protein content of crab meal can range from 27 to 44% (Johnson, 1988), although the average is approximately 32% crude protein. It generally has a low fat content (2.0%), moderately high crude fiber content (10.6%) and extremely high ash value (41%). As a consequence, the digestible energy content is low (1380 kcal/kg). The crude fiber content arises as a result of the presence of chitin which has a molecular structure very similar to that of cellulose, differing only in the substitution of an acetylamine group for the hydroxyl group on carbon 2 of the glucose unit (Pond and Maner, 1984). A representative analysis of crab meal is presented in Table 16.1.

Table 16.1. Proximate Analysis of Crab Meal (as fed)[1]

	Mean	CV[2]
Dry Matter (%)	92.3	3
D.E. (kcal/kg)	1380	-
Crude Protein (%)	31.4	9
Crude Fiber (%)	10.5	11
Ether Extract (%)	2.0	46
Ash (%)	40.8	13

[1]National Academy of Sciences, 1971.
[2]Coefficient of Variation.

The presence of the shell has a negative effect on the availability of the protein fraction. Some of the nitrogen present in crab meal is found in a non-protein form, bound to chitin, which makes it relatively unavailable to the pig. However, other protein contributed by the viscera and residual meat can be of good value to monogastrics. The relative amounts of nitrogenous material from these two major sources could result in large differences in availability of protein and amino acids. A breakdown of amino acid composition is presented in Table 16.2.

Due to the presence of the exoskeleton of the crab, the calcium content of the meal is extremely high, ranging from 9 to 21% (Johnson, 1988). The high calcium and relatively low phosphorous content of crab meal can represent a difficult problem when attempting to balance calcium /phosphorous ratios in swine diets. Usually, a non-calcium containing source of phosphorous is required to correct the imbalance between these two minerals. The high level of calcium in the crab meal is a limiting factor for its inclusion in diets. Levels of minerals are presented in Table 16.3.

Table 16.2. Amino Acid Content of Crab Meal (% as fed)[1]

	Mean	CV[2]
Arginine	1.65	8
Histidine	0.50	16
Isoleucine	1.19	17
Leucine	1.59	12
Lysine	1.40	3
Cystine	0.92	15
Phenylalanine	1.19	20
Tyrosine	1.19	23
Threonine	1.00	4
Tryptophan	0.32	14
Valine	1.49	8

[1]National Academy of Sciences, 1971.
[2]Coefficient of Variation.

Table 16.3. Mineral Content of Crab Meal (as fed)[1]

	Mean	CV[2]
Calcium (%)	15.0	17
Chloride (%)	1.52	39
Magnesium (%)	0.88	48
Phosphorous (%)	1.57	8
Potassium (%)	0.45	21
Sodium (%)	0.94	3
Sulphur (%)	0.32	-
Copper (mg/kg)	32.8	3
Iodine (mg/kg)	0.56	-
Iron (mg/kg)	4350	81
Manganese (mg/kg)	134	38

[1]National Academy of Sciences, 1971.
[2]Coefficient of variation.

FEEDING CRAB MEAL

Starter Pigs

There does not appear to be any published information on feeding crab meal to newly weaned pigs. Until such work is conducted, it may be wise to avoid the use of crab meal in diets fed to starter pigs.

Growing-Finishing Pigs

Husby and Brundage (1979) suggested that crab meal can replace 50% of the protein provided by soybean meal in barley-soybean meal mash diets with only a slight reduction in feed efficiency. Corn-soybean meal diets can include a 25% replacement of soybean meal by crab meal with no detrimental effects, provided total inclusion of crab meal does not exceed 6.3% and an energy supplement is added. Anderson and Van Lunen (1986) reported similar results (Table 16.4).

Corn-soybean meal diets containing 5, 10 and 15% crab meal on a dry basis were fed in a mash form in grower and finisher diets. Pigs were fed diets containing crab meal that were formulated to meet the requirements for grower (20-60 kg) pigs and finisher (60-100 kg) pigs (National Research Council, 1979). The study indicated that inclusion rates above the 5% level resulted in a 13% reduction in growth rate and a significant deterioration in feed conversion at both stages tested.

Table 16.4. Performance of Growing Pigs Fed Graded Levels of Crab Meal

	Level of Crab Meal (%)			
	0	5	10	15
Grower Trial				
Daily Gain (kg)	0.80	0.73	0.72	0.73
Daily Intake (kg)	2.23	2.06	2.29	2.23
Feed Efficiency	2.79	2.82	3.18	3.05
Finisher Trial				
Daily Gain (kg)	0.91	0.87	0.76	0.84
Daily Intake (kg)	2.99	3.48	3.10	3.37
Feed Efficieny	3.29	4.00	4.08	4.01

Anderson and Van Lunen, 1986.

Palatability may be a problem with crab meal. This situation is most likely due to the dustiness of the product. Some fractions of the raw product, possibly the meat and/or viscera components, when dried and ground become extremely fine and dusty. Pigs fed crab meal diets appear to dislike this consistency. Pelleting feeds containing crab meal may overcome palatability problems associated with dustiness. In recent work with rats, reduced feed intakes were reported concurrent with increased gastric lesions, believed to be due to the sharp particulate matter in some samples of crab meal (Anderson, 1989). This may represent another factor involved with reduced feed intake.

Breeding Stock

The only work reported to date on the effects of crab meal on reproductive performance of sows has been from on-farm evaluations. Johnson (1988), in an evaluation consisting of 96 sows spread over two farms, fed barley-based diets containing approximately 15% crude protein. Half of the sows were fed a diet where all protein supplementation was derived from soybean meal while the other half were fed a similar diet with 8% crab meal, on a dry basis, making up part of the protein supplementation. No differences in number of pigs born alive, average birth weight, number of piglets weaned, total litter weaning weight, average weaning weight, total litter growth rate or average piglet growth rate were found over two parities. It was also noted that feed consumption of the crab meal diet during both the dry and lactating periods was higher than that of the commercial diet. Since the crab meal diet was lower in energy content, this increased feed intake is not unexpected. Table 16.5 shows some of the results from the sow work described above.

Table 16.5. Reproductive Performance of Sows Fed Crab Meal

	Conventional Diet	Crab Meal Diet
Feed in Gestation (kg/day)	2.36	2.61
Feed in Lactation (kg/day)	4.76	4.98
Piglets Born Alive	10.81	10.87
Mean Birth Weight (kg)	1.41	1.33
Number Weaned	9.64	9.66
Litter Weaning Weight (kg)	60.19	58.10
Mean Weaning Weight (kg)	6.22	6.08
Gain per Day per Litter (kg)	1.86	1.80
Gain per Day per Piglet (kg)	0.19	0.18

Johnson, 1988.

SUMMARY

In summary, it appears that crab meal can be included in diets for growing and breeding pigs. Because of palatability concerns, and because of negative effects on performance, it is suggested that inclusion rates be restricted to 6% for growing pigs and 8% for sows. Inclusion of crab meal in hog diets presents a challenge in balancing calcium/phosphorous ratios due to its high calcium content provided by the shell portion of the raw material. When assessing the financial benefits of using crab meal, the cost and availability of non-calcium, high phosphorous supplements must be taken into account. Despite this fact, in some instances crab meal does in fact offer a good, low cost option for replacement of some of the protein supplementation of hog diets. Dietary formulations using crab meal which have been successfully used in commercial practice are shown in Table 16.6.

Table 16.6. Suggested Crab Meal Diets

	Starter	Grower	Finisher	Sow
Corn	42.0	50.0	60.3	----
Barley	23.0	20.0	15.0	82.3
Crab Meal	5.0	5.0	5.0	8.0
Soybean meal	21.0	16.2	10.9	10.0
Sodium Tripoly phosphate	1.4	1.4	1.4	---
Monoammonium phosphate	---	---	---	0.8
Salt	0.5	0.5	0.5	0.4
Premix	0.6	0.6	0.6	0.5
Tallow	3.5	4.3	3.3	---
Molasses	3.0	3.0	3.0	---
Total	100	100	100	100
Calculated Analysis				
D.E. (kcal/kg)	3390	3390	3395	3000
Crude Protein (%)	18.0	16.0	14.0	15.0
Calcium (%)	0.88	0.83	0.78	1.26
Phosphorus (%)	0.72	0.71	0.69	0.73
Lysine (%)	1.00	0.80	0.60	----

Diets from Anderson and Van Lunen, 1986 and Johnson, 1988.

REFERENCES

Anderson, D.M., 1989. High protein feedstuffs for monogastrics. Annual Report on Research and Innovative Demonstration. Nova Scotia Department of Agriculture and Marketing, Truro; Grant in Aid Program.

Anderson, D.M. and Van Lunen, T.A., 1986. Crab meal in diets for growing-finishing pigs. Can J. Anim. Sci. 66: 1170 (Abstract).

Husby, F.M. and Brundage, A.L., 1979. King crab meal as a replacement for soybean meal in growing swine diets. Proc. 29th Alaska Scientific Conference Sea Grant Report No. 79-6.

International Network of Feed Information Centres, 1980. International Feed Descriptions, International Feed Names and Country Feed Names. Publ. No. 5.

Johnson, D., 1988. Crab meal as feed for hogs. Canada/Prince Edward Island Agriculture and Food Development Sub-Agreemant Report. Technology Development J88, pp. 1-53.

National Academy of Sciences, 1971. Atlas of Nutritional Data on Canadian and United States Feedstuffs. Washington, D.C.

National Research Council, 1979. Nutrient Requirements of Domestic Animals, No. 2. Nutrient Requirements of Swine. 8th ed. National Academy of Sciences, Washington, D.C.

Pond, W.G. and Maner, J.H., 1984. Swine Production and Nutrition. AVI Publishing Company, Westport Connecticut, pp. 405-407.

CHAPTER 17

Distillers By-Products

H.W. Newland and D.C. Mahan

INTRODUCTION

Distillers feeds are produced from the de-alcoholized fermentation residues that remain after high quality cereal grains have been fermented by yeast. During the fermentation process, nearly all of the starch in the grain is converted to alcohol and carbon dioxide while the remaining nutrients such as protein, fat, fiber, minerals and vitamins, undergo a three-fold increase in concentration. The fermentation residues contain yeast cells and other unidentified nutrients which are formed during the distillation-fermentation process. Since corn is the principal grain used, references to any by-products in this chapter will be assumed to be the products obtained after the removal of alcohol from a feedstock which is predominantly corn.

Distillers feeds have traditionally been divided into three categories including distillers dried grains (DDG), distillers dried solubles (DDS) and distillers dried grains with solubles (DDGS). Distillers dried grains are obtained after the removal of ethyl alcohol from the yeast fermentation by separating the coarse grain fraction of the whole stillage and drying it. Distillers dried grains with solubles are obtained after the removal of ethyl alcohol by distillation from the yeast fermentation by condensing and drying at least three-fourths of the solids of the whole stillage. Distillers dried solubles are obtained following the removal of ethyl alcohol by distillation from the yeast fermentation by condensing the thin stillage to a semisolid.

During the past fifty years, there has been a dramatic shift in the annual production of the three types of distillers feeds. In the 1940's, distillers dried grains were the predominant product, followed by distillers dried grains with solubles and distillers dried solubles. However, more recently, distillers dried grains with solubles have become the primary by-product and production of the other two types is minimal.

NUTRIENT CONTENT OF DISTILLERS DRIED BY-PRODUCTS

The composition of distillers feeds may be influenced by the raw materials used, as well as processing procedures and the type of equipment used in distillation. Although each item may contribute to a change in composition, it is important to point out that, in any given plant, most variables are minimized because equipment and processing procedures do not undergo major changes (Carpenter, 1970).

The chemical compositions of corn distillers dried by-products are shown in Table 17.1. It should be evident that during the process of alcohol production, a concentration of certain nutrients occurs as well as a synthesis of others. As a consequence, the nutrient content of the various by-products vary considerably from the parent source, corn.

Table 17.1. Proximate Analysis of Distillers Dried By-Products (% as fed)

	DDG	DDS	DDGS	Soybean Meal	Corn
Dry Matter	94.0	91.0	91.0	90.0	88.0
Crude Protein	27.0	28.5	27.0	48.5	8.5
Fat	9.8	8.4	9.3	0.9	3.6
Fiber	12.1	4.4	9.1	3.4	2.3
Ash	2.4	7.0	4.5	---	---
D.E. (kcal/kg)	3394	3330	3640	3680	3530

National Research Council, 1988.

Relatively high crude fiber levels (4.4-12.1%) but concurrent high fat values (8.4-9.8%) result in a digestible energy (D.E.) value for distillers dried grains slightly higher than shelled corn in the case of distillers dried grains with solubles and 94% the energy of corn for distillers dried solubles.

Distillers feeds are classified as medium-high protein supplements, due largely to the loss of starch in the fermentation process. Their crude protein content is approximately 27.5%, or 57% of the level in 48.5% soybean meal. However, the quality of protein in these feedstuffs is low due to an imbalance of several amino acids (Table 17.2).

Table 17.2. Essential Amino Acid Content of Distillers Dried By-Products (% as fed)

	DDG	DDS	DDGS	Soybean Meal	Corn
Arginine	1.10	0.96	0.96	3.67	0.43
Histidine	0.60	0.66	0.64	1.20	0.27
Isoleucine	1.00	1.30	1.38	3.13	0.35
Leucine	3.00	2.31	2.21	3.63	1.19
Lysine	0.60	0.91	0.70	3.12	0.25
Methionine	0.50	0.55	0.49	0.71	0.18
Cystine	0.20	0.44	0.29	0.70	0.22
Phenylalanine	1.20	1.46	1.47	2.36	0.46
Threonine	0.90	1.00	0.92	1.90	0.36
Tryptophan	0.20	0.86	0.69	0.69	0.09
Valine	1.30	1.52	1.48	2.47	0.48

National Research Council, 1988.

Like the basal feed source corn, distillers feeds are a good source of phosphorus, but are low in calcium (Table 17.3). However, there is considerable variation in the mineral content of these three by-products. Distillers dried solubles are considerably higher in all minerals than distillers dried grains and distillers dried grains with solubles, with the exception of selenium and sulfur.

Table 17.3. Mineral Composition of Distillers Dried By-Products (as fed)

	DDG	DDS	DDGS	SBM	Corn
Calcium (%)	0.11	0.30	0.14	0.26	0.03
Chlorine (%)	0.08	0.26	0.16	0.04	0.05
Magnesium (%)	0.07	0.59	0.16	0.30	0.11
Phosphorus (%)	0.43	1.44	0.66	0.64	0.28
Potassium (%)	0.18	1.64	0.40	2.13	0.33
Sodium (%)	0.10	0.23	0.52	0.01	0.01
Sulphur (%)	0.46	0.36	0.30	0.44	0.11
Copper (mg/kg)	48.00	81.00	52.80	20.30	3.50
Iron (mg/kg)	223.00	555.00	236.00	131.00	33.00
Manganese (mg/kg)	23.00	73.00	23.00	37.20	5.70
Selenium (mg/kg)	0.48	0.36	0.35	0.10	0.07
Zinc (mg/kg)	35.00	84.00	80.00	57.00	19.00

National Research Council, 1988.

A comparative analysIs of certain vitamins contained in distillers feeds is given in Table 17.4. As a group, they are a good source of water soluble vitamins and vitamin E. Distillers dried solubles, in particular, are rich in the B-complex vitamins compared with soybean meal, corn and other naturally occurring protein and energy sources.

Table 17.4. Vitamin Composition of Distillers Dried By-Products (as fed)

	DDG	DDS	DDGS	SBM	Corn
Vitamin E (mg/kg)	---	50.5	39.1	3.3	20.9
Niacin (mg/kg)	42.0	122.0	72.0	22.0	23.0
Pantothenic Acid (mg/kg)	6.6	22.9	13.9	14.8	5.1
Riboflavin (mg/kg)	3.3	17.0	8.3	2.9	1.1
Vitamin B_{12} (μg/kg)	0.3	7.0	1.5	---	---
Biotin (mg/kg)	0.2	1.6	0.8	0.3	0.1
Choline (mg/kg)	1000	4687	2552	2753	504
Folicin (mg/kg)	1.1	1.3	0.9	0.7	0.3
Thiamine (mg/kg)	2.0	6.6	2.8	3.1	3.7
Vitamin B_6 (mg/kg)	4.0	10.9	5.8	4.8	6.2

National Research Council, 1988.

FEEDING DISTILLERS DRIED GRAINS WITH SOLUBLES

Distillers dried grains with solubles represent the major distillers feed used in swine diets. The largest portion of the annual production goes into the manufacture of swine supplements by feed companies. However, for swine producers desirous of using distillers dried grains with solubles in their own formulations, they are generally available through major feed brokers and dealers.

Presently, distillers dried grains with solubles (DDGS) represent over 95% of the distillers feed by-products used in swine diet formulations (Hatch, 1989, personal communication). This is in contrast to a few decades ago when the vitamin-rich, distillers dried solubles were used extensively in swine diets for their B-vitamins and "unidentified growth factors" (Fairbanks et al., 1944).

Distillers dried grains with solubles are relatively high in fiber (9.1% vs. corn at 2.3%) which may limit their use as an energy source in swine ration formulation. However, the counteracting effect of a high fat level (9.3%), results in a D.E. value of 3640 kcal/kg, which compares favorably with that of corn at 3530 kcal/kg.

With an approximate crude protein content of 27%, distillers dried grains with solubles is a potential source of supplemental protein for swine. However, compared with soybean meal, the quality of protein is low due to the deficiency of several essential amino acids (Table 17.2). The lack of lysine, in particular, is critical, and to a lesser extent threonine, leucine, isoleucine and tryptophan.

While the lysine content of distillers dried grains with solubles is low in comparison with soybean meal (0.70 vs 3.12%), the availability appears to be good. Harmon (1975), using 15 kg weanling pigs in both growth trials and balance studies, demonstrated the lysine in distillers dried grains with solubles to be efficiently utilized. The study showed that distillers dried grains with solubles and synthetic lysine supported gains and feed efficiency equally well when substituted into different basal diets of corn and soybean meal.

Starter Pigs

Distillers dried grains with solubles have been successfully incorporated into starter pig feeding programs at levels up to 5% of the diet (Richards, 1979). With pigs weighing 5.9 kg, 2.5% distillers dried grains with solubles with synthetic lysine was incorporated into a basal diet of sorghum and soybean meal. Average daily gains and feed intake were slightly improved with distillers dried grains with solubles and feed efficiency was similar among treatments. Orr et al. (1981) reported similar results with pigs weaned at four weeks of age. Distillers dried grains with solubles were substituted at 2.5% and 5.0% in a basal diet of corn and soybean meal. Growth rate and feed intake were slightly improved by distillers dried grains with solubles at 2.5% of the diet. However, the 5.0% level gave no additional benefit (Table 17.5).

Table 17.5. Performance of Starter Pigs Fed Corn Distillers Dried Grains with Solubles (DDGS)

	Control	2.5% DDGS	5.0% DDGS
Daily Gain (kg)	0.27	0.30	0.27
Daily Intake (kg)	0.47	0.50	0.46
Feed Efficiency	1.74	1.65	1.74

Orr et al., 1981.

Higher levels of distillers dried grains with solubles in starter diets for pigs cannot be utilized effectively. Research by Walstrom and Libal (1980), using four week old pigs, compared a basal diet of corn and soybean meal with distillers dried grains with solubles substituted into the basal diet at 10, 20 and 30%. All diets contained 1.16% lysine. Daily gains decreased linearly as the level of distillers dried grains with solubles in the diet increased. There was no significant difference in feed intake and feed efficiency due to treatment.

Growing-Finishing Pigs

Research results indicate that distillers dried grains with solubles can serve as an alternate source of energy and protein for growing-finishing swine when incorporated at levels up to 10% of the diet (Walstrom and German, 1968; Cromwell et al., 1984; Cromwell et al., 1985). These studies evaluated 0, 2.5, 5.0 and 10% dietary distillers dried grains with solubles. All diets were fortified with adequate levels of vitamins and minerals. Daily gains and feed efficiency in the pigs were similar among treatments (Table 17.6).

Table 17.6. Effect of Graded Levels of Distillers Dried Grains with Solubles on the Performance of Growing-Finishing Swine (30-100 kg)

	% Distillers Dried Grains with Solubles			
	0	2.5	5.0	10.0
Daily Gain (kg)	0.83	0.83	0.83	0.82
Daily Intake (kg)	2.59	2.61	2.59	2.53
Feed Efficiency	3.12	3.14	3.12	3.09

Cromwell et al., 1984.

Although the incorporation of 20% dietary distillers dried grains with solubles for growing-finishing pigs appears to be feasible, the results are not as consistent as at lower levels. In two trials, diets containing 20% distillers dried grains with solubles supported equal gains and feed efficiency as did corn-soybean meal basal diets (Harmon, 1975; Cromwell et al., 1983). On the other hand, Walstrom and Libal (1969) reported reduced performance in pigs fed 20% dietary distillers dried grains with solubles. Inclusion rates of 30 and 40% distillers dried grains with solubles in the diets of growing-finishing swine have been shown to depress growth rate by 4 to 9% and feed efficiency 6 to 11% (Cromwell et al., 1983).

Synthetic amino acids are effective, to some degree, in counteracting the depressing effect of high levels of distillers dried grains with solubles in diets of growing-finishing pigs. Growth depression from 20% distillers dried grains with solubles was alleviated with synthetic lysine (Walstrom and Libal, 1969), and lysine plus tryptophan showed similar positive response when added to diets containing 30% distillers dried grains with solubles (Cromwell et al., 1983). In the latter case, supplemental threonine did not give an additional response over the lysine plus tryptophan treatment groups.

Breeding Stock

Distillers dried grains with solubles apparently can furnish all of the supplementary dietary protein and portions of the energy in diets fed to the breeding herd (Thong et al., 1978). Distillers dried grains with solubles were substituted at levels of 0, 17 and 44.2% in a corn-soybean meal basal diet. Corn and soybean were adjusted to accommodate the distillers dried grains with solubles and maintain lysine levels at 0.42% in all diets. Gestation body weight gains and lactation body weight losses did not differ among treatments. In addition, number of pigs farrowed per litter, average birth weight per pig, number of pigs weaned per litter and pig weight at weaning were similar among all dietary treatments (Table 17.7).

Table 17.7. Reproductive Performance of Gilts Fed Different Levels of Distillers Dried Grains with Solubles

	% Distillers Dried Grains with Solubles		
	0	17.7	44.2
Pigs Born Alive	8.8	8.6	8.2
Birth Weight (kg)	1.4	1.4	1.4
Pigs Weaned (28 days)	7.3	7.4	7.3
Weaning Weight (kg)	6.5	6.7	6.6
Lactation Feed Intake (kg)	123	105	112
Sow Weight Loss in Lactation	5.3	3.9	1.4

Thong et al., 1978.

FEEDING DISTILLERS DRIED SOLUBLES

Starter Pigs

The supplementary effect of distillers dried solubles to natural feed ingredients for creep-fed pigs has been demonstrated (Fairbanks et al., 1944). This research compared a basal ration of corn, wheat middlings, soybean meal, tankage and fish meal with the basal diet plus 6% distillers dried solubles. Both diets were creep-fed and continued after weaning. Pigs supplemented with distillers dried solubles continued to grow normally after weaning while those without distillers dried solubles grew poorly and it was necessary to fortify their diet in order to continue the experiment.

Growing-Finishing Pigs

The use of distillers dried solubles in diets of growing-finishing swine has been researched quite extensively. Distillers dried solubles (DDS) are a rich source of the B-complex vitamins and exceed distillers dried grains and distillers dried grains with solubles in this respect as well as containing a better balance of essential amino acids. This nutritional property was recognized early as researchers noted a significant growth response when distillers dried solubles was incorporated into diets of growing pigs fed in dry lot (Bohstedt et al., 1943; Synold, 1944).

A basal ration of corn, wheat middlings, soybean meal, tankage and fish meal fed to pigs in dry lot was shown to be nutritionally inadequate (Krider et al., 1944). During a growing-finishing trial, 50% of the pigs on the basal ration died. Supplementing the basal diet with 6 and 12% corn distillers dried solubles resulted in 8.6 and 22.6% improvement in gains, respectively, compared with the pigs which remained alive on the basal ration. Feed efficiency was also improved proportionately. The basal ration was analyzed for amino acids and water soluble vitamins and was considered adequate for growing-finishing swine at that time. However, on the basis of this experiment, the authors concluded that the recommended allowance for the B-complex vitamins was too low and that the inclusion of distillers dried solubles compensated for this deficiency.

Table 17.8 summarizes 15 experiments showing the effects of supplementing various types of basal diets with distillers dried solubles for growing-finishing pigs. Irrespective of the type of basal diet used, the addition of distillers dried solubles consistently improved performance. Improvement in average daily gain ranged from 2.8 to 34.5% and feed efficiency from 2.5 to 5.4%. Percentage improvement from distillers dried solubles was not directly related to type of basal diet used or level of distillers dried solubles. Response from distillers dried solubles was consistent, whether compared with a basal diet containing animal protein or a simple corn-soybean meal basal. Diets containing 5 and 6% distillers dried solubles resulted in improved performance equal to the higher levels of 17 to 20% distillers dried solubles.

The high concentration of B-complex vitamins contained in distillers dried solubles was largely responsible for the positive response observed in the data presented in Table 17.8 (Fairbanks et al., 1945). This study compared a basal diet of corn, wheat middlings, soybean meal, tankage and fish meal, with the basal diet plus crystalline B-complex vitamins as well as the basal diet plus 6% corn distillers dried solubles for growing-finishing pigs. Daily gain improvement over the basal was 17 and 47% from distillers dried solubles and B-complex vitamins, respectively.

Table 17.8. Effect of Distillers Dried Solubles on the Performance of Growing-Finishing Swine

	% DDS	Daily Gain (kg)		Feed Efficiency	
		Basal	DDS	Basal	DDS
Basal Diet: Corn, Soybean Meal					
Wilford, 1951	5.6	0.72	0.71	3.57	3.60
" "	6	0.51	0.61	3.87	3.75
" "	10	0.74	0.77	3.52	3.56
" "	12	0.70	0.70	3.51	3.17
" "	12	0.76	0.76	3.55	3.39
" "	20	0.90	0.85	3.96	3.18
Basal Diet: Corn, Soybean Meal, Alfalfa Meal, Tankage					
Krider et al., 1944	6	0.42	0.46	4.03	3.91
" "	12	0.42	0.52	4.03	3.91
Hanson et al., 1952	6	0.36	0.39	3.93	3.63
" "	6	0.37	0.37	4.40	4.53
Basal Diet: Corn, Soybean Meal, Alfalfa Meal, Meat Meal					
Robinson, 1950	5	0.71	0.74	3.67	3.57
Robinson, 1951	17	0.78	0.83	3.62	3.48
Robinson, 1952	18.5	0.74	0.79	3.72	3.61
Basal Diet: Corn, Soybean Meal, Wheat Middlings, Tankage, Fish Meal					
Fairbanks et al., 1944	6	0.46	0.58	4.11	3.86
" "	12	0.34	0.49	4.33	4.20

Breeding Stock

The effect of including corn distillers dried solubles in sow diets was studied by Wallace and Combs (1964). A diet containing 5% corn distillers dried solubles was compared with a basal ration of corn, oats and soybean meal. Results of this extensive study are presented in Table 17.9. Distillers dried solubles resulted in a slight improvement in live pigs per litter and number of pigs weaned per litter. However, it is not known whether this response was real since a test for statistical significance was not indicated in the data. Distillers dried solubles had no positive influence on birth weight of pigs, weaning weight or survival rate.

Table 17.9. Reproductive Performance of Sows Fed Corn Distillers Dried Solubles

	Control Sows	DDS Sows
Number Litters	144	165
Live Pigs per Litter	10.70	11.06
Dead Pigs per Litter	0.35	0.73
Fetal Resorptions per Litter	0.37	0.42
Average Birth Weight (kg)	1.33	1.33
Pigs Weaned per Litter	9.38	9.67
Weaning Weight (kg)	3.42	3.34
Percent Survival	87.7	87.4

Wallace and Combs, 1964.

An obvious carry-over or residual effect was observed in pigs suckling sows supplemented with distillers dried solubles compared with those on the basal diet of corn, tankage, and soybean meal (Krider et al., 1944). In one study, only 50% of the pigs which came from sows fed the basal ration survived, compared to a 93% survival rate in pigs from distillers dried solubles-supplemented sows.

FEEDING DISTILLERS DRIED GRAINS

The use of distillers dried grains (DDG) in swine diets is almost non-existent. Although the crude protein and energy levels are similar to distillers dried solubles and distillers dried grains with solubles, the quality of protein or balance of essential amino acids is considerably lower, as is the concentration of B-complex vitamins (see Tables 17.1 and 17.3).

In one trial, 22% distillers dried grains was fed to growing-finishing swine and compared to a basal diet of corn, tankage and alfalfa meal (Table 17.10). The pigs receiving the distillers dried grains gained 26% less, consumed 8% less feed and were 25% less efficient in feed conversion than those on the basal diet (Robinson, 1939).

Table 17.10. Performance of Finishing Swine Fed Corn Distillers Dried Grains (DDG) as a Substitute for Corn (33-94 kg)

	Control	22.5% DDG
Daily Gain (kg)	0.60	0.44
Daily Intake (kg)	2.50	2.30
Feed Efficiency	4.18	5.22

Robinson, 1939.

SUMMARY

Research has proven that distillers dried grains with solubles can effectively furnish portions of dietary protein and energy for all phases of swine production, from breeding through marketing. Distillers dried grains with solubles up to 5% in starter diets, 20% in grower diets, and 40% in gestation diets, appear to be practical levels. Where higher than 20% is used in grower-finisher diets, the addition of synthetic lysine and tryptophan may improve performance. However, cost effectiveness will depend on relative market prices of ingredients.

The nutritional properties of distillers dried solubles (the B-complex vitamins and amino acids in particular) make it the most desirable of the distillers feed sources for swine. Extensive research in the past has shown improved performance from distillers dried solubles in growing pigs of all ages. Growth response appeared to be primarily from the B-complex vitamin concentration and was effective at dietary levels from 5 to 20%. Effective replacement of protein and energy was also evident. Whether the same growth response from distillers dried solubles would be manifest today, with the high level of vitamin and trace mineral fortification in swine diets, is not known. However, due to the very limited supply, only minimal amounts of this by-product are being incorporated into swine feed formulations at this time.

Distillers dried grains are the least desirable, nutritionally, of the distillers feeds for swine feeding. For this reason, coupled with the lack of production by the distillation industry, distillers dried grains are an insignificant ingredient in the swine industry today.

REFERENCES

Bohstedt, G., Grummer, R.H. and Ross, O.B., 1943. Cattle manure and other carriers of B complex vitamins in rations for pigs. J. Anim. Sci. 2: 373 (Abstract).

Carpenter, L.E., 1970. Nutrient compositon of distillers feeds. Proc. 25th Distillers Feed Conference, 25: 54-61.

Cromwell, G.L., Stahly, S.T., Monegue, H.J. and Overfield, J.R., 1983. Distillers dried grains with solubles for growing-finishing swine. Kentucky Agricultural Experimental Station, Lexington, Swine Research Report No. 274, pp. 30-32.

Cromwell, G.L., Stahly, S.T. and Monegue, H.J., 1984. Distillers dried grains with solubles for growing-finishing swine. Kentucky Agricultural Experimental Station, Lexington, Swine Research Report No. 284, pp. 15-16.

Cromwell, G.L., Stahly, S.T. and Monegue, H.J., 1985. Distillers dried grains with solubles and antibiotics for weanling swine. Kentucky Agricultural Experimental Station, Lexington, Swine Research Report No. 292, pp. 10-11.

Fairbanks, B.W., Krider, J.L. and Carroll, W.E., 1944. Distillers by-products in swine rations. I. Creep-feeding and growing fattening rations. J. Anim. Sci. 3: 29-40.

Fairbanks, B.W., Krider, J.L. and Carroll, W.E., 1945. Distillers by-products in swine rations. III. Dried corn distillers solubles, alfalfa meal, and crystalline B vitamins compared for growing-fattening pigs in dry lot. J. Anim. Sci. 4: 420-429.

Harmon, B.G., 1975. The use of distillers dried grains with solubles as a source of lysine for swine. Proc. 30th Distillers Feed Conference, 30: 23-28.

Krider, J.L., Fairbanks, B.W. and Carroll, W.E., 1944. Distillers by-products in swine rations. II. Lactation and growing-fattening rations. J. Anim. Sci. 3: 107-119.

National Research Council, 1988. Nutrient Requirements of Domestic Animals, No. 2. Nutrient Requirements of Swine. 9th ed. National Academy of Sciences, Washington, D.C.

Orr, D.E., Owsley, W.F. and Tribble, L.F., 1981. Use of corn distillers dried grains, dextrose, and fish meal. Proc. 29th Swine Short Course, Texas Agricultural Experimental Station, Lubbock, pp. 48-50.

Richards, G., 1979. Effect of protein sources in isolysine starter diets. Proc. 27th Swine Short Course, Texas Agricultural Experimental Station, Lubbock, pp. 30-34.

Robinson, W.L., 1939. Substitutes for corn for growing pigs. Ohio Agricultural Experimental Station, Wooster, Research Bulletin No. 607, pp. 42-45.

Robinson, W.L., 1950. Vitamin concentrates and dried distillers grain solubles as sources of B vitamins for pigs. Ohio Agricultural Experimental Station, Wooster, Mimeo Series for Swine No. 71.

Robinson, W.L., 1951. Antibiotics, B12, cobalt, B vitamin supplements, and fish meal with soybean oil meal for pigs in drylot. Ohio Agricultural Experimental Station, Wooster, Mimeo Series for Swine No. 70.

Robinson, W.L., 1952. Materials supplemental to corn, soybean oil meal, ground alfalfa and minerals for pigs in drylot. Ohio Agricultural Experimental Station, Wooster, Mimeo Series for Swine No. 70, pp. 6-9.

Synold, R.E., 1944. Distillers feed products in rations for growing pigs. J. Anim. Sci. 3: 455 (Abstract).

Thong, L.A., Jensen, A.H., Harmon, B.G. and Cornelius, S.G., 1978. Distillers dried grains with solubles as a supplemental protein source in diets for gestating swine. J. Anim. Sci. 46: 674-677.

Wilford, E.J., 1951. Dried distillers solubles as a protein supplement for growing-fattening hogs in drylot. Kentucky Agricultural Experimental Station, Lexington, Research Bulletin No. 577: 1-16.

Wallace, H.D. and Combs, G.E., 1964. Corn distillers dried solubles in the sow ration. Proc. 19th Distillers Feed Conference, 19: 30-36.

Walstrom, R.C. and German, C., 1968. A study of distillers by-products in growing-finishing swine rations. South Dakota Agricultural Experimental Station, Brookings, Animal Science Series No. 68-27, pp. 1-5.

Walstrom, R.C. and Libal, G.W., 1969. A study of lysine and distillers dried grains with solubles in growing-finishing swine rations. South Dakota Agricultural Experimental Station, Brookings, Animal Science Series No. 69-35, pp. 1-3.

Walstrom, R.C. and Libal, G.W., 1980. Effect of distillers dried grains with solubles in pig starter diets. South Dakota Agricultural Experimental Station, Brookings, Series for Swine No. 80-6, pp. 14-16.

CHAPTER 18

Fababeans

P.A. Thacker

INTRODUCTION

The cereal grains commonly used as feed in livestock production contain insufficient quantities of several of the indispensible amino acids such as lysine, threonine and the sulfur-containing amino acids to meet the amino acid requirements of the rapidly growing pig. Therefore, it is essential that the pig's diet contain a supplementary source of protein which will supply additional quantities of these limiting amino acids.

Soybean meal is the most commonly used source of supplementary protein for swine production and it is generally a very consistent, high quality product. However, as transportation costs increase, swine producers will be forced to maximize the use of locally produced feedstuffs. Therefore, it is important that alternative sources of supplemental protein be developed. One crop which appears to have considerable potential for use in swine diets is the fababean (*Vicia faba*).

GROWING FABABEANS

The fababean (also known as the horsebean and tickbean) is an annual belonging to the legume family. It is well adapted to the moister areas of production and thrives under cool growing conditions (Evans and Rogalsky, 1974). The plant is sensitive to moisture stress and performs poorly in areas where moisture supply is limited. From a maturity standpoint, fababeans are relatively slow maturing, requiring about 115 days to mature (Aherne, 1974). However, this late maturity can be compensated for by early seeding as fababeans are very frost tolerant during the seedling stage (Agriculture Canada, 1975).

Fababeans can be grown on most soil types but appear to prosper best on sandy loams and medium to heavy clays that have good water holding capacity (Alberta Agriculture, 1982). They generally perform poorly in shallow soils with a compacted layer near the surface.

One advantage of growing fababeans is that they condition the soil and significantly increase its nitrogen level (McGregor et al., 1956). As a result, cereals or oilseed crops that are grown on fababean stubble will yield as well as, or better than, a similar crop grown on an adjacent or nearby summerfallow. In addition, properly inoculated fababeans require little or no nitrogen fertilizer, which will reduce production costs.

Unfortunately, fababean production is not without its problems. Fababeans are poor competitors with weeds, particularly in the seedling stage (Agriculture Canada, 1975). This means that weed control is of utmost importance for successful production of this crop. In addition, fababeans are susceptible to diseases common to rapeseed, sunflowers and other legume and oilseed crops which limit their place in a crop rotation (Evans and Rogalsky, 1974). Finally, conventional harvesting equipment may prove inadequate for harvesting heavy stands of fababeans.

NUTRIENT CONTENT OF FABABEANS

The chemical composition of fababeans and other commonly used feedstuffs are compared in Table 18.1. The digestible energy content of the fababean (3263 kcal/kg) is lower than that of most of the commonly used protein supplements (National Research Council, 1988). Fababeans have a fairly low fat content (1.0%) which partially accounts for its lower digestible energy content (Marquardt et al., 1975; Eden, 1968). In addition, the relatively high crude fiber content of fababeans (8.2%) may also reduce the digestible energy level (Marquardt et al., 1975).

Table 18.1. Chemical Analysis of Fababeans (% as fed)

	Barley	Soybean Meal	Canola Meal	Faba-Beans
Dry Matter (%)	89.0	90.0	94.0	89.0
D.E. (kcal/kg)	3086	3860	2998	3263
Crude Protein (%)	11.6	48.5	35.0	26.0
Crude Fiber (%)	5.1	3.9	12.4	8.2
Ether Extract (%)	1.8	1.3	1.8	1.0

National Research Council, 1988.

The average crude protein content of fababeans is approximately 26% with a range from 24 to 30% (Bond and Toynbee-Clarke, 1968; Evans et al., 1972; Hebblethwaithe and Davies, 1971). This is lower than the level of protein found in most of the commonly used protein supplements and therefore higher levels must be utilized when supplementing a swine ration with fababeans than with soybean meal. A common replacement scheme is that two units of fababeans are equivalent in feeding value to the combination of one unit of soybean meal and one unit of barley.

A comparison of the essential amino acid content of fababeans and other feedstuffs is presented in Table 18.2. Fababean protein is relatively high in lysine but like most legume seeds, it is deficient in methionine (Clarke, 1970; Wilson and McNab, 1972; Marquardt and Campbell, 1974). A high level of cystine partially overcomes this deficiency (Blair, 1977). The balance of other amino acids appears adequate.

Table 18.2. Essential Amino Acid Content of Fababeans (% as fed)

	Barley	Canola Meal	Soybean Meal	Fababeans
Arginine	0.52	2.32	3.67	2.45
Histidine	0.24	1.07	1.20	0.67
Isoleucine	0.46	1.51	2.13	1.08
Leucine	0.75	2.65	3.63	1.98
Lysine	0.40	2.27	3.12	1.68
Methionine	0.16	0.68	0.71	0.20
Phenylalanine	0.58	1.52	2.36	1.13
Threonine	0.36	1.71	1.90	0.96
Tryptophan	0.15	0.44	0.69	0.23
Valine	0.57	1.94	2.47	1.21

National Research Council, 1988.

The vitamin content of fababeans and other commonly utilized protein supplements are shown in Table 18.3. Fababeans contain lower levels of biotin, choline, niacin, pantothenic acid and riboflavin in comparison with soybean meal or canola meal (Aherne and Lewis, 1978). However, fababeans contain a higher level of thiamine. As a result of its lower vitamin content, diets containing a high level of fababeans may require a specially formulated premix in order to supply a nutritionally balanced diet.

Table 18.3. Vitamin Composition (mg/kg) of Fababeans (as fed)

	Fababeans	Canola Meal	Soybean Meal
Biotin	0.1	0.9	0.3
Choline	1670	6700	2794
Niacin	22.0	188.0	60.0
Pantothenic Acid	3.0	12.5	13.3
Riboflavin	1.6	3.1	2.9
Thiamine	5.5	1.8	1.7

Aherne and Lewis, 1978.

An indication of the mineral content of the fababean is given in Table 18.4. It can be seen that fababeans are a relatively poor source of calcium and are low in iron and manganese (Aherne and Lewis, 1978). In fact, extra supplementation of manganese appears to be required should the diet contain a large proportion of fababeans (Clarke, 1970; Presber, 1972).

Table 18.4. Mineral Composition of Fababeans (as fed)

	Fababeans	Canola Meal	Soybean Meal
Calcium (%)	0.1	0.6	0.3
Magnesium (%)	0.1	0.5	0.3
Phosphorus (%)	0.5	1.0	0.7
Potassium (%)	1.2	1.1	2.0
Sodium (%)	0.8	---	0.3
Copper (mg/kg)	4.1	5.5	36.3
Iron (mg/kg)	70.0	155.0	120.0
Manganese (mg/kg)	8.4	47.5	29.3
Selenium (mg/kg)	---	0.9	0.1
Zinc (mg/kg)	42.0	60.0	27.0

Aherne and Lewis, 1978.

UNDESIRABLE CONSTITUENTS IN FABABEANS

Fababeans contain several anti-nutritional factors which may impair animal performance if they are included at too high a level in the diet. Fababeans contain between 0.3 and 0.5% tannins (Kadirvel and Clandinin, 1974). The presence of these tannins may lead to a slight reduction in feed intake when high levels of fababeans are fed to swine (Singleton and Kratzer, 1969). In addition, tannins may inhibit the retention of certain dietary nutrients (McLeod, 1974; Marquardt et al., 1976).

Fababeans contain a fairly high level of trypsin inhibitor (Wilson et al., 1972; Bhatty, 1974). However, the level of trypsin inhibitor in fababeans is lower than the level found in raw soybeans. It is estimated that fababeans contain 2.3 units of trypsin inhibitor activity compared with 25.5 units in raw soybeans (Marquardt et al., 1974). The trypsin inhibitor may cause a reduction in the digestibility of protein if the diet contains a high level of fababean (Chubb, 1982).

Another undesirable factor, hemagglutinin, was found to be present at a level of 2900 to 4200 rabbit RBC units per gram in four cultivars of raw fababeans (Marquardt et al., 1974). The corresponding values for soybeans, wheat and barley were 650, 50 and 5 RBC units. Haemagglutinins attach to the cells lining the intestinal wall causing a non-specific interference with the digestion of dietary nutrients (Chubb, 1982). Growth inhibition has also been reported (Liener, 1974).

The activity of several of the anti-nutritional factors present in fababeans has been shown to be reduced by autoclaving (Marquardt et al., 1974; Marquardt et al., 1976). However, feed efficiency and daily gain are not improved by heat treatment of the bean (Aherne et al., 1977; Ivan and Bowland, 1976). Since autoclaving is a relatively expensive process, it is unlikely that it would be cost effective.

FEEDING FABABEANS TO SWINE

Starter Pigs

As a result of the anti-nutritional factors present in fababeans, it is recommended that they not supply all of the supplementary protein required to provide a balanced diet for swine. Although there has not been a great deal of research conducted to determine the value of including fababeans in starter diets, the few reports available suggest that fababeans should not be used at a level in excess of 15% of the diet (Table 18.5). At higher levels of inclusion, there appears to be a problem with palatability and as a consequence of the lower feed intake, growth rates are impaired.

Table 18.5. Performance of Starter Pigs (10-25 kg) Fed Fababeans

	Fababean Level (%)				
	0	10	15	20	25
Daily Gain (kg)	0.55	0.54	0.54	0.48	0.51
Daily Intake (kg)	1.27	1.24	1.20	1.11	1.17
Feed Efficiency	2.31	2.31	2.22	2.31	2.30

Aherne et al., 1977.

Growing-Finishing Pigs

Several feeding trials have demonstrated a reduction in daily gain when fababeans constituted the sole source of supplementary protein in growing-finishing rations (Livingstone et al., 1970; Stothers, 1974; Castell, 1976; Onaghise and Bowland, 1977). The results of a feeding trial conducted at the University of Alberta clearly demonstrate the adverse effects of inclusion of high levels of fababeans in the diet of the growing-finishing pig (Aherne et al., 1977). As the level of fababean in the diet increased, daily gain and feed efficiency decreased (Table 18.6). The decrease in performance was particularly evident at levels of inclusion greater than 20% of the diet. Although fababeans contain relatively low levels of methionine, supplementation of diets containing fababeans with synthetic methionine has not proven to be beneficial (Aherne and McAleese, 1964; Cole et al., 1971).

Table 18.6. Performance of Pigs Fed Diets Containing Fababeans

	Fababean Level (%)					
	0	10	15	20	25	30
Daily Intake (kg)	1.87	1.94	1.89	1.90	1.98	1.95
Daily Gain (kg)	0.68	0.66	0.65	0.65	0.61	0.58
Feed Efficiency	2.75	2.92	2.89	2.94	3.26	3.40

Aherne et al., 1977.

Perhaps the most attractive characteristic of fababeans is that they can be utilized directly on the farm without further processing. Unlike other home grown protein supplements such as canola meal or sunflower meal which require oil extraction or dehulling before they can be utilized, raw fababeans can be mixed directly into the ration provided that they have been ground previously.

Fababean oil contains a high content of unsaturated fatty acids. The presence of these unsaturated fatty acids can lead to an early development of rancidity after the crop has been processed (Clarke, 1970). Therefore, processed fababeans should not be stored for more than a week before using. Despite the fairly high level of unsaturated fatty acids present in fababeans, rations containing up to 30% fababeans have been found to have no effect on the taste or texture of the carcass of pigs consuming them (Hansen and Clausen, 1969).

Breeding Stock

The effect of including fababeans in breeding stock diets has received little attention. However, Danish workers have reported a significant reduction in litter size, both at birth and weaning, when fababeans were included at high levels in gestation rations (Nielsen and Kruse, 1974; Table 18.7). In addition, milk yield and milk protein content have been reported to be reduced when fababeans arc included at high levels in lactation rations. Some caution has to be exercised in the feeding of fababeans to pregnant sows and young animals as the fababeans generate stomach gasses and may cause a tendency towards constipation (Presber, 1972).

Table 18.7. Effect of Fababeans on Reproductive Performance

	Fababeans (%)		
	0	17	34
Pigs Born Alive	12.0	10.5	9.7
Birth Weight (kg)	1.4	1.4	1.4
Pigs Weaned (8 Weeks)	9.2	8.3	8.1
Weaning Weight (kg)	19.1	19.0	19.6
Milk Yield (kg/Day)	6.3	5.5	5.2
Milk Dry Matter (%)	19.0	18.9	18.6
Milk Protein (%)	7.1	7.0	6.5
Milk Fat (%)	5.5	5.6	5.7

Nielsen and Kruse, 1973.

SUMMARY

Fababeans have much to offer as a protein supplement and a considerable reduction in feed costs may be achieved by their inclusion in the diet of pigs. However, inclusion at too high a level will impair animal performance. Therefore, it is recommended that fababeans not be included at levels greater than 15% in starter diets, 20% in grower-finisher rations and 10% in rations fed to breeding stock.

REFERENCES

Agriculture Canada, 1975. Growing and Using Fababeans. Canada Department of Agriculture, Ottawa, Publication No. 1540.

Aherne, F.X. and McAleese, D.M., 1964. Evaluation of tick beans (*Vicia faba*) as a protein supplement in pig feeding. Proc. Royal Soc. Dubl. B. 1: 113-121.

Aherne, F.X., 1974. Fababeans as a protein supplement. 53rd Annual Feeders Day Report, University of Alberta, Edmonton, pp. 53-54.

Aherne, F.X., Lewis, A.J. and Hardin, R.T., 1977. An evaluation of fababeans as a protein supplement for swine. Can. J. Anim. Sci. 57: 321-328.

Aherne, F.X. and Lewis, A.J., 1978. The nutritive value of fababeans and low glucosinolate rapeseed meal for swine. Adv. Exp. Med. Biol. 105: 453-471.

Alberta Agriculture, 1982. Fababean Production in Alberta. Edmonton. Agdex 142/20-7.

Bhatty, R.S., 1974. Chemical composition of some faba bean cultivars. Can. J. Plant Sci. 54: 413-421.

Blair, R., 1977. Fababeans: An improved crop for animal feeding. Feedstuffs 49:15-21.

Bond, D.A. and Toynbee-Clarke, G., 1968. Protein content of spring and winter varieties of field beans (*Vicia faba*) sown and harvested on the same dates. J. Agric. Sci. (Camb) 70: 403-404.

Castell, A.G., 1976. Comparison of faba beans (*Vicia faba*) with soybean meal or field peas (*Pisum sativum*) as protein supplements in barley diets for growing-finishing pigs. Can. J. Anim. Sci. 56: 425-432.

Chubb, L.G., 1982. Anti-nutritional factors in animal feedstuffs. In: W. Haresign, ed. Recent Advances in Animal Nutrition, Butterworths, London, pp. 21-37.

Clarke, H.E., 1970. The evaluation of the field bean (*Vicia faba*) in animal nutrition. Proc. Nutr. Soc. 29: 64-73.

Cole, D.J.A., Blades, R.J., Taylor, R. and Luscombe, J.R., 1971. Field beans (*Vicia faba*) in the diets of bacon pigs. Exper. Husb. 20: 6-11.

Eden, A., 1968. A survey of the analytical composition of field beans (*Vicia faba*). J. Agric. Sci. (Camb) 70: 299-301.

Evans, L.E., Seitzer, J.F. and Bushuk, W., 1972. Horsebeans - A protein crop for Western Canada. Can. J. Plant Sci. 52: 657-659.

Evans, L.E. and Rogalsky, J.R., 1974. Fababean Production. Manitoba Department of Agriculture, Winnipeg, Publication No. 541.

Hansen, V. and Clausen, H., 1969. Hestebonner (*Vicia faba*) som foder til slagterisvin. 374. betetning fra forsogstaboraloriet Vdgivit af Statens Husdyrbrugsudvalg, Kobenhavn.

Ivan, M. and Bowland, J.P., 1976. Digestion of nutrients in the small intestine of pigs fed diets containing raw and autoclaved fababeans. Can. J. Anim. Sci. 56: 451-456.

Kadirvel, R. and Clandinin, D.R., 1974. The effect of fababeans (*Vicia faba*) on the performance of turkey poults and broiler chicks from 0-4 weeks of age. Poult. Sci. 53: 1810-1816.

Liener, I.E., 1974. Phytohemagglutinins: Their nutritional significance. J. Agric. Food Chem. 22: 17-22.

Livingstone, R.M., Fowler, V.R. and Woodham, A.A., 1970. The nutritive value of field beans (*Vicia faba*) for pigs. Proc. Nutr. Soc. 29: 46A-47A.

Marquardt, R.R. and Campbell, L.D., 1974. Deficiency of methionine in raw and autoclaved fababeans in chick diets. Can. J. Anim. Sci. 54: 437-442.

Marquardt, R.R., Campbell, L.D., Stothers, S.C. and McKirdy, J.A., 1974. Growth responses of chicks and rats fed diets containing four cultivars of raw or autoclaved fababeans. Can. J. Anim. Sci. 54: 177-182.

Marquardt, R.R., McKirdy, J.A., Ward, T. and Campbell, L.D., 1975. Amino acid, hemagglutinin and trypsin inhibitor levels, and proximate analysis of faba beans (*Vicia faba*) and faba bean fractions. Can. J. Anim. Sci. 55: 421-429.

Marquardt, R.R., Ward, T., Campbell, L.D. and Cansfield, P., 1976. Purification of the growth inhibitor in fababeans (*Vicia faba*). Proc. West. Sec. Amer. Soc. Anim. Sci. 27: 138-141.

McGregor, W.G., MacLean, A.J. and Wallen, V.R., 1956. Field beans in Canada. Canada Department of Agriculture, Ottawa, Publication No. 843.

McLeod, M.N., 1974. Plant tannins: Their role in forage quality. Nutr. Abst. Rev. 44: 803-815.

National Research Council, 1988. Nutrient Requirements of Domestic Animals, No. 2. Nutrient Requirements of Swine. 9th ed. National Academy of Sciences, Washington, D.C.

Neilsen, H.E. and Kruse, P.E., 1974. Effect of dietary horse beans (*Vicia faba*) on colostrum and milk composition and milk yield in sows. Livest. Prod. Sci. 1: 179-185.

Onaghise, G.T. and Bowland, J.P., 1977. Influence of dietary fababeans and cassava on performance, energy and nitrogen digestibility and thyroid activity of growing pigs. Can. J. Anim. Sci. 57: 159-167.

Presber, A.A., 1972. European experience with the small faba bean (Horse bean). Canada Grains Council. Winnipeg, Manitoba.

Singleton, V.L. and Kratzer, F.H., 1969. Toxicity and related physiological activity of phenolic substances of plant origin. J. Agric. Food Chem. 17: 497-512.

Stothers, S.C., 1974. Fababeans in swine production. Proc. 1st National Fababean Conference, Winnipeg, Manitoba. pp. 90-95.

Wilson, B.J., McNab, J.M. and Bentley, H., 1972. Trypsin inhibitor activity in the field bean (*Vicia faba*). J. Sci. Food Agric. 23: 679-684.

Wilson, B.J. and McNab, J.M., 1972. The effect of autoclaving and methionine supplementation on the growth of chicks given diets containing field beans (*Vicia faba*). Brit. Poult. Sci. 13: 67-73.

CHAPTER 19

Field Peas

A.G. Castell

INTRODUCTION

Field peas represent a source of both energy and supplementary amino acids for pigs. Forage peas, also termed "whole crop peas," have been traditionally used for ruminant livestock but may have a place in pig production under certain circumstances (Aman and Graham, 1987; Graham and Aman, 1987; Hakansson and Malmlof, 1984). In addition, industrial methods can be used to separate the constituents of peas to produce starch and high-protein concentrates (Bhatty and Christison, 1984; Wright, 1985). However, intentional use of these products for pigs is unlikely.

Although human experience over many years would imply that peas should be nourishing for pigs, consumption of the raw (uncooked) ingredient may produce results which discourage high rates of dietary inclusion. Economic aspects, such as the relative cost of alternative sources of required nutrients, will also influence the usage of peas under practical conditions.

GROWING FIELD PEAS

It is generally agreed that all cultivated peas belong to the same species, *(Pisum sativum* L.), but different subspecies may not always be identified by the same common name. According to Davies (1976), the distinction initially made was between varieties grown in fields (subsp. *arvense*; used for animal feed and characterised by long vines, colored flowers, small pods and dark seeds), garden peas (subsp. *hortense*; typically with white flowers and large green or yellow seeds, used for humans or monogastric livestock) and sugar peas (subsp. *axiphium*) which have edible pods and are intended for human consumption. There are also differences in the composition of seeds produced by "spring" and "winter" cultivars and between the "round" and "wrinkled" mature seeds from specific varieties.

Although evidence suggests that the pea crop was first domesticated in the Middle East, where species of wild pea still exist, it can now be found world-wide. Gane (1985) indicated that dry pea production occupied 7,395,000 ha worldwide with the major producing countries being the USSR (57%), China (21%), India (8%), Africa (6%), Europe (3.7%) and South America (2%).

Temperate regions provide the most suitable climatic conditions but satisfactory yields can be obtained in warmer regions if peas are grown in the cool season or at higher altitudes (Gane, 1985). Yields of "garden" or "vining" types can exceed 12 tonne/ha but varieties intended for harvesting as dry seed ("combining" or "field" peas) are lower-yielding (Gane, 1985). Peas prefer well-drained soil of good texture (Gane, 1985).

NUTRIENT COMPOSITION OF PEAS

Peas are intermediate to cereal grains and soybeans in their content of crude protein. They contain more than twice as much crude fiber as corn although the pea hull (testa) represents less than 10% of the seed weight. However, the fat content of peas is low. A chemical analysis of field peas and other commonly used feedstuffs is presented in Table 19.1.

Table 19.1. Proximate Composition of Cereals, Peas and Soybean Seed and Meal (% as fed)

	Barley	Corn	Peas	Soybean Seed	Soybean Meal
Dry Matter	89	88	89	92	90
Crude Protein	11.5	8.5	23.2	39.2	48.5
Ether Extract	1.7	3.6	1.1	17.2	0.9
Crude Fiber	5.0	2.3	5.5	5.3	3.4

National Research Council, (1982) for soybean seed and (1988) for other ingredients.

Digestible energy (D.E.) levels, in kcal/kg dry matter (DM), range from about 3610 (Edwards et al., 1987) to 3980 (Grosjean and Gatel, 1986) in comparison to corresponding values for soybean meal from 3877 to 4088 (National Research Council, 1988). Marquardt and Bell (1988) concluded that a value of 3800 kcal D.E. per kg DM was realistic for peas fed to growing-finishing pigs. When comparing peas with soybean meal as energy sources, D.E. may not be an appropriate measure of relative availabilities. Taverner and Curic (1983) reported that the ileal digestibilities of energy in diets containing 30% of peas or soybean meal were 79.5 and 74.4%, respectively, but both had the same net energy value (2318 kcal per kg DM).

The available energy content of peas is affected by variety and was reported to be higher (3750-3920 vs. 3400 kcal D.E. per kg DM) in white-flowered compared to colored pea types (Hlodversson, 1987). Higher energy levels were reported (Henry and Bourdon, 1977) in round-seeded compared to wrinkle-seeded varieties and this has been attributed to the differences in carbohydrate composition between the two types (Table 19.2).

Table 19.2. Carbohydrate Constituents of Various Pea Types (% DM)

	Round vs. Wrinkled[1]		Spring vs. Winter[2]	
NFE	67.0	60.0	63.9	60.1
Starch	47.9	32.9	50.0	47.5
Total Sugars	8.0	10.2	7.0	7.0
Sucrose	3.4	4.2	---	---
Total Oligosaccharides	6.1	11.4	---	---

[1]Cerning-Beroard and Filiatre, 1977.
[2]Grosjean and Gatel, 1986.

Starch is the predominant form of storage energy in peas. Varietal differences occur in the shape of the starch grains (Cousin, 1983) and in the proportions of amylose and amylopectin in the starch (Grosjean, 1985). The presence of oligosaccharides (raffinose, stachyose and verbascose) represent a potential problem as these are a major cause of flatulence in humans who lack galactosidase enzymes. These enzymes are also deficient in swine (Aman and Graham, 1987). Crude fiber levels normally range from 5 to 8%, dry matter basis, with the pea hull containing up to 75% fiber (Wright, 1985). The crude fiber was reported to be 72% digestible, compared to 97% for the nitrogen-free extract (Grosjean, 1985).

Although fat is a minor constituent of peas, linoleic acid represents from 45 to 50% of the total fatty acids and unsaturated fatty acids are a majority (83-84%) of these (Grosjean, 1985; Ogle and Hakansson, 1988).

Analyses for crude protein content of peas reveal wide ranges, from 15.5 to 39.7% of the dry matter (Monti, 1983) and from 21.4 to 30.3% of the dry matter (Ogle and Hakansson, 1988). This variability is not only a matter of different genotypes, for it was shown by Matthews and Arthur (1985) that crude protein content is not very consistent for seeds from the same plant, or even the same pod. However, on average, seeds from "winter" peas tend to have more protein than spring peas (27 vs. 25% of DM; Grosjean and Gatel,

1986) and "wrinkle-seed" exceed "round-seed" cultivars in crude protein content (Cerning-Beroard and Filiatre, 1977; Henry and Bourdon, 1977).

The differences in content of crude protein could be a result of the form in which nitrogen is present in the mature pea. Seed proteins can be separated according to their solubility in specific solutions, such as water-soluble albumins or globulins soluble in salt solutions. The albumins, according to Boulter (1977), contain most of the metabolic proteins while the other solubility fractions consist of storage proteins; all crop-legume protein appears to consist of 10 to 20% albumin, about 70% globulin and 10 to 20% glutelin (Table 19.3).

Table 19.3. Solubility Classes of Protein from Legumes (g/100 g protein)

	Albumins	Globulins	Glutelins
Broadbean	20	60	15
Common Bean	15	75	10
Lupin	10-20	80-90	0
Pea	21	66	12
Soybean	10	90	0

Boulter, 1977.

The different storage globulins can be separated according to their rate of sedimentation to obtain typical proteins such as 11S (e.g. legumin) and 7S (e.g. vicilin) molecular species with different, but characteristic, amino acid compositions. Davies and Domoney (1983) reported differences in the rate of synthesis of legumin between round and wrinkled peas (37.7 and 18.6% legumin of globulin in mature seed, respectively) and identified the relative effects of deficiencies of potassium, phosphorus and sulphur on seed protein composition. Pea varieties do not contain a constant ratio of the major storage proteins. For example, Wright (1985) reported a range for legumin from 20 to 60% of the total (legumin + vicilin). Given that the nitrogen content of specific amino acids and proteins is not constant (Huet et al., 1987), it is obvious that determination of crude protein is not a very useful measure of the value of peas relative to other ingredients for pig diets.

It is not surprising that published analyses of the amino acid content of peas reveal a range of values although there is general agreement with respect to the ranking of peas relative to cereals and soybeans. Pea protein appears to be a relatively poor source of the sulfur amino acids and only just adequate in tryptophan when compared to National Research Council requirements. However, on average, it is at least equal to soybean protein in content of the other essential amino acids (Table 19.4).

In view of the variability in reported crude protein contents, caution is necessary in the application of mean values for amino acid content. For example, Monti (1983) reported the following ranges, as g/100 g pea protein:

lysine (6.9-8.2), methionine (0.6-1.0), cystine (0.8-1.7) and tryptophan (0.7-1.9). The genetic variability for amino acid composition is very low (Cousin et al., 1985) and levels of essential amino acids can be influenced by the prior history of the pea sample. This was clearly suggested in results presented by Reichert and MacKenzie (1982) who found cystine, lysine, methionine and threonine in a single cultivar (Trapper) were each inversely correlated to the total N content of the seed. A similar study by Davies (1984a) implied that the corresponding changes in amino acid profile were not nutritionally significant.

Table 19.4. Typical Content (g/100 g crude protein) of Some Ingredients Compared to Amino Acids Required in Grower Diets

	Barley	Corn	Pea	Soybean	NRC[1]
Arginine	4.3	4.5	6.2	7.3	1.7
Histidine	2.0	2.7	2.9	2.5	1.5
Isoleucine	3.8	3.6	5.1	5.4	3.1
Leucine	6.3	12.6	7.9	7.7	4.0
Lysine	3.3	2.6	6.8	6.2	4.4
Methionine + Cystine	3.1	4.1	2.0	2.8	2.7
Phenylalanine + Tyrosine[2]	7.8	9.0	8.0	7.8	4.4
Threonine	3.1	3.6	4.1	4.2	3.2
Tryptophan	1.3	0.8	1.0	1.4	0.8
Valine	4.8	4.6	5.6	5.3	3.2

[1]National Research Council, 1988 (for 20 to 50 kg pig).
[2]Value for pea derived from Davies, 1984a; Lund and Hakansson, 1986; Marquardt and Bell, 1988.

Most reports (Bell and Wilson, 1970; Grosjean, 1985; Henry and Bourdon, 1977; Lund and Hakansson, 1986) indicate that the apparent digestibility of crude protein in peas is within the range from 80 to 90%, which is close to the quoted values for soybeans (Marquardt and Bell, 1988). Some varieties, characterized by colored flowers or dark-hulled seed, may have less digestible pea protein (e.g. 71-80% vs. 83-90%; Ogle and Hakansson, 1988) than is found in white-flowered types.

The relative usefulness of alternative protein sources is more likely to be reflected in their contents of essential amino acids which are available to the pig. Green (1988) compared diets containing meat meal, soybean meal and pea meal (from "Amino;" a spring-sown, white-flowered, round-seeded cultivar). Apparent digestibilities of cystine (62 vs. 84%), methionine (76 vs. 84%) and phenylalanine (78 vs. 83%) were significantly lower in the pea, compared to soybean, diet but digestibilities of lysine (83 vs. 84%, respectively) and total nitrogen (76 vs. 79%) were similar. In summarizing the potential of peas as a supplementary protein source for pigs, it is evident that particular attention should be paid to the effect upon dietary levels of sulphur amino acids and possibly the change in tryptophan levels when other ingredients are replaced by peas.

Normally, the major dietary ingredients which are the sources of energy, nitrogen and amino acids in pig diets are not considered to be adequate in terms of providing all the required minerals and vitamins. Although use of peas in the diet will make a contribution to the total requirement for these nutrients, mineral and vitamin supplements are usually included in order to meet these needs. Average levels, from the National Research Council (1988), are shown in Table 19.5.

Table 19.5. Calculated Mineral and Vitamin Composition of Pea Dry Matter

Macro Minerals	**% of DM**	**Vitamins**	**mg/kg DM**
Calcium	0.13	Biotin	0.2
Magnesium	0.13	Choline	615
Phosphorus	0.46	Folacin	0.2
Potassium	1.13	Niacin	35
Sodium	0.04	Pantothenic acid	21
		Riboflavin	2.0
Trace Minerals	**mg/kg DM**	Thiamine	5.2
		Vitamin B_6	2.2
Iron	73	Vitamin E	3.4
Manganese	3.2		
Zinc	26		

National Research Council, 1988.

UNDESIRABLE CONSTITUENTS

There are several natural constituents which can interfere with utilization of legume seeds or disrupt physiological processes in the pig. Apparently, the major concerns for peas are the presence of lectins (haemagglutinins), protease (trypsin and chymotrypsin) inhibitors and tannins (which are more characteristic of dark-hulled varieties). Levels are normally low and could be rendered innocuous by the application of heat, although this is unlikely to be economical or necessary for pig diets. Varietal differences in the levels of these constituents have been reported (Griffiths, 1984) and it has been suggested (Davies and Domoney, 1983) that there is a positive correlation of protein content and lectin, or trypsin inhibitor, levels.

Legume haemagglutinins (lectins) agglutinate red blood cells of several animal species (Leiner, 1969) and consist of proteins with an affinity for specific sugar molecules (Chubb, 1982). Valdebouze et al. (1980) reported maximum haemagglutinating activity levels of 400 HU/mg in eight samples of peas (including round and wrinkled, spring and winter types). Corresponding values for a commercial sample of defatted soybean meal ranged from 1600 to 3200 HU/mg. No evidence could be found by Marquardt and Bell (1988) that lectins from peas had adversely influenced pig performance.

Protease inhibitors have been identified in many legume species and studied comprehensively in soybeans where there are two main groups, the Kunitz and Bowman-Birk inhibitors. The former, which consists of a mixture of proteins with molecular weights from 20,000 to 25,000 has a specifity towards trypsin whereas the Bowman-Birk mixture of smaller molecules (6,000-10,000) has an ability to inhibit both trypsin and chymotrypsin (Chubb, 1982). In some species, consumption of legumes leads to pancreatic hypertrophy as a result of trypsin inhibition but this type of response is less common in pigs (Chubb, 1982). Published reports suggest that the trypsin inhibitor activity of peas is less than one-fifth of that found in soybeans. The trypsin inhibitor content of spring seeded varieties is generally between 2.3 and 5.5 TIA (trypsin inhibiting activity) units/kg DM while winter varieties generally average between 8.9 and 15.9 TIA units/mg DM. It was concluded by Marquardt and Bell (1988), that such levels would have only minor effects on animal performance.

Pea tannins, which appear to be present in the hull of seeds from predominantly colored-flower varieties, have an inhibitory effect on protein solubility and on trypsin (Griffiths, 1983). The presence of tannins was considered to be a major cause of reduction in pig performance when high levels of dietary peas were fed (Hlodversson, 1987; Thomke, 1986).

Some concerns have been expressed about the presence of other constituents such as saponins (cyanogenetic glycosides) which have a bitter taste and relatively greater effect on poultry than pigs (Chubb, 1982). However, there is no definitive evidence that other compounds in peas pose any consistent problem.

FEEDING PEAS TO PIGS

Starter Pigs

European experiments (Fekete et al., 1984; Grosjean, 1985; Grosjean and Gatel, 1986) have shown that the peas grown in France should be limited to 15% of diets for weaned pigs less than 25 kg liveweight. Using peas as the only supplementary source was not recommended since problems related to anti-proteases and amino acid balance could occur.

Growing Pigs

Australian results (Davies, 1984b) indicated that satisfactory results were obtained with growing-finishing pigs fed to scale diets containing up to 28% peas but growth rates were less when a diet containing 53% peas was consumed. A subsequent report from England (Edwards et al., 1987) suggested that up to 30% peas could be used in diets for finishing pigs (33-85 kg liveweight) without adverse effects on live performance or carcass grades. Two previous reports (Davies, 1984a; Stone and McIntosh, 1977) emphasized the potential for calcium deficiency with pea-containing diets.

In North America, the use of inclusion rates of peas in excess of 30% also appears to be practical. Bell and Wilson (1970) used field peas to replace the protein supplement (fishmeal-soybean meal) in grower-finisher diets without detriment to pig performance or carcass quality and also obtained satisfactory performance from diets containing 40% cull peas. No benefit was observed from adding 0.3% methionine to the diet containing 24% peas. However, positive responses to methionine and to tryptophan supplements were observed by Grosjean (1985).

Castell et al. (1988) fed isonitrogenous diets (16% CP) in which up to 33% pea screenings replaced barley and soybean meal and observed no decline in performance of pigs fed ad libitum from 26 to 94 kg liveweight or in carcass grades. Inclusion rates as high as 55% peas were fed by England et al. (1986) without detrimental effects on performance of pigs marketed at 100 kg liveweight.

Results of feeding Scandinavian cultivars of peas reveal some evidence of adverse effects when high levels of peas are used in the diet. Madsen and Mortensen (1985) indicated that inclusion rates above 20% required appropriate supplementation with amino acids (methionine, threonine or tryptophan) to prevent deterioration in pig performance. These authors also noted an inverse relationship between the level of peas in the diet and percentage of meat in the carcass. In a review of Swedish experiments, Thomke (1986) concluded that live performance was not adversely influenced when peas replaced barley and soybean meal on an equal lysine basis in diets fed from 20 to 105 kg, but carcass quality might be reduced. Ogle and Hakansson (1988), summarizing Norwegian research, concluded that when the level of dietary peas increased to the point when adverse effects appeared, supplementation with methionine usually rectified the problem.

Breeding Stock

There are suggestions from Scandinavian studies, reported by Ogle and Hakansson (1988), that moderate levels (<20%) of dietary peas in gestation and lactation diets may reduce litter size at birth and weaning. However, from an experiment extending over more than four parities, Gatel et al. (1988) found no differences in reproductive performance between control sows and those which received equivalent diets containing 16% peas during gestation and 24% peas in lactation.

SUMMARY

Peas have considerable potential as an ingredient for pig diets. Pea protein is an excellent source of lysine and the availability of nutrients is quite similar to that in soybean meal. The low content of fat and of anti-nutritional factors in many varieties reduce any requirement for processing prior to feeding peas and favour maximum use when warranted by relative prices of ingredients. Palatability is unlikely to restrict consumption but use of peas as the only source of supplementary protein may not be wise in view of reported variability in the feeding value of different pea varieties.

On the basis of current information, which suggests peas are not uniform in composition or feeding value, it would be unwise to expect recommendations to be valid without taking into account any available knowledge of the pea variety to be fed and the dietary ingredients to be replaced. However, erring on the side of caution, an acceptable maximum in diets for young pigs (<25 kg) would be 15%. For growing-finishing pigs, levels up to 30% should be satisfactory, while 15% peas appears to be a safe limit for sow diets. In all cases, it is assumed that any necessary adjustments will be made to the formula of the pea diet to ensure that it provides the recommended levels of nutrients.

REFERENCES

Aman, P. and Graham, H., 1987. Whole-crop Peas. I. Changes in botanical and chemical composition and rumen in vitro degradability during maturation. Anim. Feed Sci. Technol. 17: 15-31.

Bell, J.M. and Wilson, A.G., 1970. An evaluation of field peas as a protein and energy source for swine rations. Can. J. Anim. Sci. 50: 15-23.

Bhatty, R.S. and Christison, G.I., 1984. Composition and nutritional quality of pea (*Pisum sativum* L.), faba bean (*Vicia faba* L. spp. minor) and lentil (*Lens culinaris Medik.*) meals, protein concentrates and isolates. Qual. Plant Foods Human Nutr. 34: 41-51.

Boulter, D.A., 1977. Quality problems in "Protein Plants" with special attention paid to the proteins of legumes. In: Protein Quality From Leguminous Crops. Kirchberg, Luxembourg, pp. 11-47.

Castell, A.G., Neden, L.R. and Mount, K., 1988. Potential of field pea (*Pisum sativum*) screenings as feed for market pigs. Can. J. Anim. Sci. 68: 577-579.

Cerning-Beroard, J. and Filiatre, A., 1977. Characterization and distribution of soluble and insoluble carbohydrates in legume seeds: Horse beans, peas, lupins. In: Protein Quality From Leguminous Crops. Kirchberg, Luxembourg, pp. 65-79.

Chubb, L.G., 1982. Anti-nutritive factors in animal feedstuffs. In: W. Haresign, ed. Recent Advances in Animal Nutrition, Vol. 16, Butterworths, London, pp. 21-37.

Cousin, R., 1983. Breeding for yield and for protein content in pea. In: R. Thompson and R. Casey, eds. Perspectives for Peas and Lupins as Protein Crops. Martinus Nijhoff, The Hague, pp. 146-164.

Cousin, R., Messager, A. and Vingere, A., 1985. Breeding for yield in combining peas. In: P.D. Hebblethwaite, M.C. Heath and T.C.K. Dawkins, eds. The Pea Crop - A Basis for Improvement. Butterworths, London, pp. 115-129.

Davies, D.R., 1976. Peas. In: N.W. Simmonds, ed. Evolution of Crop Plants. Longman, New York, pp. 172-174.

Davies, D.R. and Domoney, C., 1983. Storage proteins of round and wrinkled peas in vivo and in vitro. In: R. Thompson and R. Casey, eds. Perspectives for Peas and Lupins as Protein Crops. Martinus Nijhoff, The Hague, pp. 256-271.

Davies, R.L., 1984a. Field peas (*Pisum sativum*) as a feed for growing and finishing pigs. 1. Nutrient levels in commercial crops. Aust. J. Exp. Agric. Anim. Husb. 24: 350-353.

Davies, R.L., 1984b. Field peas as a feed for growing and finishing pigs. 2. Effects of substituting peas for meat meal or fish meal in conventional diets. Aust. J. Exp. Agric. Anim. Husb. 24: 507-511.

Edwards, S.A., Rogers-Lewis, D.S. and Fairbairn, C.B., 1987. The effects of pea variety and inclusion rate in the diet on the performance of finishing pigs. J. Agric. Sci. (Camb.) 108: 383-388.

England, D.C., Chitko, C., Dickson, R. and Cheeke, P.R., 1986. Utilization of yellow peas in swine grower-finisher diets. Oregon Agricultural Experimental Station, Corvallis, Special Report No. 771, pp. 1-7.

Fekete, J., Castaing, J., Lavorel, O., Leuillet, M. and Quemere, P., 1984. Utilisation des pois proteagineux par le porcelet sevre: Bilan des essais realises en France. Journee Recherche Porcine en France 16: 393-400.

Gane, A.J., 1985. The pea crop - Agricultural progress, past, present and future. In: P.D. Hebblethwaite, M.C. Heath and T.C.K. Dawkins, eds. The Pea Crop - A Basis for Improvement. Butterworths, London, pp. 3-15.

Gatel, F., Grosjean, F. and Leuillet, M., 1988. Utilization of white-flowered smooth-seeded Spring peas (*Pisum sativum hortense*, cv. Amino) by the breeding sow. Anim. Feed Sci. Technol. 22: 91-104.

Graham, H. and Aman, P., 1987. Whole crop peas. II. Digestion of early- and late-harvested crops in the gastrointestinal tract of pigs. Anim. Feed Sci. Technol. 17: 33-43.

Green, S., 1988. A note on amino acid digestibility measured in pigs with pre- or post-valve ileo-rectal anastomoses, fed soya-bean, pea and meat meals. Anim. Prod. 47: 317-320.

Griffiths, D.W., 1983. The polyphenolic content of field peas and their possible significance on nutritive value. In: R. Thompson and R. Casey, eds. Perspectives for Peas and Lupins as Protein Crops. Martinus Nijhoff, The Hague, pp. 322-327.

Griffiths, D.W., 1984. The trypsin and chymotrypsin inhibitor activities of various pea (*Pisum* spp.) and field bean (*Vicia faba*) cultivars. J. Sci. Food Agric. 35: 481-486.

Grosjean, F., 1985. Combining peas for animal feed. In: P.D. Hebblethwaite, M.C. Heath and T.C.K. Dawkins, eds. The Pea Crop - A Basis for Improvement, Butterworths, London, pp. 453-462.

Grosjean, F. and Gatel, F., 1986. Peas for pigs. Pig News and Information 7: 443-448.

Henry, Y. and Bourdon, D., 1977. Utilization of legume seeds (field beans and peas) by the pig. In: Protein Quality from Leguminous Crops. Kirchberg, Luxembourg, pp. 252-272.

Hlodversson, R., 1987. The nutritive value of white- and dark-flowered cultivars of pea for growing-finishing pigs. Anim. Feed Sci. Technol. 17: 245-255.

Huet, J.C., Baudet, J. and Mosse, J., 1987. Constancy of composition of storage proteins deposited in Pisum sativum seeds. Phytochem. 26: 47-50.

Leiner, I.E., 1969. Toxic Constituents of Plant Foodstuffs. Academic Press, New York. 500 pp.

Lund, S. and Hakansson, J., 1986. Nutritional and growth studies with pea-crop meals and peas for growing-finishing pigs. Anim. Feed Sci. Technol. 16: 119-128.

Madsen, A. and Mortensen, H.P., 1985. Aerter til slagtesvin [Peas for bacon pigs]. Beretning fra Statens Husdyrbrugsforsog, No. 581. Rolighedsvej 26, 1958 Kobenhavn V. 45 pp.

Marquardt, R.R. and Bell, J.M., 1988. Future potential of pulses for use in animal feeds. In: R.J. Summerfield, ed. World Crops: Cool Season Food Legumes. Kluwer, London, pp. 421-444.

Matthews, P. and Arthur, E., 1985. Genetic and environmental components of variation in protein content of peas. In: P.D. Hebblethwaite, M.C. Heath and T.C.K. Dawkins, eds. The Pea Crop - A Basis for Improvement. Butterworths, London, pp. 369-381.

Monti, L.M., 1983. Natural and induced variability in peas for protein production. In: R. Thompson and R. Casey, eds. Perspectives for Peas and Lupins as Protein Crops. Martinus Nijhoff, The Hague. pp. 23-39.

National Research Council, 1982. United States-Canadian Tables of Feed Composition, 3rd ed. National Research Council, Washington, D.C. 148 pp.

National Research Council, 1988. Nutrient Requirements of Domestic Animals, No. 2. Nutrient Requirements of Swine. 9th ed. National Academy of Sciences, Washington, D.C. 93 pp.

Ogle, R.B. and Hakansson, J., 1988. Nordic research with peas for pigs. Pig News and Information 9: 149-155.

Reichert, R.D. and Mackenzie, S.L., 1982. Composition of peas (*Pisum sativum*) varying widely in protein content. J. Agric. Food Chem. 30: 312-317.

Stone, B.A. and McIntosh, G.H., 1977. The influence of calcium, phosphorus and vitamin D supplementation of a pea-barley ration for growing pigs. Aust. J. Agric. Res. 28: 543-550.

Taverner, M.R. and Curic, D.M., 1983. The influence of hind-gut digestion on measures of nutrient availability in pig feeds. Proc. Inter. Network Feed Information Centres, Commonwealth Agricultural Bureaux, Slough, England, pp. 295-298.

Thomke, S., 1986. Swedish experiments on energy density in pig diets and with domestically grown protein feedstuffs. A review. World Rev. Anim. Prod. 22: 89-95.

Valdebouze, P., Bergeron, E., Gaborit, T. and Delort-Laval, J., 1980. Content and distribution of trypsin inhibitors and hemagglutinins in some legume seeds. Can. J. Plant Sci. 60: 695-701.

Wright, D.J., 1985. Combining peas for human consumption. In: P.D. Hebblethwaite, M.C. Heath and T.C.K. Dawkins, eds. The Pea Crop - A Basis for Improvement. Butterworths, London, pp. 441-451.

CHAPTER 20

Fish Silage

T.A. Van Lunen

INTRODUCTION

Fish silage is a liquid product, made up of whole fish or fish processing offal, that is mechanically ground and liquefied by the action of endogenous enzymes present in the digestive tract of the fish. These enzymes, in the presence of organic or inorganic acids, break down the fish protein into smaller units of increased solubility. In addition, the presence of acid decreases pH, which inhibits the growth of molds and bacteria in the silage, making long term storage possible.

The process of making fish silage offers the potential of utilizing wastes from the fishing industry in areas where the quantity of waste is insufficient to justify the production of fish meal. Swine producers with operations located in the vicinity of these processing plants can obtain a high quality protein supplement at a relatively low cost and thereby increase the efficiency and profitability of their operations.

NUTRIENT CONTENT OF FISH SILAGE

The potential feeding value of fish silage is determined to a large extent by the quality of the material being ensiled. Fish silage produced using a high percentage of whole fish will have a higher nutritional value than will silage produced using offal. In addition, the type of fish used will also affect the quality of the silage produced. For example, fish silage based on white fish is different from that based on herring and it is important to differentiate between these in discussing their nutritive value.

White fish silage has a dry matter content of about 20%. On a dry matter basis, it contains approximately 70% crude protein, 3% ether extract and 16% ash. In contrast, herring fish silage has a dry matter content of about 35% and on a dry matter basis, it contains only 43% crude protein and 8% ash. However, it has an ether extract content of over 42%. Although fish silage is considered primarily as a protein supplement, it also contains energy, provided by the oil fraction. The energy value will vary with oil content.

Fish silage can be a good source of minerals, although silage made from viscera will be lower in mineral content than silage made from whole fish or heads and frames. Fish silage is an excellent source of both calcium and phosphorus since it contains relatively high levels of these two nutrients and the acidic nature of the product increases their availability. Additional data on the nutrient content of various fish silages are presented in Table 20.1.

Table 20.1. Nutrient Composition of Fish Silages (% DM)

Type of Fish Silage	Crude Protein	Crude Fat	Dry Matter	Calcium	Phosphorus
Cod Offal[1]	64.5	2.1	20.0	3.5	2.2
White Fish Offal[2]	71.1	2.4	---	---	---
White Fish Offal[3]	70.3	3.1	21.2	3.8	1.9
Whole Herring[1]	46.8	---	31.0	---	---
Herring Offal[3]	42.7	43.9	33.2	2.1	1.6
Herring Offal[4]	48.3	28.2	27.3	3.5	1.8
Deoiled Herring Offal[4]	66.3	3.6	22.4	4.8	2.6

[1]Agriculture Canada, Nappan samples (unpublished).
[2]Tatterson and Windsor, 1974.
[3]Strom et al., 1981.
[4]Whittemore and Taylor, 1976.

The amino acid profile of fish silage is similar to that of fish meal made from the same type of raw material (Table 20.2). This means that the level of some of the limiting amino acids in cereal-based diets, such as lysine, threonine and the sulfur containing amino acids, are present at fairly high levels. As a consequence, fish silage would appear to be an excellent protein source for use as a supplement to the cereal grains.

Fish silage protein is in a degraded form, resulting in high levels of free amino acids. However, these amino acids appear to be stable in fish silage. Reports indicate that less than 8% of the amino nitrogen is released as ammonia in fish silages stored for as long as 220 days. The amino acids most ammonia in fish silages stored for as long as 220 days. The amino acids most likely to decompose are tryptophan, methionine and histidine. This degradation occurs most often at temperatures exceeding 25°C and therefore, does not appear to be a real concern under normal storage conditions. Some less soluble amino acids present in fish silage remain in the "sludge" fraction (heavy fraction at bottom of the tank). Therefore, care must be taken when utilizing "de-sludged" fish silage to avoid amino acid deficiencies.

Table 20.2. Amino Acid Composition of Fish Silages and Fish Meal (%)

	White[1] Fish	Deoiled[2] Herring	Mackerel[3]	Fish[2] Meal
Arginine	5.1	5.1	3.7	4.3
Histidine	1.8	1.7	1.2	1.1
Isoleucine	2.8	2.8	2.5	2.4
Leucine	4.5	5.2	4.3	4.2
Lysine	4.8	6.2	5.1	4.4
Methionine	1.8	1.8	1.4	1.7
Phenylalanine	2.3	3.5	1.9	2.3
Threonine	3.1	3.0	2.0	2.5
Tyrosine	2.1	1.6	1.5	2.0

[1]Smith and Adamson, 1976.
[2]Whittemore and Taylor, 1976.
[3]Green et al., 1988.

UNDESIRABLE CONSTITUENTS IN FISH SILAGE

Although fish silage can be fed to all classes of swine, there are a number of constraints and limiting factors which must be taken into account before including fish silage in swine diets. According to Raa and Gildberg (1982), the most severe restriction to the use of fish silage in diets for pigs is the fishy off-flavor which may result in the meat. Several suggestions have been made as to the cause of these off-flavors and it appears most likely that their cause is flavors carried in the lipid fraction of the silage. This off-flavor problem appears to be of special concern for smoked pork products such as bacon and ham.

In addition to the off-flavor concern, fish lipids also represent a risk in terms of premature oxidation of pork fat. Since fish lipids contain relatively high levels of long chain, unsaturated fatty acids, they are subject to premature oxidation or rancidity. If these same fatty acids are passed on to the pig, the same oxidation risk will exist. Recent work conducted in Sweden (Ruderus, personal communication) has shown a direct linear relationship between the level of fish fat in the diet and the length of time frozen pork can be stored. As the amount of fish-source lipid in the pig's diet increases, the length of time the pork can be stored in a conventional home freezer, before it becomes rancid, decreases. Taken to the extreme, 10% fish lipid in the diet can result in oxidized pork fat within days of slaughter.

Oxidizing oils present in fish silage may also cause the destruction of vitamins A and E in the diet which could result in a vitamin deficiency. It is therefore imperative that antioxidants be added during processing. In addition, fish silage also contains high levels of the enzyme thiaminase which may act to destroy the vitamin thiamine (Anglesa and Jackson, 1985). Therefore, it may be advisable to supplement vitamins at a higher than normal level to ensure that a vitamin deficiency does not occur when fish silage is included in the diet.

Another concern regarding fish silage is the possible presence of high levels of mercury (Batterham et al., 1983). Fish accumulate mercury in their body tissues and the possibility exists that pigs fed silage could produce a carcass unacceptable for human consumption due to mercury contamination. Fish silages containing in excess of 0.5 mg/kg mercury should not be used in swine rations.

FEEDING FISH SILAGE

In the past, fish silage was not utilized to any great extent in swine diets, primarily because of its high water content but also due to competition for raw product with fish meal plants. However, recently there has been renewed interest in this product due to increasing fish meal production costs and environmental problems associated with the disposal of fish offal. In addition, pork producers are becoming more interested in fish silage as they search for lower cost protein alternatives and products compatible with liquid by-product feeding.

Starter Pigs

The results of a feeding trial conducted at the University of Georgia, using weanling pigs fed either 0, 3, 6 or 9% fish silage for six weeks are presented in Table 20.3. It can be seen that the performance of weaner pigs

Table 20.3. Performance of Weaner Pigs Fed Diets Containing Fish Silage

	Level of Fish Silage (%)			
	0	3	6	9
Daily Gain (kg)	0.42	0.40	0.43	0.39
Daily Intake (kg)	0.87	0.89	0.91	0.80
Feed Efficiency	2.07	2.22	2.12	2.07

Tibbetts et al., 1981.

fed diets containing 3 or 6% fish silage was not significantly different from the control group while those fed diets containing 9% fish silage gained at a slower rate. The major factor responsible for the reduction in growth rate appeared to be a reduction in feed intake.

Growing-Finishing Pigs

Growing-finishing pigs perform well on fish silage. Numerous reports exist which describe the feeding value of fish silage for growing pigs (Green et al., 1983). In general, it can be said that fish silage is a good source of protein for this category of swine. Levels as high as 10% (DM) of the diet can be fed with no negative effects on growth, feed consumption, feed conversion or carcass grade. Table 20.4 illustrates the performance of growing pigs fed various levels of low-fat fish silage. It should be noted that results shown are from two distinct studies which are not directly comparable. However, general trends are evident. As can be seen, the 15% level of fish silage results in a reduction in growth rate as compared to lower levels.

Table 20.4. Performance of Growing Pigs Fed Diets Containing Fish Silage

	Fish Silage Level (% DM)					
	0[1]	3[1]	6[1]	9[1]	10[2]	15[2]
Daily Intake (kg)	2.3	2.5	2.6	2.1	2.0	2.2
Daily Gain (kg)	0.73	0.73	0.74	0.68	0.68	0.62
Feed Efficiency	3.1	3.4	3.5	3.8	3.0	3.2
Backfat (cm)	3.3	3.5	3.6	3.4	5.1	5.0

[1]Tibbetts et al., 1981.
[2]Van Lunen, 1984.

There are some limitations to the use of fish silage in grower rations. As mentioned previously, research has demonstrated that the inclusion of fish silage in the diet of pigs can result in "off-flavors" in pork. In order to avoid these off-flavors, it is recommended that fish silage be removed from the diet at least 20 days prior to slaughter. Some reports suggest withdrawal times of as long as 30 days prior to slaughter. In addition, because of the fatty acid structure of fish source lipids and their effect on pork fat, it is recommended that the lipid level of fish silage be restricted to 1% on a wet basis.

Breeding Stock

There would appear to be some problems when it comes to feeding fish silage to breeding stock. The results of one study in which fish silage was fed to sows during gestation are shown in Table 20.5. Preweaning mortality has been shown to be significantly higher when diets containing 6% fish silage are fed to sows during gestation. The reason for this increase in mortality has not been determined.

Table 20.5. Reproductive Performance of Sows Fed Fish Silage

	Conventional Diet	Fish Silage
Pigs Born Alive	11.4	11.1
Birth Weight (kg)	1.4	1.3
Pigs Weaned	9.7	8.2
Weaning Weight (kg)	4.5	4.4
Mortality (%)	14.9	26.1

Tibbetts et al., 1981.

Constraints to the Use of Fish Silage

There are a number of constraints to the use of fish silage in swine diets. The first of these limitations is the requirement for specialized storage, mixing and feed delivery systems capable of handling liquid or semi-liquid products such as fish silage. At present, the systems most able to perform such tasks are computerized liquid feeding systems which can represent a significant capital commitment.

Another major problem with feeding fish silage is finding an acceptable method of feeding the product. Because of the high moisture content of the silage, diets containing fish silage must be mixed on a daily basis or else the cereal portion of the diet might spoil. In addition, rations containing high levels of fish silage tend to bridge if fed in traditional feeders. Therefore, unless an acceptable method of feeding fish silage is developed, its use will be limited to small scale producers who mix and feed by hand.

Another constraint to the economical use of fish silage for pork production is the cost of transportation. Since fish silage typically contains 75 to 80% water, the cost of transportation per unit of dry nutrients is high. For example, it could cost approximately five times as much per unit of dry protein to transport fish silage protein the same distance as soybean meal protein.

Associated with transportation is the constraint of consistency of supply. Many species of fish are caught on a seasonal basis resulting in a glut of fish silage on the market at certain times of the year while at other times no fish silage is available. As well, the consistency of nutrient content of the fish silage available could change with seasonal species changes.

Pork producers considering the inclusion of fish silage in their hog diets must compare the per unit dry protein cost of this product as compared to the price of other protein sources such as soybean meal. For example, soybean meal characteristically contains 48% crude protein and 10% water. On a dry basis then, soybean meal contains approximately 53% crude protein (48 + 0.9). If that soybean meal is selling for $400 per tonne, its value per unit of dry protein is $7.55 ($400 + 53%). Assuming the fish silage in question contains 80% water and 65% crude protein on a dry basis, its price must not exceed $490 (65% x $7.55) per dry tonne. Therefore, its maximum value per wet tonne would be $98 ($490 x 0.2) in order to remain competitive with soybean meal. This may be difficult, considering the present high price of the acids required to produce fish silage. However, if fish silage can be produced and delivered to the farm at a price significantly lower than that of the competition, it will no doubt become a common feed ingredient in hog diets.

SUMMARY

In summary, it can be said that fish silage offers good potential as a protein supplement for use in swine diets. It contains 42 to 71% crude protein, in an available form. All classes of pigs can utilize fish silage to some degree. However, limitations do exist in the level of fish silage inclusion and the type of fish silage used. Weaner pigs, grower pigs and breeding animals can all be fed levels of up to 10% of low fat fish silage on a dry basis with no detrimental effects on growth, feed consumption or carcass quality. It is recommended that high fat fish silage be restricted to weaner pigs and breeding animals because of possible meat quality problems associated with the fatty acid content of fish source fat. It is also recommended that fish silage be removed from the diet three weeks prior to slaughter, to avoid possible off-flavors in the pork.

REFERENCES

Anglesa, J.D. and Jackson, A.J., 1985. Thiaminase activity in fish silage and moist fish feed. Anim. Feed Sci. Technol. 13: 39-46.

Batterham, E.S., Gorman, T.B. and Chvojka, R., 1983. Nutritive value and mercury content of fish silage for growing pigs. Anim. Feed Sci. Technol. 9: 169-180.

Green, S., Wiseman, J. and Cole, D.J.A., 1988. Examination of stability and its effects on the nutrient value of fish silage in diets for growing pigs. Anim. Feed Sci. Technol. 21: 43-56.

Green, S., Wiseman, J. and Cole, D.J.A. 1983. Fish Silage in Pig Diets. Pig News and Information. 4: 269-273.

Parry, D.A., Hillyer, G.M. and Fraser, J.C., 1982. The evaluation of liquid de-oiled herring silage in diets for growing pigs: palatability studies. J. Sci. Food Agric. 33: 11-15.

Pond, W.G. and Maner, J.H., 1984. Swine Production and Nutrition. AVI Publishing Company, Westport, Conneticut 731 pp.

Raa, J. and Gildberg, A., 1982. Fish silage: A review. CRC Critical Reviews in Food Science and Nutrition. April, 1982.

Smith, P. and Adamson, A.H., 1976. Pig feeding trials with white fish and herring liquid protein (fish silage). Proc. Torry Research Station Symposium on Fish Silage. Aberdeen, Scotland.

Strom, T. and Eggum, B.O., 1981. Nutritional value of fish viscera silage. J. Sci. Food Agric. 32: 115-120.

Tatterson, I.N. and Windsor, M.L., 1974. Fish silage. J. Sci. Food Agric. 25: 369-379.

Tibbetts, G.W., Searley, R.W., Campbell, H.C. and Vezey, A.S., 1981. An evaluation of an ensiled waste fish product in swine diets. J. Anim. Sci. 52: 93-100.

Van Lunen, T.A., 1984. Evaluation of fish silage as a feed source for swine. Research Summmary, Nappan Experimental Farm, Nappan, Nova Scotia pp. 33-39.

Whittemore, C.T. and Taylor, A.G., 1976. Nutritive value to the growing pig of deoiled liquified herring offal preserved with formic acid (fish silage). J. Sci. Food Agric. 27: 239-243.

CHAPTER 21

Lentils

A.G. Castell

INTRODUCTION

Lentils (*Lens culinaris*) are a pulse crop grown primarily for the human food market. On occasion, they are made available to pork producers at a competitive price due to adverse market conditions, frost damage or aschocyta blight. When available for swine production, lentils are most often utilized to replace a portion of the protein supplement in the diet. However, they can also be used to replace part of the cereal grain.

GROWING LENTILS

The lentil is an annual legume which is well adapted to a variety of soil types and growing conditions. Depending upon the severity of seasonal conditions, lentils are grown either as a winter (e.g. in India) or spring-sown (e.g. in North America) crop. The young plants do not compete well with weeds but, once established, have a moderate resistance to drought. They can also tolerate high temperatures but a long cool growing season is favored for maximum yields. Effective innoculation of the seed with the appropriate strain of Rhizobium promotes subsequent root nodulation and nitrogen fixation, which reduces the requirement for fertilizer input. More specific information about appropriate methods of crop management is provided by Webb and Hawtin (1981).

The average yield of lentils is estimated to be 668 kg/ha, but there is a wide range (from 200 to 4143 kg/ha) among countries (FAO, 1985). Seeded acreage is about 2,469,000 ha with worldwide production of lentils estimated at approximately 1,650,000 metric tonnes (FAO, 1985). Approximately 88% of the total production in 1985 was produced in the developing countries where lentils play an important dietary role for humans. In other regions, such as North America, where relatively higher yields are obtained but per capita consumption is relatively low, the area in production appears to be related to the potential for export.

NUTRIENT CONTENT OF LENTILS

As observed for many crops, the composition of lentils is influenced by the cultivar seeded and the growing conditions prior to harvesting. Bhatty et al. (1976) reported the following ranges (% of DM) for six cultivars grown under similar conditions: crude protein 26.1 to 31.3, ether extract 0.5 to 0.9, crude fiber 3.5 to 4.9 and ash 2.9 to 3.3. The hull usually represents less than 5% of the mature seed (Champ et al., 1986) and has only about 1.7% lignin. The proximate composition of lentils and several other commonly used feedstuffs is presented in Table 21.1.

Table 21.1. Proximate Composition of Lentils Compared with Cereals and Soybeans (% as fed)

	Barley	Corn	Lentil	Soybean Seed	Soybean Meal
Dry Matter	89.0	88.0	90.0	92.0	90.0
Crude Protein	11.5	8.5	25.7	39.2	48.5
Ether Extract	1.7	3.6	0.6	17.2	0.9
Crude Fiber	5.0	2.3	4.0	5.3	3.4

Bhatty et al., 1976; National Research Council, 1982; 1988.

Proximate analyses reveal that lentils (25.7% CP) are intermediate to cereals and soybeans (whole or defatted) as a source of crude protein. Lentil seed proteins can be separated according to their solubility in dilute acid (glutelins), ethanol (prolamines), dilute salt (globulins) and water (albumins). Average values for these, for the six cultivars analyzed by Bhatty et al. (1976), were 14.9, 3.1, 47.7 and 3.8% of the total protein, respectively. Subsequently, Bhatty (1982) reported the distribution of seed N between the different fractions in several grain legumes (Table 21.2) and noted the difference in amino acid compositions of the non-storage (albumins) and storage (globulins) proteins for each legume.

Table 21.2. Nitrogen Distribution in Fababeans, Lathyrus, Lentils and Pea Seeds[1]

	Fababean	Lathyrus	Lentil	Pea
Total N (% of seed)	5.0	4.2	4.6	3.5
Distribution of Total N (%)[2]				
Nonprotein Fraction	11.6	21.3	20.7	17.6
Albumin Fraction	8.6	13.0	8.1	14.1
Globulin Fraction	42.0	33.3	34.3	30.7
Insoluble N[3]	30.0	22.8	28.8	30.0

[1]Bhatty, 1982.
[2]Percentages recovered: 92.2, 90.4, 91.9 and 92.4, respectively.
[3]Remaining after salt-soluble fractions removed.

With respect to the pig's requirements, the content of the sulphur amino acids and threonine place lentils second to soybean meal as a supplementary protein source (Table 21.3). However, appropriate blends with cereals can supply a satisfactory balance of amino acids. In addition, both Abu-shakra and Tannous (1981) and Bhatty (1986) noted the significant presence of free amino acids in lentils, although the identities and roles of these are not fully known. The amino acid composition of wild species (*L. orientalis, L. ervoides* and *L. nigricans*) was found to be similar to that of *L. culinaris* (Bhatty, 1986).

Table 21.3. Essential Amino Acids (g per 100 g crude protein) in Dietary Sources Compared to the Requirements for Grower Pigs (<50 kg)[1]

	Barley	Corn	Lentil	Soybean	ARC[2]	NRC[3]
Arginine	4.3	4.5	6.9	7.3	---	1.7
Histidine	2.0	2.7	2.1	2.5	2.3	1.5
Isoleucine	3.8	3.6	3.5	5.4	3.8	3.1
Leucine	6.3	12.6	6.4	7.7	7.0	4.0
Lysine	3.3	2.6	6.3	6.2	7.0	5.0
Methionine + Cystine	3.1	4.1	1.9	2.8	3.5	2.7
Phenylalanine + Tyrosine	7.8	9.0	6.4	7.8	6.7	4.4
Threonine	3.1	3.6	3.1	4.2	4.2	3.2
Tryptophan	1.3	0.8	0.9	1.4	1.0	0.8
Valine	4.8	4.6	4.0	5.3	4.9	3.2

[1]National Research Council, 1982; Bhatty, 1986.
[2]Ideal protein basis (Agricultural Research Council, 1981).
[3]Calculated from National Research Council, 1988.

The relatively poor protein quality in raw lentil has been confirmed by determination of its protein efficiency ratio (PER), biological value (BV) and digestibility. El-Nahry et al. (1980) reported a range in PER from 0.57 to 0.79, compared to a value of 2.5 for casein, and BV's from 38.8 to 42.3. Bhatty and Christison (1984) confirmed the low PER and found poorer protein digestibility (77.8 vs. 87.4% for casein) using mice. In vitro digestibility of the total nitrogen was 49.6% compared to 98.9% for casein in a study by Shekib et al. (1986) using treatment with pepsin and pancreatin. However, Sarwar and Peace (1986) suggested that crude protein digestibility is not a good predictor of amino acid availability, since they found the true digestibility of methionine and cystine were lower (58 and 66%, respectively) than for total N (85%).

The energy fraction was reported to be more digestible than crude protein (78 vs. 72%) in damaged lentils by Bell and Keith (1986), who estimated the digestible energy (D.E.) content to be 3405 kcal per kg dry matter (DM). In comparison, soybean meal contains from 3878 to 4088 kcal/kg DM (National Research Council, 1988). The nitrogen free extract of six cultivars of lentil analyzed by Bhatty et al. (1976) averaged 63.1% of the dry weight and the contents of starch and amylose ranged from 61.1 to 65.5% and from 33.0 to 37.5%, respectively. Total, ethanol-soluble oligosaccharide content of lentils was 4.2% of the dry seed (Eskin et al., 1980), similar to levels which have been found in fababean, field pea and soybean. Although pigs lack galactosidases which are required for metabolism of these carbohydrates, there does not appear to be any incidence of flatulence or similar associated problems experienced by humans (Abu-Shakra and Tannous, 1981).

Although ether extractives amount to less than 2% of the lentil seed, linoleic (C18:2) and linolenic (C18:3) account for 44 and 12%, respectively, of the total fatty acids (Eskin and Henderson, 1977). The levels of lipoxygenase present were considered by these workers as a potential for rapid development of rancidity after mechanical disruption (e.g. grinding) of lentil seeds.

Lentils are not considered as a major source of minerals or vitamins in human nutrition although they contribute to these nutrient requirements. Genotype and availability of nutrients during plant growth and development of the seed can influence ultimate concentrations (Table 21.4). According to Wassimi et al. (1978), lentils are a better source of iron (120 mg/kg) and zinc (240 mg/kg) than of copper (6 mg/kg) or manganese (14 mg/kg). The cooking quality of lentils was found to be related to the ratio of (Ca + Mg):P by Bhatty (1984).

The vitamin data appear restricted to ranges for the levels of niacin (20-30 mg/kg), riboflavin (2-3 mg/kg), thiamine (4-7 mg/kg) and ascorbic acid (10-70 mg/kg) which were obtained from El-Nahry et al. (1980), Hsu et al. (1980) and Kylen and McCready (1975).

Table 21.4. Range, Mean and Coefficient of Variability (CV) for Macrominerals in 101 Samples of Canadian-Grown Lentils (% as fed)

	Ca^{2+}	P	Mg^{2+}	Na^{+}	K^{+}
Range	0.04-0.16	0.28-0.63	0.08-0.14	0.02-0.18	0.88-1.44
Mean	0.07	0.45	0.10	0.04	1.16
CV	28.57	15.56	10.00	75.00	8.62

Bhatty, 1984.

UNDESIRABLE CONSTITUENTS IN LENTILS

In common with many other legume seeds, raw lentils contain some undesirable constituents although the levels of these are not likely to be of concern in swine feeding. Weder (1981) reported the presence of several inhibitors of proteases (chymotrypsin, plasmin and trypsin) in lentils. Marquardt and Bell (1988) also identified lectins (haemagglutinins), phytic acid, saponins and tannins as potential problems but could find no evidence that these had adversely affected performance of pigs fed lentils. It has been shown (Shekib et al., 1986) that cooking improves the nutritive value of lentils for humans but the effects of consumption of the raw material by nonruminants have not been well documented.

FEEDING LENTILS TO SWINE

Material available for livestock feeding will normally consist of only that portion of the crop which cannot be assigned for consumption by humans, e.g. the whole plant has been used for grazing or as a fodder for livestock. The lentil straw and pod walls, with the following average composition: moisture 10.2%, fat 1.8%, protein 4.4%, carbohydrate 50.0%, fiber 21.4% and ash 12.2% (Nygaard and Hawtin, 1981), represent a nutrient source for ruminants. Occasionally the starch is extracted from lentils and the residue, containing nearly 40% protein, is a useful feedstuff.

There is a paucity of published information regarding the feeding value of lentils for swine but recent Canadian studies provide some information. Bell and Keith (1986) conducted studies to determine the digestibility and potential of frost- and blight-damaged lentils using growing-finishing pigs. Dietary levels up to 30% did not adversely affect the pigs' live performance (Table 21.5) but the higher apparent digestibility of energy compared with protein implied some impairment of protein utilization for the lentil diets.

Table 21.5. Effects of Cull Lentils on Digestibility and Pig Performance

	Content of Lentils (%)			
	0	10	20	30
Apparent Digestibility (%)				
Dry Matter	79	79	78	79
Energy	80	80	78	79
Nitrogen	78	78	75	76
Performance (from 23 to 100 kg)				
Daily Gain (kg)	0.82	0.83	0.86	0.86
Feed Efficiency	3.09	2.99	2.98	3.00

Bell and Keith, 1986.

Subsequent studies (Castell and Cliplef, 1988), involving rates of inclusion up to 40% of isonitrogenous diets fed to growing-finishing pigs, suggested the optimum rate was in the range of 10 to 20%. Higher levels appeared to result in a reduction in meat quality, as judged by a trained taste panel. In a later (unpublished) study, these authors found the addition of 0.1% methionine to a 40% lentil diet improved the performance and carcass characteristics of pigs to a similar level to pigs receiving an isonitrogenous control (barley-soybean meal) diet.

It was shown by Sarwar and Peace (1986) that the digestibilities of the sulphur amino acids were about 20 percentage units less in lentil than soybean protein. Furthermore, there is some evidence (Roth and Kirchgessner, 1987) that the relative amounts of methionine and cystine, as well as the dietary level of lysine, influence the requirements for maximum performance in pigs. Therefore, it would seem wise, when high levels of lentils are used as a replacement ingredient for alternative sources of dietary amino acids, to consider supplementation with methionine.

SUMMARY

Lentils which are rejected for human consumption, or available at a competitive price to other ingredients, represent a potential source of energy and protein for pigs. Lentil protein, relative to soybean meal, is equal in lysine but poorer in threonine and the sulphur amino acids. Energy content of lentils is lower than that of soybean meal but the levels of anti-nutritive factors are low enough that their removal is not necessary.

There is insufficient information from published studies to suggest optimum rates of inclusion for lentils in diets for weaner pigs or breeding stock. In view of the variability in nutrient composition and levels of potentially anti-nutritional factors, using lentils as the sole source of supplementary protein cannot be recommended. Inclusion rates up to 40% of balanced diets for growing-finishing pigs may be satisfactory but more studies will be required to confirm the most appropriate dietary level for these and the other classes of pigs.

REFERENCES

Abu-Shakara, S. and Tannous, R.I., 1981. Nutritional value and quality of lentils. In: C. Webb and G. Hawtin, eds. Lentils. Commonwealth Agricultural Bureaux, England. pp. 191-202.

Agricultural Research Council, 1981. The Nutrient Requirements of Pigs. Commonwealth Agricultural Bureaux, England. 307 pp.

Bell, J.M. and Keith, M.O., 1986. Nutritional and monetary evaluation of damaged lentils for growing pigs and effects of antibiotic supplements. Can. J. Anim. Sci. 66: 529-536.

Bhatty, R.S., 1982. Albumin proteins of eight edible grain legume species: Electrophoretic patterns and amino acid composition. J. Agric. Food Chem. 30: 620-622.

Bhatty, R.S., 1984. Relationships between physical and chemical characters and cooking quality in lentil. J. Agric. Food Chem. 32: 1161-1166.

Bhatty, R.S., 1986. Protein subunits and amino acid composition of wild lentil. Phytochemistry 25: 641-644.

Bhatty, R.S. and Christison, G.I., 1984. Composition and nutritional quality of pea (*Pisum sativum* L.), faba bean (*Vicia faba* L. spp. minor) and lentil (*Lens culinaris Medik.*) meals, protein concentrates and isolates. Qual. Plant Foods Human Nutr. 34: 41-51.

Bhatty, R.S., Slinkard, A.E. and Sosulski, F.W., 1976. Chemical composition and protein characteristics of lentils. Can. J. Plant Sci. 56: 787-794.

Castell, A.G. and Cliplef, R.L., 1988. Live performance, carcass and meat quality characteristics of market pigs self-fed diets containing cull-grade lentils. Can. J. Anim. Sci. 68: 265-273.

Champ, M., Brillouet, J.M. and Rouau, X., 1986. Nonstarchy polysaccharides of *Phaseolus vulgaris, Lens esculenta* and *Cicer arietinum* seeds. J. Agric. Food Chem. 34: 326-329.

El-Nahry, F.I., Mourad, F.E., Abdel-Khalik, S.M. and Bassily N.S., 1980. Chemical composition and protein quality of lentils (Lens) consumed in Egypt. Qual. Plant Foods Human Nutr. 30: 87-95.

Eskin, N.A.M. and Henderson, H.M., 1977. A study of lipoxygenase from lentils and lupins. Annals Technol. Agric. 26: 139-149.

Eskin, N.A.M., Johnson, S., Vaisey-Genser, M. and McDonald, B.E., 1980. A study of oligosaccharides in a select group of legumes. J. Inst. Can. Sci. Technol. Aliment. 13: 40-42.

FAO, 1985. Production Yearbook. Vol 39. Food and Agriculture Organization of the United Nations, Rome, 335 pp.

Hsu, D., Leung, H.K., Finney, P.L. and Morad, M.M., 1980. Effect of germination on nutritive value and baking properties of dry peas, lentils and faba beans. J. Food Sci. 45: 87-92.

Kylen, A.M. and McCready, R.M., 1975. Nutrients in seeds and sprouts of alfalfa, lentil, mung beans and soybeans. J. Food Sci. 40: 1008-1009.

Marquardt, R.R. and Bell, J.M. 1988. Future potential of pulses for use in animal feeds. In: R.J. Summerfield, ed. World Crops: Cool Season Food Legumes. Kluwer, Dordrecht, pp. 421-444.

National Research Council, 1982. United States-Canadian Tables of Feed Composition, 3rd ed. National Academy of Sciences, Washington, D.C. 148 pp.

National Research Council, 1988. Nutrient Requirements of Domestic Animals, No. 2. Nutrient Requirements of Swine. 9th ed. National Academy of Sciences, Washington, D.C.

Nygaard, D.F. and Hawtin, G.C., 1981. Production, trade and uses. In: C. Webb and G. Hawtin, eds. Lentils. Commonwealth Agricultural Bureaux, England, pp. 7-14.

Roth, F.X. and Kirchgessner, M., 1987. Biological efficiency of dietary methionine or cystine supplementation with growing pigs. Z. Tierphysiol. Tiererahrg. u. Futtermittelkde 58: 267-280.

Sarwar, G. and Peace, R.W., 1986. Comparisons between true digestibility of total nitrogen and limiting amino acids in vegetable proteins fed to rats. J. Nutr. 116: 1172-1184.

Shekib, L.A.H., Zoueil, M.E., Yossef, M.M. and Mohamed, M.S., 1986. Amino acid composition and in vitro digestibility of lentil and rice proteins and their mixture (Koshary). Food Chem. 20: 61-67.

Wassimi, N., Abu-Shakra, S., Tannous, R. and Hallab, A.H., 1978. Effect of mineral nutrition on cooking quality of lentils. Can. J. Plant Sci. 58: 165-168.

Webb, C. and Hawtin, G. 1981. Lentils. Commonwealth Agricultural Bureaux, England. 216 pp.

Weder, J.K.P., 1981. Protease inhibitors in the Leguminosae. In: R.M. Polhill and P.H. Raven, eds. Advances in Legume Systematics. British Museum of Natural History, London, pp. 533-560.

CHAPTER 22

Linseed Meal

J.P. Bowland

INTRODUCTION

Flax (*Linum usitatissimum* L.) is one of the oldest crops known to man and the botanical name means "most useful" (Peterson, 1958). World production of seed flax has remained virtually constant over the past 50 years varying from 2.6 to 3.3 million tonnes per year (Prentice, 1989). However, since the demand for other oilseeds has increased significantly, the importance of flax relative to other crops has declined. In the 1984-87 period, Canada accounted for 32% of the world flaxseed supply with Argentina, India, U.S.S.R. and U.S.A. producing 20, 16, 8 and 7%, respectively.

The oil content of flaxseed ranges from 40 to 45%, depending on cultivar and growing conditions. Linseed meal is the by-product that remains following oil extraction from flaxseed. Linseed meal is viewed primarily as a crushing industry by-product which can be disposed of relatively easily, but at a price determined by the prices of its many substitutes (Churchward and Prentice, 1987). During the 1980s, annual linseed meal utilization has averaged approximately 1.3 million tonnes with over one half of this consumption taking place in the European Economic Community (Agriculture Canada, 1988; The Flax Council of Canada, personal communication). Most linseed meal is used in diets of ruminants and horses with very little being fed to swine. The limited use of linseed meal for swine is demonstrated by the fact that the meal is not included in the feedstuff composition tables in the Nutrient Requirements of Swine (National Research Council, 1988).

GROWING FLAX

Flax is an herbaceous annual plant which belongs to the genus *Linum*, one of the 10 genera of the family *Linaceae*. The genus contains over 100 annual and perennial species. The species *L. usitatissimum* contains two types of cultivars, one grown for seed and the other for fiber. Flax is adapted to growth in warm or cool temperate climates with seed flax being more tolerant to warm, dry climates than fiber flax (Peterson, 1958). The life cycle of the flax plant consists of a 45 to 60 day vegetative period, a 15 to 25 day flowering period and a maturation period of 30 to 40 days (Saskatchewan Agriculture, undated). Water stress, high temperature or disease can shorten any of these periods and have an adverse effect on yield.

Historically, fusarium wilt caused by the fungus *Fusarium oxysporum* and rust disease caused by the fungus *Melampsora lini* have been limiting factors in flax production. Breeding of resistant cultivars and control measures such as crop rotation can keep these diseases and other minor diseases under control. Treatment of flax seed with a recommended fungicide is essential to reduce seed decay and seedling blight (Saskatchewan Agriculture, undated). As is true of all crops, flax may be infested from the time of emergence to maturity by various insect pests. Because flax does not shade the ground as much as cereal grains, weeds have an excellent chance to develop if left uncontrolled.

One advantage of flax production is that it can be sown and harvested with conventional equipment used for cereal production. Flax usually grows well on types of land suitable for cereals such as wheat and yields of 950 kg/ha are typical under Canadian conditions.

NUTRIENT CONTENT OF LINSEED MEAL

Most of the research conducted using linseed meal as a protein supplement in swine diets has used expeller-processed meal, the so-called "old process." The more common process now in use for extracting oil from flaxseed is commonly referred to as solvent extraction but should more correctly be called prepress-solvent extraction.

Flaxseed is high in oil content (40-45%) so that the flakes formed are too fragile to extract directly. Therefore, the flax is prepressed to remove more than half the oil and reduce the residual oil content to 10 to 20%. The material, after solvent extraction, contains 1 to 2% oil (Peterson, 1958). This residual oil is highly unsaturated (Iodine Number, 160 to 190), has a low melting point (remaining liquid below 18°C) and is readily digestible.

There is no evidence that there are marked consistent differences between expeller and solvent-processed linseed meals for swine. The proximate analyses of linseed meal processed by either the expeller or the solvent process are given in Table 22.1. The major difference in proximate analyses between expeller-processed and solvent-processed meals is in the ether extract which averages 5.6% for expeller meals compared with 1.4% for solvent meals. This difference in oil content is reflected to a minor degree in crude protein, crude fiber and digestible energy levels. There is, however, greater variability between meals within each of the processing procedures than there is between expeller and solvent-processed meals.

The crude fiber content of linseed meal is relatively high but not as high as that found in cottonseed, peanut, rapeseed or sunflower meals (National Academy of Sciences, 1971). Total ash content is comparable to most oilseed meals. D.E. values are slightly lower compared with soybean meal.

Table 22.1. Nutrient Composition of Linseed Meal (as fed)

	Expeller Processed Linseed Meal[1]	Solvent Processed Linseed Meal[1]	Soybean Meal (solvent)[2]
Crude Protein (%)	35.6	36.3	44.0
Crude Fiber (%)	9.2	9.3	7.3
Ether Extract (%)	5.6	1.4	0.8
Ash (%)	6.3	6.3	6.8
D.E. (kcal/kg)	3340	3330	3350

[1]Morrison, 1956; National Academy of Sciences, 1971; Fonnesbeck et al., 1984.
[2]National Research Council, 1988.

Linseed meal is intermediate in protein content when compared with other oilseed meals, averaging 36%. However, meals may vary from 34 to 42% protein (Saskatchewan Agriculture, undated). Schneider (1947) and National Academy of Sciences (1971) reported an average digestibility by swine for crude protein in linseed meal of 90%. Linseed meal is seriously limiting in lysine (Table 22.2). Therefore, it does not effectively make good the amino acid deficiencies in cereal grains when it is fed in swine diets (Janick et al., 1969; Pond and Maner, 1984).

Valid comparisons with other protein supplements must be made on a protein-equivalent basis but even on this type of comparison linseed meal contains only 55% as much lysine as is present in soybean meal. Comments in review articles and texts list either methionine (Aherne and Kennelly, 1984) or tryptophan (Cunha, 1957) as also being limiting amino acids in linseed meal. However, linseed meal on an equivalent protein basis contains approximately 10% more methionine or combined methionine and cystine than is present in soybean meal. Similarly, linseed meal contains approximately 10% more tryptophan than soybean meal. Assuming that amino acid digestibilities and utilization are similar in the two meals, it appears unlikely that either methionine (total sulfur amino acids) or tryptophan are limiting for swine. Most of the other essential amino acids are lower in linseed meal than in soybean meal, on an equivalent protein basis, but these amino acids are usually not severely limiting in cereal grains. Therefore, it seems evident that for pig feeding the major essential amino acid deficiency in linseed meal, whether expeller or solvent-processed, is lysine. With the current availability of synthetic L-lysine at a competitive price, this may not be as serious a deficiency as it was formerly.

Table 22.2. Essential Amino Acid Content of Linseed Meal (% as fed)

	Expeller Processed Linseed Meal[1]	Solvent Processed Linseed Meal[1]	Soybean Meal (solvent)[2]
Arginine	3.2	3.2	3.2
Histidine	0.7	0.8	1.1
Isoleucine	1.8	1.8	2.0
Leucine	2.1	2.2	3.4
Lysine	1.2	1.3	2.9
Methionine	0.7	0.6	0.5
Cystine	0.7	0.7	0.6
Phenylalanine	1.6	1.6	2.1
Threonine	1.3	1.4	1.5
Tryptophan	0.6	0.6	0.6
Valine	1.8	1.9	2.2

[1]National Academy of Sciences, 1971; Fonnesbeck et al., 1984.
[2]National Research Council, 1988.

The levels of the major macro minerals in linseed meal are in a range similar to those present in other oilseed meals although calcium, phosphorus and magnesium are all higher than the levels in soybean meal (Table 22.3). Micro minerals vary widely. Linseed meal contains low levels of iodine (National Academy of Sciences, 1971). However, selenium, which averages 0.9 mg/kg, is much higher in linseed meal than in soybean meal. Arthur (1971) reported average selenium values of 1.05 mg/kg in linseed meal of Canadian origin with a range of 0.65 to 1.51 mg/kg. It must be recognized that the concentration of selenium in plant materials varies widely with geographic location, soil composition, growing conditions and treatment during the manufacturing process.

In experiments with chicks (Jensen and Chang, 1976) and with rats (Halverson et al., 1955; Levander et al., 1970) it was demonstrated that linseed meal contains a protective factor against selenosis. Wahlstrom et al. (1956) found that when 14% linseed meal replaced soybean meal in pig diets, the linseed meal protected against visible signs of selenium toxicity but did not alter the rate of gain of pigs fed the selenized diets. It is generally recognized that protein may minimize the severity of selenium poisoning but linseed meal seems to be the most effective supplement.

Table 22.3. Mineral Content of Linseed Meal (as fed)

	Expeller Processed Linseed Meal[1]	Solvent Processed Linseed Meal[1]	Soybean Meal (solvent)[2]
Calcium (%)	0.41	0.41	0.26
Phosphorus (%)	0.90	0.85	0.64
Magnesium (%)	0.59	0.62	0.30
Potassium (%)	1.26	1.43	2.13
Sodium (%)	0.11	0.14	0.01
Chlorine (%)	0.04	0.04	0.04
Sulfur (%)	0.39	0.43	0.44
Iron (mg/kg)	180	340	131
Copper (mg/kg)	31.4	26.6	23.0
Manganese (mg/kg)	36.5	38.9	37.2
Zinc (mg/kg)	52.0	----	57.0
Selenium (mg/kg)	0.89	0.91	0.10

[1]Morrison, 1956; National Academy of Sciences, 1971; Fonnesbeck et al., 1984.
[2]National Reseach Council, 1988.

The B-complex vitamin content of linseed meal is similar to soybean meal and most other oilseed meals (Table 22.4). Linseed meal contains an average vitamin E content of 8.5 mg/kg meal (Fonnesbeck et al., 1984) and traces of carotene (National Academy of Sciences, 1971).

Table 22.4. Vitamin Content (mg/kg) of Linseed Meal (as fed)

	Expeller Processed Linseed Meal[1]	Solvent Processed Linseed Meal[1]	Soybean Meal (solvent)[2]
Biotin	0.36	----	0.32
Choline	1847	1736	2753
Folacin	3.1	1.1	0.7
Niacin	39.7	39.2	22.0
Pantothenic Acid	15.4	16.4	14.8
Pyridoxine	6.1	9.5	4.8
Riboflavin	3.5	3.5	2.9
Thiamine	4.9	7.0	3.1

[1]National Academy of Sciences, 1971; Fonnesbeck et al., 1984.
[2]National Research Council, 1988.

UNDESIRABLE CONSTITUENTS IN LINSEED MEAL

Flaxseed, particularly immature flaxseed, may contain a cyanogenic glucoside called linamarin which, when acted upon by an associated enzyme, linase, in the seeds, yields hydrocyanic acid (Boucque and Fiems, 1988; Peterson, 1958). This reaction occurs at certain temperatures (optimum 40-50°C), conditions of acidity (optimum pH 5) and in the presence of moisture. The concentration of the glucoside can be reduced by appropriate processing and seed selection. The enzyme, linase, is normally destroyed by the heat to which ground flaxseed is subjected in oil extraction. However, when the oil is removed under low-temperature processing conditions, a part of the linamarin and linase remain unchanged in the meal and such meal has caused the death of some animals (Peterson, 1958).

The amount of linamarin present varies with variety, maturity and oil content of the flaxseed. Mature seed contains little or none. A study of linseed meal samples collected from various countries showed the amount of linamarin to be equivalent to hydrocyanic acid levels of 100 to 300 mg/kg of meal. When dry meal containing linase is consumed by non-ruminants, the enzyme is destroyed by acidic digestive juices before it can react with linamarin (Peterson, 1958).

A vitamin B_6 antagonist has been isolated and synthesized from linseed meal (Klostermann et al., 1967). The antagonist is a dipeptide called linatine that yields 1-amino-D-proline and glutamic acid when hydrolyzed. Samples of linseed meal may differ in their content of linatine, with a range of 20 to 100 mg/kg meal being reported (Peterson, 1958).

Bishara and Walker (1977) reported that pigs fed diets containing 30% of linseed meal showed greater gain, nitrogen retention, blood packed cell volume and hemoglobin levels if supplemental pyridoxine (5.4 mg/kg) was added as compared with those receiving only the basal diet. The results suggest that, under some conditions, diets containing linseed meal may be marginally deficient in vitamin B_6.

UTILIZATION OF LINSEED MEAL

The market potential for linseed meal in general is dependent upon future market growth in the demand for linseed oil. Linseed oil has been used traditionally as an industrial oil and the oil is not currently suitable for use as an edible oil because of its high concentration of linolenic acid (45-65%). This polyunsaturated fatty acid is readily oxidized and imparts the rapid drying properties to linseed oil. Cultivars capable of producing edible quality oil would be of agricultural benefit because the market for edible vegetable oils is large and expanding whereas that for industrial-quality linseed oil is declining owing to the increased use of substitutes (The Flax Council of Canada, personal communication).

Starter Pigs

Danielsen and Nielsen (1983) conducted studies in Denmark with a total of 20 litters of piglets in which diets contained 0 or 3% linseed meal with or without 5% animal fat. The linseed meal and fat were substituted for barley and soybean meal. The results of these trials were published in Danish and the comments here are based on a personal communication (Danielsen, 1989). The studies were conducted during a pre-weaning period when the pigs were 21 to 35 days of age and a post-weaning period from 36 to 70 days of age. Daily gain for all diets averaged 325 g/day during the pre-weaning period and 488 g/day during the post-weaning period. No interactions of linseed meal and animal fat were observed and responses to linseed meal were very small and insignificant. The results suggest that low levels of at least 3% linseed meal may be included in creep feed or early weaning diets if relative price relationships justify such use.

Growing-Finishing Pigs

Linseed meal has a lower protein content and a much lower lysine level compared with soybean meal. Therefore, linseed meal is grossly deficient in lysine when used as a supplement to cereal grains to meet the nutrient requirements of pigs (Agricultural Research Council, 1981; National Research Council, 1988). If used as the only protein supplement to corn or other grains in swine feeding, linseed meal is much less efficient than supplements that supply protein of better quality. For this reason, if it is to be used in swine diets it should be combined with a complementary protein source. Morrison (1956) in his summary of experiments, prior to that date, indicated that linseed meal is an excellent protein supplement for swine, when it is fed in combination with such animal by-products as tankage, fishmeal or skim milk. Early popularity of linseed meal as a protein supplement was related in part to its palatability (Morrison, 1956) even when it was more expensive per unit of protein than other supplements.

Linseed meal produces fairly satisfactory results when fed as the sole protein supplement to growing and finishing pigs on good pasture, but the gains are less rapid than when a better balanced protein source is fed. Morrison (1956) reported that, in nine experiments, 27 kg pigs fed corn and linseed meal on alfalfa or rape pasture gained 0.58 kg per head daily while others fed corn and tankage gained 0.61 kg daily.

Prior to 1965, research workers have fed linseed meal as the major protein supplement up to 25% of the diet for swine with reasonable results but experiments were generally not conducted as comparative studies of linseed meal with other supplements. For example, Crampton et al. (1955) used linseed meal as the basal protein in their studies with malt sprouts and Wahlstrom et al. (1956) found that linseed meal was protective against selenium toxicity.

Bethke et al. (1928) conducted some of the earlier comparative studies with linseed meal and another protein supplement, in this case, cottonseed meal. They fed diets containing 25% linseed meal to pigs with no adverse effects. They also observed that not all species of animals respond alike to linseed meal and cottonseed meal which indicates a danger in extrapolating between species. Cunha (1957) noted that wheat and barley are higher in lysine and tryptophan than corn, hence linseed meal should be more satisfactory as a protein supplement when used with wheat or barley diets as compared with corn. Morrison (1956) had previously summarized experiments that indicated that linseed meal is more satisfactory as the only protein supplement to barley and wheat than as a supplement to corn. He also noted that experiments in 1919 found that the proteins from flaxseed and rye complemented each other.

If too much flaxseed is fed to pigs, "soft pork" may be produced (Morrison, 1956). "Soft pork" is the general term used to refer to pork fat that is more highly unsaturated than normal and which tends to be liquid at room temperature. There are no reports of this problem occurring with pigs fed expeller-processed linseed meal which usually contains over 5% oil.

In summary, there is no research available comparing varying levels of linseed meal in diets to growing and finishing pigs. As linseed meal has been fed to growing and finishing pigs at levels of up to 25% of the diet with no evident adverse effects, it appears that the meal may be fed as a major portion of the protein supplement if the lysine deficiency is corrected.

Breeding Stock

A unique property of linseed meal is the content of mucilage, a water dispersible polysaccharide. Mucilage constitutes 2 to 7% of the weight of the dry seed and 3 to 10% of the defatted meal (Wolf, 1983). The outer coating of the seed, or epiderm, is a layer of mucilage about 0.1 to 0.2 mm thick. This complex mixture of carbohydrate disperses readily in water to form a thick slime (Peterson, 1958). The mucilage is considered to be almost completely indigestible by non-ruminant animals (Boucque and Fiems, 1988). However, the mucilage is able to absorb large amounts of water. Linseed meal tends to have a laxative effect and is sometimes added to sow diets at the time of parturition to prevent constipation (Pond and Maner, 1984). The mucilage content of the meal is undoubtedly largely responsible for this characteristic of the meal.

There is no literature in which the comparative value of linseed meal for gestating and lactating sows has been studied but it appears that because of its laxative properties and palatability, it could form at least 10% of the diet if properly balanced with other supplements to provide adequate lysine. The laxative properties of linseed meal might become a disadvantage if very high levels were fed to lactating sows on high feed intake.

SUMMARY

Although linseed meal has been widely used in the past as a portion of the protein supplement for pigs, it is unlikely to be recommended at present.

Very little linseed meal is currently used in swine diets because of its relatively low protein content and low lysine level which make the meal non-competitive with other oilseed meals. If linseed meal is used in swine diets it should be combined with a complementary protein source.

Most of the seed selection and hybridization being done on flax has been for the purpose of increasing yield by developing resistance to disease, parasites and adverse climate (Peterson, 1958). Research on the amount and quality of oil and protein has lagged. Green (1986a, 1986b) working in Australia has reported on genetic removal of linolenic acid from linseed oil to a level more consistent with that required for an edible oil. New cultivars of flax are being developed in Canada by crossing mutated strains from Australia, which are low in linolenic acid, with Canadian cultivars (The Flax Council of Canada, personal communication). Tests on the new flax oil show it to be similar to sunflower oil. If such oil enters the edible oil market it will allow research on the meal with the hope of eliminating negative or limiting factors as was done with rapeseed meal in the development of canola meal. Industry research suggests that a new flax crop, to be called linola, could be in widespread production in 10 years. Therefore, although not widely used at present, the future of linseed meal as a protein supplement for pigs appears to offer promise.

REFERENCES

Agriculture Canada, 1988. Fats and Oils in Canada Annual Review 1988. Prepared by Agriculture Canada Grains and Oilseeds Branch, Ottawa, Canada.

Agricultural Research Council, 1981. The Nutrient Requirements of Pigs. Commonwealth Agricultural Bureaux, Slough, England.

Aherne, F.X. and Kennelly, J.J., 1984. Oilseed meals for livestock feeding. In: D.J.A. Cole and W. Haresign, eds. Recent Developments in Pig Nutrition. Butterworths, London pp. 278-316.

Arthur, D., 1971. Selenium content of some feed ingredients available in Canada. Can. J. Anim. Sci. 51: 71-74.

Bethke, R.M., Bohstedt, G., Sassaman, H.L., Kennard, D.C. and Edington, B.H., 1928. The comparative nutritive value of the proteins of linseed meal and cottonseed meal for different animals. J. Agric. Res. 36: 855-871.

Bishara, H.N. and Walker, H.F., 1977. The vitamin B_6 status of pigs given a diet containing linseed meal. Brit. J. Nutr. 37: 321-331.

Boucque, C.V. and Fiems, L.O., 1988. Feedstuffs 4. Vegetable by-products of agro-industrial origin. Livest. Prod. Sci. 19: 97-136.

Churchward, S. and Prentice, B.E., 1987. Flaxseed, linseed oil and linseed cake/meal market profile: Prospects for growth in the Canadian market. The Flax Council of Canada, Winnipeg, Manitoba.

Crampton, E.W., MacKay, V.G. and Lloyd, L.E., 1955. The biological value of the protein and the apparent digestibility of the crude fiber of malt sprouts. J. Anim. Sci. 14: 688-692.

Cunha, T.J., 1957. Swine Feeding and Nutrition. Interscience Publishers. New York, 352 pp.

Danielsen, V. and Nielsen, H.E., 1983. Horfro og fedt i foderblandinger til smogrise. Statens Husdyrbrugsforsog Meddelelse N.R. 468.

Fonnesbeck, P.V., Lloyd, H., Obray, R. and Romesburg, S., 1984. International Feedstuffs Institute Tables of Feed Composition. Utah State University, Logan, Utah.

Green, A.G., 1986a. A mutant genotype of flax (*Linum usitatissimum* L.) containing very low levels of linolenic acid in its seed oil. Can. J. Plant Sci. 66: 499-503.

Green, A.G., 1986b. The development of edible-oil flax, a potential new polyunsaturated oilseed crop. CSIRO Division of Plant Industry Report, Canberra, Australia.

Halverson, A.W., Hendrick, C.M. and Olson, O.E., 1955. Observations of the protective effect of linseed oil meal and some extracts against chronic selenium poisoning in rats. J. Nutr. 56: 51-60.

Janick, J., Schery, R.W., Woods, F.W. and Ruttan, V.W., 1969. Plant Science. W. H. Freeman and Company, San Francisco, 447 pp.

Jensen, L.S. and Chang, C.H., 1976. Fractionation studies on a factor in linseed meal protecting against selenosis in chicks. Poult. Sci. 55: 594-599.

Klostermann, H.J., Lamoureux, G.L. and Parsons, J.L., 1967. Isolation, characterization, and synthesis of linatine, a vitamin B_6 antagonist from flaxseed (*Linum usitatissimum*). Biochem. (Easton) 6: 170-177.

Levander, D.A., Young, M.L. and Meeks, S.A., 1970. Studies on the binding of selenium by liver homogenates from rats fed diets containing either casein or casein plus linseed meal. Toxicol. Appl. Pharmacol. 16: 79-87.

Morrison, F.B., 1956. Feeds and Feeding. 22nd Edition. The Morrison Publishing Company, Ithaca, New York, 1165 pp.

National Academy of Sciences, 1971. Atlas of Nutritional Data on United States and Canadian Feed. National Academy of Sciences, Washington, D.C.

National Research Council, 1988. Nutrient Requirements of Domestic Animals, No. 2. Nutrient Requirements of Swine, 9th ed. National Academy of Sciences, Washington, D.C.

Peterson, S.W., 1958. Linseed oil meal. In: A.M. Altschul, ed. Processed Plant Protein Foodstuffs. Academic Press, New York, pp. 593-615.

Pond, W.G. and Maner, J.H., 1984. Swine Production and Nutrition. AVI Publishing Company, Westport, Connecticut, 731 pp.

Prentice, B.E., 1989. Canada's Flax Industry. Presented to the Sixth International Feed and Oilseed Program. Canadian International Grains Institute, Winnipeg, Manitoba.

Saskatchewan Agriculture - Agriculture Development Fund. Not dated. Growing Flax. The Flax Council of Canada, Winnipeg, Manitoba.

Schneider, B.H., 1947. Feeds of the World. Their Digestibility and Composition. Agricultural Experiment Station, West Virginia University, Morgantown, West Virginia.

Wahlstrom, R.C., Kamstra, L.D. and Olson, O.E., 1956. The effect of organic arsenicals, chlortetracycline and linseed oil meal on selenium poisoning in swine. J. Anim. Sci. 15: 794-799.

Wolf, W.J., 1983. Nutritional supplements for animals: Oilseed proteins. In: M. Rechiegl, ed. CRC Handbook of Nutritional Supplements. Vol. II: Agricultural Use. CRC Press, Boca Raton, Florida, pp. 163-175.

CHAPTER 23

Leucaena Leaf Meal

T.E. Ekpenyong

INTRODUCTION

The use of leguminous trees to supply livestock with a source of nitrogen and protein during the dry season has recently increased in the tropics and sub-tropics. Of the numerous legumes available, one of them, *Leucaena (L. glauca)*, has been given pride of place as a versatile legume adapted to areas with an average annual rainfall varying from 600-1800 mm (National Academy of Sciences, 1977). The meal obtained by grinding the dried leaves of the Leucaena tree has a protein content of nearly 30%. Therefore, it may be possible to use leucaena leaf meal as a means of meeting the amino acid requirements of pigs fed in the tropics.

GROWING LEUCAENA

Leucaena (also called ipil-ipil, koa-haole, lamtoro, guaje and subabul) originated in Central America (Southern Mexico, Guatemala, El Salvador and Honduras) and some varieties were spread by the Maya and Zapotec civilizations. Spanish conquistadors took the "Acapulco" variety from the port of Acapulco during their trade expeditions into the Philippines. From there, Leucanena spread into Indonesia, Papua New Guinea, Malaysia and other parts of Southeast Asia and later into countries like Hawaii, Australia,

Caribbean Islands, India, Fiji as well as East and West Africa where it has established itself as a persistent, vigorous and rugged crop (National Academy of Sciences, 1977; Pound and Martinez Cairo, 1983).

The genus Leucaena is a member of the family Leguminosae and of the subfamily Mimosoideae. There are about 100 strains. Some are multi-branched shrubs that average 5 m in height at maturity while others are single-trunk trees which attain a height in excess of 20 m at maturity. Like most legumes, leucaena forms a symbiotic relationship with soil bacteria, especially Rhizobium. These bacteria fix nitrogen from the air into compounds which the plant eventually utilizes for protein formation. The leucaena-Rhizobium partnership can fix more than 500 kg N/ha/annum (Blom, 1980). The plant has a strong taproot which penetrates deep into the soil to utilize water and other nutrients (Laudencia, 1972). Good varieties, when planted on good soils, produce annual yields of dry matter ranging from 12 - 20 tonnes/ha (6 - 10 tons/acre) which is equivalent to 800 - 4300 kg of protein/ha (800 - 4300 lbs/acre).

NUTRIENT CONTENT OF LEUCAENA LEAF MEAL

Various reports and reviews have been published in several tropical and sub-tropical countries on the nutrient and mineral composition of leucaena (Ylangan and Salud, 1957; Castillo et al., 1963, 1964; Oakes, 1968; National Academy of Sciences, 1977; Adeneye, 1979; D'Mello and Fraser, 1981; Akbar and Gupta, 1984; Ekpenyong, 1986). It would appear that there is considerable variation in nutrient content between varieties, between young and mature leaves and also between fresh and dried leaves. A chemical analysis of leucaena leaf meal is shown in Table 23.1.

Table 23.1. Chemical Analysis of Four Samples of Dried Leucaena Leaf Meal (as fed)

	Source and Type of Leucaena Leaf Meal			
	Malawi		Thailand	
	Sun-Dried	Sun-Dried	Oven-Dried	Sun-Dried
Crude Protein (%)	29.41	29.10	28.13	22.44
Crude Fiber (%)	7.33	8.91	8.84	12.36
Ether Extract (%)	3.40	4.77	4.53	3.98
Ash (%)	10.41	7.00	6.84	9.78
Gross Energy (Mcal/kg DM)	4.52	4.67	4.64	4.60

D'Mello and Fraser, 1981.

Leucaena leaf meal contains about 29% crude protein and therefore fairly high levels would have to be included in the diet if leucaena leaf meal were used as the sole source of supplementary protein. The gross energy concentrations of four cultivars of dried leucaena leaves have been reported to be 4.52, 4.67, 4.64 and 4.60 Mcal/kg DM (D'Mello and Fraser, 1981). However, it should be noted that much of this energy is probably unavailable to non-ruminant animals (D'Mello and Thomas, 1978). The lignin content of leucaena leaf meal has been reported to be as high as 140 g/kg (Arora and Joshi, 1984).

A few workers have determined the amino acid composition of various cultivars of leucaena. All of these reports (National Academy of Sciences, 1977; D'Mello and Acamovic, 1988; D'Mello and Fraser, 1981; Ekpenyong, 1986) show that leucaena protein is of high nutritional quality and has amino acids in well-balanced proportions (Table 23.2).

Table 23.2. Amino Acid Analysis (g/kg DM) of Four Samples of Dried Leucaena Leaf Meal

	Source and Type of Leucaena Leaf Meal			
	Malawi		Thailand	
	Sun-dried	Sun-dried	Oven-dried	Sun-dried
Arginine	16.4	15.1	15.8	10.2
Histidine	6.6	5.4	5.7	4.0
Isoleucine	15.3	13.7	14.1	12.4
Leucine	22.6	21.7	24.2	16.0
Lysine	15.5	17.6	16.9	12.8
Methionine	5.4	4.6	4.6	2.3
Cystine	2.1	2.0	1.6	1.6
Phenylalanine	15.8	14.8	14.2	10.7
Tyrosine	12.1	12.5	11.7	8.1
Threonine	13.5	12.1	11.9	8.7
Tryptophan	3.3	3.8	3.3	2.4
Valine	15.2	14.4	13.0	10.1
Mimosine	10.2	25.5	14.7	14.1

D'Mello and Fraser, 1981.

Leucaena leaves contain appreciable levels of calcium and phosphorus. Other reports (Akbar and Gupta, 1984) have shown the leaves to be poor in sodium and with a very wide calcium to phosphorus ratio. With age, the minerals accumulate. The values reported for calcium, potassium and sodium by Adeneye (1979) and Ekpenyong (1986) in leaves were higher in Nigeria than those reported in Malawi by D'Mello and Thomas (1978). The mineral composition of leucaena leaf meal is presented in Table 23.3.

Leucaena, besides its role as a protein supplement, also serves as a source of vitamin A. The vitamin A requirement of gestating sows is met by supplementing the diet with 4.5% leaf meal (Eusebio et al., 1961). Leucaena is rich in carotene and this is of interest in the tropics, especially during drought when leucaena leaves retain the green pigment in the leaf better and much longer than any other tropical forage.

Table 23.3. Mineral Composition of Four Samples of Dried Leucaena Leaf Meal

	Source and Type of Leucaena Leaf Meal			
	Malawi		Thailand	
	Sun-Dried	Sun-Dried	Oven-Dried	Sun-Dried
Major Minerals (g/kg DM)				
Calcium	23.3	18.1	18.5	23.7
Phosphorus	2.5	2.5	2.9	1.9
Sodium	0.4	0.0	0.1	0.2
Potassium	19.9	8.0	8.1	18.0
Magnesium	4.7	5.1	5.6	4.2
Trace Minerals (mg/kg DM)				
Copper	7.0	9.9	10.6	9.7
Iron	407.1	239.9	191.0	181.2
Zinc	21.0	21.2	29.9	18.1
Manganese	60.0	42.1	46.0	49.2

D'Mello and Fraser, 1981.

UNDESIRABLE CONSTITUENTS IN LEUCAENA

A major constraint inhibiting the widespread utilization of leucaena as an animal feed has been the presence of a rare amino acid, mimosine (beta-(N-3-hydroxy-4-ozopyridyl)-a-amino propionic acid). Mimosine occurs in the seeds and all green parts of the plant at levels up to 12% (Jones, 1979).

Mimosine has general cytotoxic activity, causing alopecia, low liveweight gain, excessive salivation, enlarged thyroid glands, incoordination of gait, loss of appetite and reproductive failure (Owen, 1958). Crounse et al. (1962) suggested that mimosine may interfere with tyrosine metabolism. Circulating dihydroxypyridine prevents iodination of tyrosine, the first step in the synthesis of thyroxine, resulting in goitre and reduced levels of thyroxine in the serum associated with loss of appetite, drooling and alopecia (Jones et al., 1976). Castillo et al. (1964) showed that rations containing 1 to 6% unprocessed leucanena caused ruffled hair, skin eruptions and alopecia.

Leucaena contains tannins which have adverse effects on growth and digestibility in animals (Ford and Hewitt, 1979). D'Mello and Fraser (1981) showed variation in tannin content from 11.6 to 43.6 and 21.7 to 33.6 g/kg DM, respectively, using two different methods. Tannins reduce protein quality of feeds by inhibiting protease activity and/or binding with food proteins (Reddy et al., 1985) and also decrease iron and starch availability (Rao and Prabhavathi, 1982).

Leucaena also contains saponins which are surface triterpenoid and sterols glycosides (Basu and Rastogi, 1967; Oakenfull, 1981). Some authors have shown that the consumption of saponin-containing diets by monogastric animals has adverse effects on growth and also affects cholesterol metabolism (Oakenfull, 1981; Cheeke, 1976). Saponins have been shown to be effective in reducing cholesterol in blood of animals consuming diets which produce hypercholesterolemia. The bitter taste of leucaena has been attributed partly to the presence of tannins (D'Mello and Fraser, 1981; Fraser and D'Mello, 1981) and partly to saponins.

Leucaena leaves also contain procyanidins which react with leaf proteins during leaf protein extraction and form indigestible complexes. In addition, D'Mello and Acamovic (1988) have shown that other antinutritional factors, such as protease inhibitors and glactomannan gums are present, and these also reduce performance.

ELIMINATION OF TOXICITY

Despite concern about the negative effects of leucaena in monogastric animals, with a better understanding of the mode of action of its toxic factors it is possible to process the meal so as to minimize the impact of these factors. Various methods for the elimination of toxicity are summarized below.

1. Under practical conditions, swine do not die spontaneously and the symptoms when observed early enough should lead to withdrawal of leucaena meal from the diet. It is inappropriate to give fresh leucaena leaves to swine of any age.

2. The toxicity of leucaena can be eliminated by supplementation with metal ions which chelate with mimosine and prevent its absorption in the gastrointestinal tract. Ferrous sulfate and aluminum sulfate serve as chelating agents (Yoshida, 1945). Mimosine inhibits the activity of metal-containing enzymes such as alkaline-phosphatase (Chang, 1960), succinoxidase (Tsai, 1961) and L-dopadecarboxylase (Lin et al., 1963) in different animals. In pigs, an inhibition by mimosine on aspartate-glutamate transaminase in the heart has been reported (Lin et al., 1962).

3. Heat treatment during drying reduces the mimosine content. Temperatures for drying should be about 70°C. In local situations,

the fresh leaves should be dried in the sun for 3 to 4 days. The leaves are then threshed off from the branches. Dried leaves may be milled before storage and later mixed in practical swine diets.

4. It has been demonstrated that the addition of polyethylene glycol can reduce the toxic effects of tannins (Acamovic and D'Mello, 1981).

5. If seeds are fed, they should be boiled in alkali for 30 minutes before incorporating in diets for pigs. This will lessen the impact of the condensed tannins.

FEEDING LEUCAENA LEAF MEAL TO PIGS

Starter Pigs

In weaner pigs, 1.5 to 6% leucaena leaf meal has been used to correct vitamin A deficiencies (Castillo et al., 1964). The effect of including leucaena leaf meal in the starter diet on pig performance is presented in Table 23.4.

Table 23.4. Effect of Leucaena Meal on Growth of Weaner Pigs (11-30 kg)[1]

	Level of Leucaena Meal (%)					
	0	10	20	30	40	50
Daily Gain (kg)[2]	0.28[b]	0.40[a]	0.38[a]	0.29[ab]	0.21[bc]	0.15[c]
Daily Intake (kg)	1.20	1.21	1.15	1.44	1.07	1.00
Feed Efficiency	4.48[b]	3.04[a]	3.02[a]	4.98[b]	5.12[b]	6.70[c]

[1]Malynicz, 1974.
[2]Means with different superscripts are different ($P<0.05$).

Growing-Finishing Pigs

When pigs are fed moderate levels of leucaena during the growing and finishing phases, little toxicity is observed. Brewbaker and Hutton (1979) observed no side effects when leucaena was fed at the 10% level. Patricio (1956) fed pigs in the Philippines with rations containing 0, 5 or 10% leaf meal. Weight gains, although not significantly different, appeared to increase as the level of leucaena in the diet increased (213, 270 and 299 g/day from 0, 5 and 10% leucaena, respectively). At the 10% level of inclusion, feed efficiency also tended to improve.

In Taiwan, Chen et al. (1981) used 0 to 16% dehydrated, finely-ground leucaena meal with 15 mg/g mimosine content in the leaf. There were no statistical differences in weight gain or feed efficiency between treatments.

Also, there was no mimosine in the feces, liver or meat of pigs fed leucaena meal. In another study carried out in Papua New Guinea, Malynicz (1974) used higher levels (0, 10, 20, 30, 40 and 50%) of leaf meal in a commercial grower ration with the addition of 2 g/kg ferrous sulphate to the feed. Their results showed declining performance at levels beyond 20%. This was attributed to either leucaena toxicity or an excessive crude fiber content in the ration. It was speculated by the author that the addition of iron salt at 2 g/kg to the diet was insufficient to counteract the effect of mimosine at high levels of leucaena inclusion.

Gonzalez and Wyllie (1982a) determined the effects of iron salts and polyethylene glycol (PEG) on feed intake, growth rate and feed conversion efficiency of pigs fed leucaena leaf meal. From the results, the authors concluded that although performance was similar to the control diet when leucaena made up 20% of the diet, treatment of leucaena meal with ferrous sulphate and PEG had some beneficial effects.

In another set of experiments, Chen et al. (1981) used rations for pigs at levels of 0, 4, 8, 12 and 16% dehydrated finely ground leucanena leaf meal containing 15 mg/g mimosine. Another ration contained 13% leucaena leaf meal plus 0.25% $FeS0_4$ plus 0.15% $CuS0_4$. All the rations had 15% crude protein and 3310 kcal/kg of digestible energy. Results showed no statistical differences in weight gain and backfat thickness of live pigs and carcasses. There were no deleterious effects and no mimosine was found in the feces, liver or muscles at 16% level of supplementation for growing-finishing pigs.

Gonzalez and Wyllie (1982b) used Landrace x Large White female pigs to determine nitrogen balances and the digestibility coefficients of leucaena based diets. They found that the mean digestibility values for dry matter, crude protein and gross energy were significantly lower compared to the basal diets, due partly to a negative effect of the higher fiber content.

Breeding Stock

Several reports have shown the negative effects of mimosine on reproductive performance of male and female pigs. Willett et al. (1945) and Willett et al. (1947) in two separate studies showed that gilts failed to conceive in three and four estrous cycles when fed fresh leucaena leaves free choice. Wayman and Iwanaga (1957) showed that at 15% level, dried leucaena leaf reduced conception, average litter size and piglet weight. Wayman et al. (1970), in five separate studies, showed that 12.5% fresh leucaena forage or 15% dehydrated leaf meal caused embryonic or fetal death followed by resorption. When leucaena leaves were fed free choice to pregnant sows from the 4th week of pregnancy, neither reproductive performance nor litter size were affected.

FEEDING LEUCAENA SEEDS TO PIGS

Leucaena leaf meal is the most common product used in animal nutrition. However seeds are also processed and incorporated into animal rations.

Boiling the seeds in water or alkali for 30 minutes reduces the tannin content by more than 70% (Komasi and Stamet, 1986). Mimosine is present in the seeds but other antinutritional factors present may also contribute to the toxicity. These factors include protease inhibitors and galactomannan gums (D'Mello and Acamovic, 1989).

Dehulled leucaena seeds supplemented with fish or yeast have been used as a protein supplement for growing finishing pigs with satisfactory results (Lee and Yi, 1963). Feeding breeding boars hulled processed seeds showed that average sperm number per cubic centimeter and the total sperm in ejaculate of the boars were lower but the difference was not statistically significant (Lee, 1963).

SUMMARY

Leucaena leaf meal and dehulled seed meal have unique nutritional properties in the tropics and sub-tropics as feedstuffs for pigs. In several countries, both the leaves and seeds have been used in protein-deficient areas, especially during the dry season. Leucaena contains high levels of protein and high carotene with well balanced amino acids in its protein.

Leucaena contains a rare and toxic amino acid, mimosine, which is harmful to pigs when consumed in its unprocessed form or when consumed over a prolonged period. Other undesirable constituents include tannins, protease inhibitors and saponins. Leucaena can be used in pig feeding if properly processed to eliminate the undesirable constituents. Common means to render leucaena nontoxic to pigs include sun-drying or supplementing the diet with metal ions. For practical feeding, weaners should be fed not more than 5% of leucaena leaf meal in the diet, fatteners up to 15%, while for breeders the level should not exceed 15% or complete withdrawal of leucanena leaf meal from the diet 2 to 4 weeks prior to breeding. On no account should the level of leucaena exceed 20%.

REFERENCES

Acamovic, T. and D'Mello, J.P.F., 1981. The effect of iron supplemented leucaena diets on the growth of young chicks. Leucaena Res. Rep. 2: 60-61.

Adeneye, J.A., 1979. A note on the nutrient and mineral composition of Leucaena leucocephala in Western Nigeria. Anim. Feed Sci. Technol. 4: 221-225.

Akbar, M.A. and Gupta, P.A., 1984. Nutrient composition of different cultivars of leucaena leucocephala. Leucaena Res. Rep. 5: 14-15.

Arora, S.K. and Joshi, U.N., 1984. Chemical composition of leucaena seeds. Leucaena Res. Rep. 5: 16-17.

Basu, N. and Rastogi, R.P., 1967. Triterpenoid saponins and sapogenins. Phytochemistry 6: 1249-1270.

Blom, P.S., 1980. Leucaena, a promising versatile leguminous tree for the tropics. Abstract on Tropical Agriculture 6: 9-17.

Brewbaker, J.L. and Hutton, E.M., 1979. Leucaena: Versatile tropical tree legume. In: G.M. Ritchie ed. New Agricultural Crops. AAAS Westview Press, Boulder Colorado, pp. 207-257.

Castillo, L.S., Gerpacio, A.L., Javier, T.R., Gloria, L.A. and Gerpacio, C.P., 1963. Quantitative changes in nutrient composition of ipil-ipil leaf during storage. Philipp. Agric. 46: 681-700.

Castillo, L.S., Aglibut, F.B., Gerpacio, A.L., Javier, T.R., Puyaoan, R.B. and Gatmaitan, O.M., 1964. Leucaena glauca Benth for poultry and livestock. In: Levels of leaf meal in repleting vitamin A deficient chicks and pigs. Vol. 48 (No. 6-7): 269-286.

Chang, L.T., 1960. The effect of mimosine on alkaline phosphatase of mouse kidney. J. Formosan Med. Assoc. 59: 882.

Cheeke, P.R., 1976. Nutritional and physiological properties of saponins. Nutr. Rep. Int. 13: 315-324.

Chen, M.T., Cheng, T. and Lin, L.S., 1981. Feeding efficiency of leucaena leaf meal used in pig rations. Leucaena Res. Rep. 2: 46.

Crounse, R.G., Maxwell, J.D. and Blank, H., 1962. Inhibition of growth of hair by mimosine. Nature 194: 694.

D'Mello, J.P.F. and Thomas, D., 1978. The nutritive value of dried leucaena leaf meal from Malawi: Studies with young chicks. Trop. Agric. (Trinidad) 55: 45-50.

D'Mello, J.P.F. and Fraser, K.W., 1981. The composition of leaf meal from Leucaena leucocephala. Tropical Science 23: 75-78.

D'Mello, J.P.F. and Acamovic, T., 1988. The toxicity of leucaena leaf meal for poultry: A critical assessment of recent evidence concerning the mode of action. Leucaena Res. Rep. 9: 97-98.

D'Mello, J.P.F. and Acamovic, T. 1989. Leucaena leucocephala in poultry nutrition - A review. Anim. Feed Sci. Technol. 26: 1-28.

Eusebio, J.A., 1961. A preliminary study on the effects of white corn in the ration for gestating sows and their vitamin A requirement. Philipp. Agric. 44: 439-452.

Ekpenyong, T.E., 1986. Nutrient and amino acid composition of Leucaena leucocephala (Lam.) de Wit. Anim. Feed Sci. Technol. 15: 183-187.

Ford, J.E. and Hewitt, D., 1979. Protein quality in cereals and pulses. 3. Bioassays with rats and chickens on sorghum (*Sorghum vulgare* Pers.), barley and field beans (*Vicia faba*, L.). Influence of polyethylene glycol on digestibility on the protein in high-tannin grain. Brit. J. Nutr. 42: 325-340.

Fraser, K.W. and D'Mello, J.P.F., 1981. Methods of determination of tannins in Leucaena leucocephala. Leucaena Res. Rep. 2: 64-65.

Gonzalez, V.D. and Wyllie, D., 1982a. Treated dried leucaena leaf meal in diets for growing pigs. Leucaena Res. Rep. 3: 74-75.

Gonzalez, V.D. and Wyllie, D., 1982b. Nutritive value of leucaena for growing pig. Leucaena Res. Rep. 3: 76.

Jones, R.J., Blunt, C.G. and Holmes, J.H.G., 1976. Enlarged thyroid glands in cattle grazing Leucaena pastures. Trop. Grasslands 10: 113-116.

Jones, R.J., 1979. The value of Leucaena leucocephala as a feed for ruminants in the tropics. World Anim. Rev. 31: 1-11.

Komari, L. and Dewi Sabita Stamet, 1986. Effect of alkali on tannin content of Leucaena seeds. Leucaena Res. Rep. 7: 77-78.

Laudencia, P.N., 1972. Soil and water conservation through the improvement of soil cover. Philipp. Geog. J. 16: 42-52.

Lee, P.K., 1963. Effect of Leucaena glauca seed instead of soybean cake in ration upon the semen production of breeding boars. Expt. Reports in Animal Nutrition of Taiwan Livestock Institute. Dec. 1963, pp. 32-34.

Lee, P.K. and Yi, J.J., 1963. An experiment on the feeding value of dried white fish, seaweed and Leucaena glauca seed as a substitute for soybean cake in hog ration. Expt. Reports on Animal Nutrition of Taiwan Livestock Research Institute. Dec. 1963, pp. 28-31.

Lin, J.Y., Shih, Y.M. and Ling, K.H., 1962. Effect of mimosine on the activity of glutomic-aspartic transaminase, in vitro. J. Formosan Med. Assoc. 61: 1004.

Lin, J.Y., Lin, K.T. and Ling, K.H., 1963. Studies on the mechanism of toxicity of mimosine B-N-3- hydroxypyridone X-amino propionic acid. 3. The effect of mimosine on the activity of L-dopa decarboxylase in vitro. J. Formosan Med. Assoc. 62: 587.

Malynicz, G., 1974. The effect of adding Leucaena leucocephala meal to commercial rations for growing pigs. Papua New Guinea Agric. J. 25: 12-14.

National Academy of Sciences, 1977. Leucaena: Promising Forage and Tree Crop for Tropics. National Academy of Sciences, Washington, D.C. 122 pp.

Oakenfull, D., 1981. Saponins in food: A review. Food Chem. 6: 19-40.

Oakes, A.J., 1968. Leucaena leucocephala: Description, culture, utilisation. Advancing Frontiers of Plant Science (New Delhi, India) 20: 1-114.

Owen, L.N., 1958. Hair loss and other toxic effects of Leucaena glauca ("Jumbey"). Vet. Rec. 70: 454-457.

Patricio, V.B., 1956. Ipil-ipil leaf meal as a supplement to rations for growing and fattening pigs. Philipp. Agric. 40: 212.

Pound, B. and Martinez Cairo, L., 1983. Leucaena: Its Cultivation and Uses. Overseas Development Administration, London.

Rao, B.S.N. and Prabhavathi, T., 1982. Tannin content of food commonly consumed in India and its influence on ionizable iron. J. Sci. Food Agric. 33: 89-96.

Reddy, N.R., Pierson, M.D., Sathe, S.K. and Salunkhe, D.K., 1985. Dry beans tannins: A review of nutritional implications. J. Amer. Chem. Soc. 62: 541-549.

Tsai, K.C., 1961. The effect of mimosine on several enzymes. J. Formosan Med. Assoc. 60: 58.

Willett, E.L., Henke, L.A. and Maruyama, C.I., 1945. Roughage for brood sows. Hawaii Agric. Experimental Station, Honolulu, Report 1942-1944.

Willett, E.L., Quisenberry, J.H., Henke, L.A. and Maruyama, C.I., 1947. Kao haole as a roughage for non-ruminants. Hawaii Agric. Experimental Station, Honolulu, Report 1944-1946.

Wayman, O. and Iwanaga, I.I., 1957. The inhibiting effect of L. glauca (Kao haole) on reproduction performance in Swine. Proc. West. Sec. Amer. Soc. Anim. Prod. 8: 5.

Wayman, O., Iwanaga, I.I. and Hugh, W.I., 1970. Fetal resorption in swine caused by Leucaena leucocephala (Lam.) de Wit. in the diet. J. Anim. Sci. 30: 583-588.

Ylangan, M.M. and Salud, S.S., 1957. Carotene and nitrogen content of ipil-ipil leaves. Philipp. Agric. 41: 238-239.

Yoshida, R.K., 1945. Chemical and physiological study of the nature and properties of Leucaena glauca (Koa haole). Proc. Hawaii Acad. Sci. 19: 5.

CHAPTER 24

Lupins

R.H. King

INTRODUCTION

Lupins have been cultivated as a grain crop for over 3,000 years, primarily in the Mediterranean, parts of the Middle East and in South America. However, the extreme bitterness of the seed has generally made the crop unsuitable for human or animal consumption without prior treatment to remove toxic alkaloids (Gladstones, 1972). The breeding of low alkaloid or "sweet" varieties of lupins, first achieved in Germany in the 1920s, revolutionized lupins as a grain crop and this vital breakthrough provided a firm base for all subsequent breeding programs. More recent varietal improvements, such as the incorporation of shatter resistance, soft seededness and other characteristics, have further improved the harvesting properties of lupins (Gladstones, 1972).

Although there are over 1,700 named species of lupins, only two (*Lupinus albus* and *L. angustifolius*) are widely grown at present (Hill, 1977). Other species which may have potential as grain crops include *L. luteus* (yellow lupin) and *L. mutabilis* (South American pearl lupin). *L. albus* (white lupin) is a broad-leafed lupin which has large, white, permeable, flat seeds and commercial cultivars include Kiev, Ultra, Neuland, Hamburg and Kalina. *L. angustifolius* (narrow leaf lupin) has smaller, round white seeds and commercial cultivars include Unicrop, Uniharvest, Uniwhite, Marri and Illyarrie.

GROWING LUPINS

Lupins are annual, winter-grown legumes which can be grown on a wide range of soils provided that they are well drained. Average annual rainfall should be at least 450 mm and may be as high as 1000 mm (Reeves, 1974). Lupins can be used in a crop rotation as a nitrogen-fixing legume and to provide a useful break in the buildup of disease in cereals. In areas that have not previously grown lupins, the seed should be inoculated with the recommended strain of Rhizobium bacteria.

Lupins are very susceptible to competition from weeds. However, the development of pre-emergence herbicides has allowed effective control of a wide range of weeds in lupins (Reeves and Lumb, 1974). Another advantage of lupins is that they can be sown and harvested with conventional equipment.

NUTRIENT CONTENT OF LUPINS

The proximate analyses of the seeds of the two main species of lupins are shown in Table 24.1. The average composition of soybean meal and wheat are provided for comparison.

Table 24.1. Nutrient Composition of Lupinseed Meal (as fed)[1]

	Narrow Leaf Lupin[2]	Broad Leaf Lupin[3]	Soybean Meal	Wheat
Crude Protein (%)	29.0	34.9	47.8	12.4
Crude Fiber (%)	13.1	10.4	5.3	2.6
Ether Extract (%)	5.5	8.1	2.1	2.1
Ash (%)	2.8	3.2	6.8	1.6
D.E. (kcal/kg)	3440	3680	3660	3420

[1]Adapted from the Standing Committee on Agriculture, 1987.
[2]*L. angustifolius*
[3]*L. albus*

The crude protein content of *L. angustifolius* may range from 25 to 34%. The level of protein found in *L. albus* is usually higher and ranges from 31 to 40% (Hill, 1977). Lupin protein is relatively high in lysine and threonine (Table 24.2) but like most legumes, the methionine content of lupins is low.

The protein and essential amino acids in lupins are well digested and absorbed from the small intestine; the true ileal digestibility of the essential amino acids of lupins is about 90% for pigs which is similar to that of soybean meal (Taverner et al., 1983). However, the utilization of the amino acids in lupins by the pig appears to be much lower. Batterham et al. (1984) tested the availability of lysine in lupins for pigs using the slope-ratio growth assay and found that the availability of lysine in four samples of lupins ranged from

37 to 65%, yet the true digestibility of lysine at the terminal ileum of one of the lupin samples was 86%. Later, Batterham et al. (1986b) confirmed the low availability of lysine in lupins when they found that the availability of lysine, assessed by the slope-ratio assay, ranged from 44 to 57% in three samples of lupins. The average availability of lysine in *L. angustifolius* and *L. albus* is considered to be 57% and 51% respectively (Standing Committee on Agriculture, 1987). The low availability of lysine in lupins for pigs is not associated with low digestibility, but may reflect either the presence of an unidentified growth inhibitor or that the lysine is in a form that is digested but inefficiently utilized.

Table 24.2. Essential Amino Acid Content of Lupinseed Meal (% as fed)

	Narrow Leaf Lupin[1]	White Lupin[2]	Soybean Meal[3]	Wheat[3]
Arginine	3.04	3.45	3.67	0.65
Histidine	0.75	0.73	1.20	0.30
Isoleucine	1.16	1.39	2.13	0.53
Leucine	2.08	2.33	3.63	0.87
Lysine	1.36	1.64	3.12	0.40
Methionine	0.14	0.20	0.71	0.22
Cystine	0.46	0.55	0.70	0.30
Phenylalanine	1.01	1.29	2.36	0.71
Threonine	1.01	1.29	1.90	0.37
Tryptophan	0.26	0.31	0.69	0.17
Valine	1.07	1.39	2.47	0.58

[1]Means obtained from data presented by Batterham, 1979; Barnett and Batterham, 1981; King, 1981; Batterham et al., 1986a.
[2]Means obtained from data presented by Batterham, 1979; King, 1981; Aguilera et al., 1985.
[3]National Research Council, 1988.

Lupins have a high level of crude fiber which is contained largely in the seed coat (Hill, 1977). It is well known that the chemical composition of the cell wall influences its physical structure and thereby its biological degradation. Lignin accounts for only 2.1% of the acid-detergent fiber fraction in lupin seed (Aguilera et al., 1985) and the small degree of lignification in lupins may explain the relatively high digestibility of structural carbohydrates in lupins. An analysis of the data presented by Taverner (1975) reveals that the apparent digestibility of crude fiber in *L. angustifolius* is 76%. Aguilera et al. (1985) also reported digestibilities in excess of 80% for the fiber fractions in *L. albus*. The relatively high level of fiber in lupins also does not appear to markedly depress the digestibility of other important nutrients (Pearson and Carr, 1976).

Because of the well digested fiber fraction and high oil content of the seed, the digestible energy content of lupins is high (Table 24.1). Nevertheless, there are many unusual features respecting the site and extent of dry matter and energy digestion from lupins by the pig. In contrast to the high absorption of amino acids in lupins from the small intestine, only about 46% of the dry matter and 51% of the energy in *L. angustifolius* are absorbed prior to the proximal end of the small intestine (Taverner et al., 1983). Much of the carbohydrate in lupins is digested by microbial fermentation in the cecum and proximal colon. Furthermore, energy which is absorbed from the hind gut is less efficiently utilized by the pig (Just, 1981) and consequently the net energy of lupins will be lower than anticipated from its gross and digestible energy contents.

An indication of the mineral content of lupins is given in Table 24.3. The mineral composition of *L. albus* is similar to that of *L. angustifolius* except that *L. albus* accumulates much higher concentrations of manganese in its seeds than other Lupinus species. Levels in the whole seed of *L. albus* ranging from 164 to 3397 mg/kg have been recorded (Hill, 1977). Fortunately, the tolerance of the pig for manganese is thought to be high (ARC, 1967). King (1981) reported that growing pigs could tolerate up to 1330 mg/kg dietary manganese, which is likely to exceed the upper level of manganese attained in pig diets containing *L. albus*.

Table 24.3. Mineral Composition of Lupinseed Meal (as fed)

	Narrow Leaf Lupin[1]	White Lupin[1]	Soybean Meal[2]	Wheat[2]
Calcium (%)	0.19	0.20	0.26	0.04
Magnesium (%)	0.15	---	0.30	0.12
Phosphorus (%)	0.28	0.37	0.64	0.37
Potassium (%)	0.66	---	2.13	0.43
Sodium (%)	0.03	0.02	0.0	0.02
Iron (mg/kg)	50	---	220	80
Copper (mg/kg)	7	5	18	6
Manganese (mg/kg)	17	393	39	49
Selenium (mg/kg)	0.10	---	0.20	0.10
Zinc (mg/kg)	42	26	53	26

[1]Standing Committee on Agriculture, 1987.
[2]National Research Council, 1988.

UNDESIRABLE CONSTITUENTS IN LUPINS

Plant breeders have developed sweet varieties of lupins which lack the toxic alkaloid components of the bitter varieties (Gladstones, 1972). However, because of possible contamination with bitter seed varieties (Hill, 1977) and certain agronomic and seasonal conditions (Godfrey et al., 1985), not all of the new sweet lupin varieties are free of alkaloids (Ruiz, 1978).

Pearson and Carr (1977) reported poor growth and food consumption by pigs given diets containing *L. albus* (cv. Neuland). However, food intake and growth performance was restored in response to reducing the alkaloid content of the seed from 0.9 g/kg to 0.2 g/kg (Pearson and Carr, 1977). Godfrey et al. (1985) recently examined a range of lupin alkaloid concentrations in pig diets and found that growing pigs could tolerate up to 0.2 g/kg of dietary lupin alkaloids before feed intake was reduced. Nevertheless, the alkaloid content in the present sweet varieties of lupins is low and there have been very few reports of toxicity or feed intake depression in pigs given diets containing up to 30-40% *L. angustifolius* or *L. albus*.

Many legume seeds contain heat-labile proteins that inhibit the activity of trypsin and other pancreatic enzymes. However, there is no evidence that trypsin inhibitors occur in *L. angustifolius* or *L. albus* (Hove and King, 1979). Furthermore, the lack of a positive response when either *L. angustifolius* (Batterham et al., 1986a) or *L. albus* (Batterham et al., 1986b) are subjected to heat treatment prior to feeding indicates that there are no heat-labile anti-nutritional factors present in lupins.

PROCESSING LUPINS

The major advantage lupins have over more common protein sources is that they can be stored for considerable time without appreciable loss of nutritive value. They also only require crushing before being included in pig diets. Lupins are also easily dehulled and the resultant lupin kernel meal has been shown by Godfrey and Payne (1987) to be a valuable and superior protein and energy source compared to lupinseed meal. Another method of processing lupin seeds, prior to feeding to pigs, is extrusion. However, extrusion of lupins has failed to provide any beneficial effects on the growth performance of weaner pigs (Hew Lap Im, personal communication). Simply crushing lupins with either a hammermill or roller mill remains the easiest and often most appropriate treatment of lupins prior to their inclusion in pig diets.

UTILIZATION OF LUPINSEED MEAL

The most extensive and thorough evaluations of lupinseed meal for pigs have been conducted in Australia and New Zealand. Often the results from other studies have been confounded by the possible effects of alkaloid in lupins. For example, the presence of alkaloids has been implicated as the likely reason for the depressed feed intake and poor performance of pigs given quite low levels of lupins (Castell and Tsukamoto, 1984; Hale and Miller, 1985).

Starter Pigs

Lupins have considerable potential for use in diets fed to weaner pigs. The high level of fiber in lupins does not appear to restrict their inclusion in starter diets provided that they are used in association with a cereal base of low fiber content. Barnett and Batterham (1981) replaced soybean meal in wheat-based diets which were equal in energy content and equal in lysine content with *L. angustifolius* and found that weaner pigs between 6 and 20 kg could tolerate up to 43% lupinseed meal without adversely affecting growth (Table 24.4). Earlier, Pearson and Carr (1976) found that lupins could completely replace blood meal or skim milk powder in starter diets and be included up to a dietary level of 26% in these diets.

Table 24.4. Performance of Weaner Pigs Fed Lupinseed Meal (6-20 kg)

	Level of Lupinseed Meal (%)		
	0	18.5	43.0
Daily Gain (g)	418	406	394
Daily Intake (g)	698	682	634
Feed Efficiency	1.67	1.68	1.61

Barnett and Batterham, 1981.

Growing-Finishing Pigs

The critical evaluation of lupins for growing pigs began in the 1970's with growth studies which investigated the response of pigs to *L. angustifolius* cultivars. Taverner (1975) reported that the cultivar Uniwhite could constitute the sole source of supplementary protein in wheat-based diets for growing pigs. Similarly, Pearson and Carr (1976) found that Uniwhite could be included at levels up to 37% and could completely replace more conventional protein concentrates without detrimental effects on the growth of grower/finisher pigs.

The cultivars of *L. albus* have slightly higher crude protein, lysine and DE contents than *L. angustifolius* cultivars. However, some studies have shown that the performance of pigs given *L. albus* is inferior to that of pigs given *L. angustifolius*. Batterham (1979) found that pigs fed either *L. angustifolius* or soybean meal as the sole protein concentrate exhibited similar growth. However, during the early grower phase, daily gain and feed conversion efficiency were significantly depressed when *L. albus* was the sole protein concentrate. King (1981) also observed that the growth of pigs given *L. albus* tended to be poorer than that of pigs offered *L. angustifolius*. Furthermore, reduction in growth became evident at levels of *L. albus* inclusion of 20% or greater in the diet for grower/finisher pigs (King, 1981; Table 24.5).

Batterham et al. (1986b) suggested that although growing pigs could tolerate *L. albus* at levels of inclusion of up to 37%, the slower growth observed in the above studies with *L. albus* was due to the lower lysine availability in lupinseed meal. However, if allowance is made for this in dietary formulation, these meals are capable of supporting acceptable growth in pigs. Lupins are also characteristically low in methionine and methionine deficiencies may occur at high inclusion levels. Diets containing lupins as the sole protein concentrate often benefit from supplementation with synthetic methionine, particularly in the grower phase (Leibholz, 1984).

Table 24.5. Performance of Growing Pigs Fed Lupinseed Meal (22-70 kg)

	Level of Lupinseed Meal (%)			
	0	10.3	20.7	31.0
Daily Gain (kg)	0.69	0.69	0.65	0.62
Daily Intake (kg)	2.03	2.00	2.00	2.00
Feed Efficiency	2.92	2.90	3.04	3.22
Backfat (mm)	21.6	18.4	19.1	18.3
Dressing Percentage	84.3	83.5	81.2	80.2

King, 1981.

Apart from the low availability of amino acids in lupins the effect of lupins on dressing percentage should also be taken into account when assessing the relative value of lupins for diets for growing pigs. The dressing percentage of pigs decreases by approximately 0.8 (Pearson and Carr, 1976) to 1.4 (King, 1981) percentage units for each 10% increment in dietary levels of lupinseed meal, when lupins are given to pigs over the entire grower/finisher phase.

Breeding Stock

The effect of including lupins in diets for pregnant and lactating sows has received little attention. However, lupins have been recommended at levels of up to 20% in dry sow and lactating sow diets and these levels have been used extensively throughout Australia without adversely affecting reproductive efficiency of sows.

SUMMARY

Lupinseed meal has unique nutritional properties and considerable potential as a feedstuff for pigs. This potential has been realized in Australia and New Zealand where lupinseed meal constitutes one of the major protein concentrates in pig diets. The seeds of the most common commercial cultivars of *L. albus* and *L. angustifolius* do not contain any major anti-nutritional factors but the low availability of lysine (50-60%) in lupins, together with the adverse effect of lupins on dressing percentage, must be taken into account when determining the relative value of lupins as a source of protein in pig diets.

The two species of lupins, *L. albus* and *L. angustifolius*, provide the major source of the present low alkaloid or "sweet" varieties of lupins. The seed of *L. albus* (white lupin) contains about 35% CP and 3680 kcal DE/kg whereas the seed of *L. angustifolius* (narrow leaf lupin) contains about 29% CP and 3440 kcal DE/kg. Lupins can be stored for a considerable time and the most common treatment of lupins prior to feeding to pigs is simply crushing with either a hammermill or roller mill to produce lupinseed meal.

The common commercial cultivars of *L. albus* and *L. angustifolius* together with recommended maximum inclusion levels are shown in Table 24.6. However, under practical feeding situations, it is unlikely that these levels of lupinseed meal, particularly those of *L. angustifolius*, will be reached in commercial pig diets.

Table 24.6. Recommended Maximum Inclusion Levels of Lupins in Pig Diets

Diet Type	White Lupin	Narrow Leaf Lupin
Starter Diet	10%	20%
Grower/Finisher Diet	15%	30%
Sow Diet	15%	20%

REFERENCES

Agricultural Research Council, 1967. Nutrient Requirements of Farm Livestock. No. 3. Pigs. Agricultural Research Council, London.

Aguilera, J.F., Molina, E. and Prieto, C., 1985. Digestibility and energy value of sweet lupinseed (*Lupinus albus* var. multiopolupa) in pigs. Anim. Feed Sci. Technol. 12: 171-178.

Barnett, C.W. and Batterham, E.S., 1981. *Lupinus angustifolius* (cv. Unicrop) as a protein and energy source for weaner pigs. Anim. Feed Sci. Technol. 6: 27-34.

Batterham, E.S., 1979. *Lupinus albus* cv. Ultra and *Lupinus angustifolius* cv. Unicrop as protein concentrates for growing pigs. Aust. J. Agric. Res. 30: 369-375.

Batterham, E.S., Andersen, L.M., Burnham, B.V. and Taylor, G.A., 1986a. Effect of heat on the nutritional value of lupin (*Lupinus angustifolius*) - seed meal for growing pigs. Brit. J. Nutr. 55: 169-177.

Batterham, E.S., Andersen, L.M., Lowe, R.F. and Darnell, R.E., 1986b. Nutritional value of lupin (*Lupinus albus*) - seed meal for growing pigs: Availability of lysine, effect of autoclaving and net energy content. Brit. J. Nutr. 56: 645-659.

Batterham, E.S., Murison, R.D. and Andersen, L.M., 1984. Availability of lysine in vegetable protein concentrates as determined by the slope-ratio assay with growing pigs and rats and by chemical techniques. Brit. J. Nutr. 51: 85-99.

Castell, A.G. and Tsukamoto, J.Y., 1984. Responses of growing gilts to dietary inclusion of a Manitoba lupin. Can. J. Anim. Sci. 64: 1095 (Abstract).

Gladstones, J.S., 1972. Lupins in Western Australia. Department of Agriculture, Western Australia, Perth, Bulletin No. 3834

Godfrey, N.W., Mercy, A.R., Emms, Y. and Payne, H.G., 1985. Tolerance of growing pigs to lupin alkaloids. Aust. J. Exp. Agric. 25: 791-795.

Godfrey, N.W. and Payne, H.G., 1987. The value of dietary lupin kernel meal for pigs. In: Manipulating Pig Production. Australasian Pig Science Association, Werribee, Victoria.

Hale, O.M. and Miller, J.D., 1985. Effects of either sweet or semi-sweet blue lupin on performance of swine. J. Anim. Sci. 60: 989-997.

Hill, G.D., 1977. The composition and nutritive value of lupin seed. Nutr. Abst. Rev. 47: 511-529.

Hove, E.L. and King, S., 1979. Trypsin inhibitor contents of lupin seeds and other grain legumes. New Zealand J. Agric. Res. 22: 41-42.

Just, A., 1981. Energy and protein utilization of diets for pigs. Proc. 4th Aust. Poultry and Stock Feed Convention, Perth, pp. 86-91.

King, R.H., 1981. Lupin-seed meal (*Lupinus albus* cv. Hamburg) as a source of protein for growing pigs. Anim. Feed Sci. Technol. 6: 285-296.

Leibholz, J., 1984. A note on methionine supplementation of pig grower diets containing lupin-seed meal. Anim. Prod. 38: 515-517.

National Research Council, 1988. Nutrient Requirements of Domestic Animals, No. 2. Nutrient Requirements of Swine. 9th ed. National Academy of Sciences, Washington, D.C.

Pearson, G. and Carr, J.R., 1976. Lupin-seed meal (*Lupinus angustifolius* cv. Uniwhite) as a protein supplement to barley-based diets for growing pigs. Anim. Feed Sci. Technol. 1: 631-642.

Pearson, G. and Carr, J.R., 1977. A comparison between meals prepared from the seeds of different varieties of lupin as protein supplements to barley-based diets for growing pigs. Anim. Feed Sci. Technol. 2: 49-58.

Reeves, T.G., 1974. Lupins - a new grain legume crop for Victoria. J. Agric. Vic. 72: 285-289.

Reeves, T.G. and Lumb, J.M., 1974. Selective chemical control of annual ryegrass (*Lolium rigidum*) in oilseed rape, field peas and lupins. Aust. J. Exp. Agric. Anim. Husb. 14: 771-776.

Ruiz, L.P., 1978. Alkaloid analysis of "sweet" lupin seed by GLC. New Zealand J. Agric. Res. 21: 241-242.

Standing Committee on Agriculture, 1987. Feeding Standards for Australian Livestock, Pigs, Editorial and Publishing Unit, CSIRO, East Melbourne, Australia.

Taverner, M.R., 1975. Sweet lupin seed meal as a protein source for growing pigs. Anim. Prod. 20: 413-419.

Taverner, M.R., Curic, D.M. and Rayner, C.J., 1983. A comparison of the extent and site of energy and protein digestion of wheat, lupin and meat and bone meal by pigs. J. Sci. Food Agric. 34: 122-128.

CHAPTER 25

Minor Oilseed Meals

V. Ravindran

INTRODUCTION

There is a growing awareness that the cost and scarcity of traditional protein supplements is one of the most difficult aspects of world food production. This problem is likely to become more critical in view of the increasing demand for livestock products, arising from an expanding world population and rising standards of living. Therefore, it is timely that all potential protein sources be explored as animal feeds. In this context, the use of minor oilseed meals in feed formulations may be expected to increase in the future, particularly in the developing world.

A wide variety of minor oilseed meals are available in both the temperate and tropical regions (Salunkhe and Desai, 1986) and these could be used to increase the supply of feed for livestock. Some of these minor oilseed meals are now emerging as important nontraditional feedstuffs and their potential feed value is reviewed in this chapter. Only a small proportion of the total quantities of these oilseed meals are currently utilized for animal feeding. Some important reasons for this underutilization are highlighted in Table 25.1.

Table 25.1. Reasons for the Underutilization of Minor Oilseed Meals

1. Presence of anti-nutritional factors and hence the high cost of processing.
2. Lower nutritive quality (e.g. poor amino acid balance, high fiber) compared with traditional oilseed meals.
3. Seasonality in supply and scattered production.
4. Lack of knowledge regarding their potential feeding value.
5. Lack of appropriate guidelines for their effective utilization.

PALM KERNEL MEAL

The oil palm (*Elaeis guineensis*) is a tree that grows up to 20 m in height. The fruit grows in bunches and consists essentially of a soft, reddish-orange colored outer skin and a fibrous pulp covering the nut, composed of a shell and a kernel. Of all the oil-bearing plants, the oil palm is the highest yielding (Hartley, 1977). For this reason, an upward trend in palm oil production is expected. The major producers of palm oil and palm kernel are Malaysia, Nigeria, Indonesia, China, Zaire and Cameroon (FAO, 1986). The fruit yields two kinds of oil, namely the palm oil from the fleshy pulp and the palm kernel oil from the kernel (Hartley, 1977).

Palm kernel meal, the residue remaining after oil extraction, contains only about 19 to 21% protein (Table 25.2), but serves as a valuable locally extracted protein source in areas of production. Palm kernel meal has a relatively good amino acid balance, with the exception of lysine (Table 25.3). Despite the lysine deficiency, palm kernel meal is reported to have a better essential amino acid index than coconut meal and is well supplied in sulphur-containing amino acids. However, significant improvements in growth have

Table 25.2. Chemical Composition of Some Minor Oilseed Meals (% DM)

	Crude Protein	Crude Fiber	Ether Extract	Ash	NFE	Calcium	Phos-phorus
Palm Kernel Meal	19.0	16.0	2.0	4.2	58.8	0.23	0.31
Kapok Seed Meal	33.8	22.1	6.4	7.9	29.8	0.45	1.25
Castor Seed Meal	39.0	18.0	4.0	7.5	31.0	0.76	0.87
Babassu Meal	25.0	15.0	6.0	6.0	48.0	0.13	0.49
Niger Seed Meal	37.0	18.0	7.0	11.0	27.0	0.09	0.82
Crambe Seed Meal	44.0	13.0	1.3	7.0	54.7	----	1.00

Gohl, 1981; Weiss, 1983; Devendra, 1988.

been observed by supplementing palm kernel meal with methionine (Babatunde et al., 1975), indicating that the availability of methionine may be relatively poor. This is in agreement with the report of Nwokolo et al. (1976) who found the availability of the amino acids in palm kernel meal to be lower than those in soybean meal.

The high crude fiber content of palm kernel meal make it less suitable for feeding non-ruminants. The meal is also dry and gritty, and is not readily accepted by pigs. Solvent-extracted meal is more unpalatable than the expeller-extracted meal. However, in mixed feeds, its unpalatability is of less importance (Gohl, 1981). Palatability can be improved by mixing with ingredients such as molasses (Hutagalung, 1981).

Palm kernel meal has been successfully fed to swine in Nigeria and Malaysia (Devendra, 1988). In the studies of Fetuga et al. (1973), good weight gains were obtained with rats fed rations containing palm kernel meal as the only source of protein. But with weaner and growing pigs, there were poor gains and utilization when palm kernel meal contributed 50% of the dietary protein (Babatunde et al., 1975). Methionine supplementation significantly improved the performance. Fetuga et al. (1977a) reported an improved utilization of palm kernel meal-based rations by the addition of blood meal, a lysine-rich protein source. In a subsequent study (Fetuga et al., 1977b), the addition of molasses was found to improve fiber digestibility and nitrogen retention of growing-finishing swine fed palm kernel meal-based rations.

Table 25.3. Amino Acid Composition of Minor Oilseed Meals (% of Protein)

	Palm Kernel Meal	Kapok Seed Meal	Castor Seed Meal	Babassu Kernel Meal	Niger Seed Meal	Crambe Seed Meal
Arginine	13.3	9.6	10.0	14.0	10.8	6.4
Histidine	1.6	1.6	1.7	1.8	3.0	2.3
Isoleucine	4.0	6.0	4.7	3.9	4.6	3.7
Leucine	6.4	6.2	5.7	6.2	7.0	6.1
Lysine	3.4	3.2	3.0	4.3	4.7	5.3
Methionine	2.1	2.0	1.5	2.3	2.1	1.6
Cystine	1.9	1.4	1.0	1.4	1.3	2.6
Phenylalanine	4.3	4.3	4.7	5.9	4.1	3.7
Threonine	3.1	3.4	3.2	3.1	3.7	4.1
Tryptophan	1.0	1.2	1.1	1.1	1.4	1.2
Valine	5.4	3.8	5.4	5.2	5.7	4.7

Gohl, 1981; Weiss, 1983.

The available evidence indicates that, with careful attention to amino acid balance and energy level of the ration, palm kernel meal may replace 50% or more of the protein supplement in the rations of growing-finishing swine.

KAPOK SEED MEAL

Kapok (*Ceiba pentandra*), also known as the silk cotton tree, is grown in Asia and tropical America for its silky floss. The seeds, a by-product of the fiber industry, are processed in many countries to produce oil and meal. Although kapok seed meal contains over 30% protein (Table 25.2), it is a poor source of lysine (Table 25.3). The high fiber content, presence of tannins and low palatability also restrict the use of kapok seed meal in monogastric rations (Kategile et al., 1978; Thanu et al., 1983). The presence of residual oils containing cyclopropenoid acids may further lower the nutritional value of the meal (Deutschman et al., 1964). These fatty acids with cyclopropene rings are known to react with the sulfhydryl groups of physiologically active proteins (Phelps et al., 1965) and may cause toxic effects in livestock (Sahai and Kehar, 1968; Siriwardene and Manamperi, 1979).

Kapok seed meal is a good ruminant feed, though somewhat inferior to cottonseed meal (Godin and Spensley, 1971). However, results of feeding trials with pigs have been less encouraging. Ganegoda and Siriwardene (1978) fed growing pigs with rations containing 0, 10, 20, and 30% kapok seed meal and reported depressions in feed intake, gains and nutrient digestibility with increasing levels of meal. Evidence of toxicity was observed in animals fed 20 and 30% kapok seed meal. Recent work with growing-finishing swine (Ravindran, 1986), showed that kapok seed meal can be included at 5% of the diet with no adverse effects on performance and that the growth depressing effects at the 10% level of inclusion can be partially overcome by supplementing with methionine and lysine. On the basis of limited available data, the use of kapok seed meal in fattening pig rations at levels higher than 5% is not recommended.

CASTOR SEED MEAL

Castor (*Ricinus communis*) is a warm-season crop which is grown as an annual in temperate regions and as a perennial in the tropics (Weiss, 1971). Brazil is the largest producer of castor seed and oil, followed by India, China, Soviet Russia and Thailand (FAO, 1986). The meal obtained from the oil extraction process contains about 39% crude protein and 18% crude fiber (Table 25.2). However, meal from undecorticated seeds may contain as high as 32% crude fiber (Gohl, 1981). Castor seed meal is deficient in lysine and methionine (Table 25.3).

Castor seed meal contains three toxic substances, ricin, a highly toxic protein, ricinine, a relatively harmless alkaloid and the castor bean allergen, an extremely potent allergen (Spies et al., 1962). These toxic principles considerably restrict the use of the meal and its current value is mainly as a fertilizer.

Use of castor seed meal as animal feed has often been considered, provided that a reliable method of detoxification could be found. Ricin can be easily detoxified by moist-heat treatment (Jenkins, 1963; Roberts, 1975). In India, steam treatment at 5 kg per cm^2 pressure for 15 minutes is recommended for detoxification. The allergen is difficult to inactivate but it seems to cause little harm to animals (Gohl, 1981; Weiss, 1983).

No published data is available on the feeding value of detoxified castor seed meal for swine. However, studies with poultry show that detoxified castor seed meal can be safely used at levels up to 40% in balanced chick rations (Vilhajalmsdottir and Fisher, 1971). If supplements are provided to balance the amino acid deficiencies, low level usage of detoxified castor seed meal in swine rations should be profitable.

BABASSU PALM MEAL

Babassu palm (*Obrignya martiana*) is a wild fan-shaped palm, which is widespread in Brazil, Mexico and British Honduras (National Academy of Sciences, 1975). The seeds are collected and the oil-rich kernels are pressed for oil. The composition of the by-product meal (Table 25.2) is remarkably similar to that of coconut meal. However, babassu palm meal is a better source of lysine and methionine (Table 25.3) than coconut meal.

In areas of production, babassu palm meal can be used in the same way as coconut meal (Gohl, 1981). Early reports of Squibb and Salazar (1951) showed babassu palm meal to be a useful protein supplement, even at 30% level of inclusion in growing swine rations. Feeding of more than 30% level is reported to cause scouring (Gohl, 1981), which has been attributed to its high magnesium content (0.97%).

NIGER SEED MEAL

Niger (*Guizotia abyssinica*) is grown extensively in India, Pakistan and Ethiopia for the edible oil obtainable from its small black seeds, which are also exported as feed for caged birds (Weiss, 1983). The meal remaining after oil extraction is very dark in color, sometimes almost black. Niger seed meal contains about 37% protein and 18% fiber (Table 25.2). Compared to soybean meal, the niger seed protein is deficient in lysine (Mehansho et al., 1973) but its methionine content is higher.

Roychoudhury and Mandal (1984) used niger seed meal to replace 0, 50, 75 and 100% of the protein provided by peanut meal in growing-finishing swine rations and observed depressed performance with increasing levels of niger seed meal. The poor performance was related to the amino acid imbalance and the high fiber content. However, niger seed meal has been successfully incorporated at up to 30% in balanced rations for layers. Although definitive research data is lacking, it appears that limited inclusion of niger seed meal in growing-finishing swine rations is possible.

CRAMBE SEED MEAL

Crambe (*Crambe abyssinica*) is an herbaceous annual, about 1 m in height. Although the crop is distributed widely in Mediterranean and

Southwest Asian regions, it is not produced commercially except in East Germany and the United States (Weiss, 1983). Crambe seed meal contains about 44% crude protein and 13% fiber (Table 25.2). The lysine content is adequate but it is deficient in methionine (Van Etten et al., 1965). The protein content of crambe meal can be increased to almost 55% by dehulling. Removal of hulls will also reduce the fiber content.

The major disadvantage of crambe meal as a stock feed is its content of glucosinolates (8-10% thioglucosides) which can be toxic to non-ruminants. The presence of sinapine imparts a bitter taste to the meal. Initial feeding trials with untreated crambe meal have produced growth depression and death in poultry (Weiss, 1983). However, modern processing methods can now produce crambe meal devoid of thioglucosides and sinapine (Baker et al., 1977; Mustakas et al., 1978). The detoxified meal can safely be fed to non-ruminants.

SUMMARY

In conclusion, attention is drawn to the fact that the list of minor oilseed meals available is longer than discussed in this chapter. For example, the Indian subcontinent alone is reported to have 86 different types of potential minor oilseed meals (Ranjhan, 1980). In general, the utilization of these oilseed meals in swine feeding is restricted due to the presence of anti-nutritional factor(s) and to their relatively poor protein quality. Added costs of detoxification and supplements often discourage their use, relative to traditional protein sources. Nevertheless, the role of these feedstuffs is likely to increase in location-specific feeding systems in areas of production, provided appropriate guidelines can be recommended for their effective utilization.

REFERENCES

Babatunde, G.M., Fetuga, B.L., Odumosu, G. and Oyenuga, V.A., 1975. Palm kernel meal as the major protein concentrate in the diets of pigs in the tropics. J. Sci. Food Agric. 26: 1279-1291.

Baker, E.C., Mustakas, G.C., Gumbman, M.R. and Gould, D.H., 1977. Biological evaluation of crambe meals detoxified by water extraction. J. Amer. Oil Chem. Soc. 54: 392-396.

Devendra, C., 1988. Non-traditional Feed Resources in Asia and the Pacific. 2nd ed. FAO Regional Office for Asia and the Pacific. Bangkok, Thailand.

Deutschman, A.J., Jr., Berry, J.W., Kircher, R.W. and Sakir, J.M., 1964. Catalytic elimination of cycloprophane containing acids. J. Amer. Oil Chem. Soc. 41: 175-176.

FAO, 1986. Production Yearbook. Vol. 39. Food and Agriculture Organization, Rome.

Fetuga, B.L., Babatunde, G.M. and Oyenuga, V.A., 1973. Protein quality of some Nigerian Feedstuffs. II. Biological evaluation of protein quality. J. Sci. Food Agric. 24: 1515-1523.

Fetuga, B.L., Babatunde, G.M. and Oyenuga, V.A., 1977a. The value of palm kernel meal in finishing diets for pigs. 1. The effect of varying the proportion of protein contribution from blood meal and palm kernel meal on the performance and carcass quality of finishing pigs. J. Agric. Sci. (Camb.) 88: 655-662.

Fetuga, B.L., Babatunde, G.M. and Oyenuga, V.A., 1977b. The value of palm kernel meal in finishing diets for pigs. 2. The effects of addition of cane molasses on the utilization of high level palm kernel meal diets. J. Agric. Sci. (Camb.) 88: 663-669.

Ganegoda, G.A.P. and Siriwardene, J.A. de S., 1978. Effect of feeding kapok seed meal to growing pigs and broiler chickens. Ceylon Vet. J. 26: 53.

Godin, V.J. and Spensley, P.O., 1971. Oils and Oilseeds. Crop and Product Digest No. 1. Tropical Products Institute, London.

Gohl, B., 1981. Tropical Feeds. Food and Agriculture Organization, Rome.

Hartley, C.W.S., 1977. The Oil Palm. Tropical Agriculture Series. 2nd ed. Longman, London.

Hutagalung, R.J., 1981. The use of tree crops and their by-products for intensive animal production. In: A.J. Smith and R.G. Gunn, eds. Intensive Animal Production in Developing Countries, D. & J. Croal Ltd., Haddington, U.K. pp. 151-184.

Jenkins, P.P., 1963. Allergenic and toxic principles of castor bean: A review of literature and studies of inactivation of these compounds. J. Sci. Food Agric. 14: 773-780.

Kategile, J.A., Ishengoma, M. and Katule, A.M., 1978. The use of kapok seed cake as a source of protein in broiler rations. J. Sci. Food Agric. 29: 317-322.

Mehansho, H., Peng, Y., Vanich, M.G. and Kemmerer, A.R., 1973. Limiting amino acids in Noog (*Guizotia abyssinica*). J. Nutr. 103: 1512-1518.

Mustakas, G.C., Kirk, L.D. and Griffin, E.L., 1978. Crambe seed processing - Improved feed meal by soda ash treatment. J. Amer. Oil Chem. Soc. 45: 53-57.

National Academy of Sciences, 1975. Underexploited Tropical Plants with Promising Economic Value. Washington, D.C.

Nwokolo, E.N., Bragg, D.B. and Kitts, W.D., 1976. The availability of amino acids from palm kernel, soyabean, cotton seed and rapeseed meal for the growing chick. Poult. Sci. 55: 2300-2304.

Phelps, R.A., Shenstone, F.S., Kemmerer, A.R. and Evans, R., 1965. A review of cyclopropenoid compounds: Biological effects of some derivatives. Poult. Sci. 44: 358-394.

Ranjhan, S.K., 1980. Animal Nutrition in the Tropics. 2nd ed. Vikas Publishing House, New Delhi, India.

Ravindran, V., 1986. Perspectives on the use of non-conventional feed resources. Seminar on Utilization of Fibrous Feeds in Animal Feeding. Digana, Sri Lanka, 10 pp.

Roberts, R., 1975. Nutritive value of oilseed cakes and meals. In: Proceedings of the Conference on Animal Feeds of Tropical and Sub-tropical Origin. Tropical Products Institute, London. pp. 171-177.

Roychoudhury, A. and Mandal, L., 1984. Utilization of deoiled niger (*Guizotia abyssinica*) cake in the rations of growing-finishing pigs. Indian Vet. J. 61: 608-611.

Salunkhe, D.K. and Desai, B.B., 1986. Postharvest Biotechnology of oilseeds. CRC Press, Inc., Boca Raton, Florida.

Sahai, B. and Kehar, N.D., 1968. Investigations on subsidiary feeds. Kapok (*Ceiba pentandra*) seed as a feed for livestock. Indian J. Vet. Sci. Anim. Husb. 38: 670-673.

Siriwardene, J.A. de S. and Manamperi, H.B., 1979. Effect of feeding kapok seed meal on growth of broiler chickens. Ceylon Vet. J. 27: 26-28.

Spies, J.R., Coulstan, E.J., Bernton, H.S., Wells, P.A. and Stevens, H., 1962. The chemistry of allergens. Inactivation of the castor bean allergens and ricin by heating with aqueous calcium hydroxide. J. Agric. Food Chem. 10: 140-144.

Squibb, R.L. and Salazar, E., 1951. Value of corozo palm nut and sesame oil meals, bananas, A.P.F. and cow manure in rations for growing and fattening pigs. J. Anim. Sci. 10: 545-550.

Thanu, K., Kadirvel, R. and Ayyaluswami, P., 1983. The effect of nutrient supplementation on the feeding value of kapok seed for poultry. Anim. Feed Sci. Technol. 9: 263-269.

Van Etten, C.H., Daxenbichler, M.E., Peters, J.E., Wolf, I.A. and Booth, A.N., 1965. Seed meal from crambe abyssinica. J. Agric. Food Chem. 13: 24-27.

Vilhajalmsdottir, L. and Fisher, H., 1971. Castor bean meal as a protein source for chickens. Detoxification and determination of limiting amino acids. J. Nutr. 101: 1185-1189.

Weiss, E.A., 1971. Castor, Sesame and Safflower. Leonard Hill, London.

Weiss, E.A., 1983. Oilseed crops. Tropical Agriculture Series. 1st ed. Longman, London.

CHAPTER 26

Mung Beans

C.V. Maxwell

INTRODUCTION

The mung bean (*Phaseolus aureus, Phaseolus mungo, Vigna radiata*) is a large-seeded legume that is a native crop of southern Asia and is an important source of dietary protein for people in tropical and subtropical countries. In Asia, the mung bean is used as a supplemental protein source for the rice diet (Asian Vegetable Research and Development Center, 1976) while in the United States, the mung bean is produced primarily as a source of bean sprouts used in oriental cuisine.

In harvesting and processing mung beans for the food industry, the undersized or split beans are of little economic value. As a consequence, they have traditionally been utilized in livestock feeds. Mung beans may also be available for use in swine feeds due to overproduction for the food industry.

GROWING MUNG BEANS

Mung beans are an annual belonging to the legume family. Oklahoma is the leading state in mung bean production in the United States with 20,000 to 28,000 hectares produced annually. Mung beans are a short season crop and flourish under hot summer weather, with yields and plant height almost twice as high in summer trials as those obtained in either fall or spring trials

(Asian Vegetable Research and Development Center, 1978). In addition, the mung bean plant has good drought resistance qualities (Staten, 1942). The mung bean plant appears to be well adapted to dry hot climates, particularly those considered too dry for soybean production.

Problems associated with mung bean production include lack of uniformity of pod maturity (which makes mechanical harvesting difficult) and variability in yield (Asian Vegetable Research and Development Center, 1976). Mung beans are also susceptible to a number of diseases even though the short growing season tends to reduce this problem.

NUTRIENT CONTENT OF MUNG BEANS

Mung beans contain almost three times as much protein as cereal grains but are approximately 40% lower in protein than soybean meal (Table 26.1). Variation in protein content ranges from 24.5 to 31.2% protein (Sood and Wagle, 1982; Hymowitz and Collins, 1975). Mung beans are low in ether extract and have a low crude fiber content. Sood and Wagle (1982) reported calcium levels from 0.09 to 0.20% in mung beans and found little variation in phosphorus content among nine varieties of mung beans.

Table 26.1. Proximate Analysis of Mung Beans and Soybean Meal (%)

	Mung Bean	Soybean Meal[1]
Dry Matter	91.0[2]	90.0
Crude Protein	26.5[2,3]	44.0
Crude Fiber	4.9[2]	7.3
Ether Extract	1.7[2]	1.1
Calcium	0.14[2]	0.30
Phosphorus	0.46[2]	0.75

[1]National Research Council, 1988.
[2]Sood and Wagle, 1982.
[3]Hymowitz and Collins, 1975.

The amino acid composition of mung beans (Table 26.2) indicates that mung beans, like many legume seeds, are high in lysine, containing between 1.64 to 1.74% lysine. The high lysine, which is comparable to soybean meal on an equal protein basis (percent lysine of crude protein), makes mung beans particularly attractive as a protein source for swine since lysine is the first limiting amino acid for most grain-soybean meal based swine diets. Lysine as a percent of protein in mung beans is approximately 6.4%, which is comparable to the 6.5% lysine in soybean meal (National Research Council, 1989).

Table 26.2. Essential Amino Acid Content (% Dry Matter) of Mung Beans

	Mung Beans				
	A[1]	B[2]	C[3]	D[4]	SBM[5]
Isoleucine	1.05	1.20	1.19	---	2.00
Leucine	1.94	1.80	2.08	---	3.37
Lysine	1.75	1.70	1.64	---	2.90
Methionine	0.30	0.20	0.37	0.27	0.52
Cystine	0.11	0.10	0.44	---	0.66
Phenylalanine	1.44	1.20	1.66	---	2.10
Threonine	0.80	0.80	0.77	---	1.70
Tryptophan	---	---	0.20	0.48	0.64
Valine	1.25	1.50	1.56	---	2.02

[1]Oklahoma State University - Average of two sources of mung beans (unpublished).
[2]Commonwealth Scientific and Industrial Research Organization, 1987.
[3]Adapted from Chatterjee and Abrol, 1975.
[4]Sood and Wagle, 1982.
[5]National Research Council, 1988.

The sulfur amino acids (methionine and cystine) are considered the limiting amino acids in mung beans. The Asian Vegetable Research and Development Center (1978) reported that Protein Efficiency Ratio (PER) was increased from 0.09 to 2.1 by supplementing the diet with methionine. A supplementation of 0.20% methionine provided the maximum PER. Tryptophan appears to be variable with estimates ranging from 0.22 to 0.59% tryptophan among nine varieties analyzed by Sood and Wagle (1982). The low value is comparable to the tryptophan level in mung beans reported by Chatterjee and Abrol (1975). Other amino acids in mung beans appear to be adequate.

UNDESIRABLE CONSTITUENTS IN MUNG BEANS

Mung beans appear to be relatively low in anti-nutritional factors. A trypsin inhibitor has been reported in mung beans (Borchers and Ackerson, 1947; Gupta and Wagle, 1978). Gupta and Wagle (1978) reported trypsin inhibitor levels of 8.72 to 12.16 units/mg protein among three varieties of mung beans. These values are considerably lower than the 25.5 units reported in raw soybeans (Marquardt et al., 1976). Similarly, hemagglutinating activity has been reported in mung beans (De Muelenaere, 1965) but the level is much lower (78 HU/g) than observed in soybeans (30,000 HU/g). Hankins and Shannon (1978) have also reported the presence of phytohemagglutin from mung beans.

Since heat treatment to eliminate the possible effects of anti-nutritional factors reported to be present in mung beans is too expensive to consider, the use of mung beans in swine diets may be limited to the level of raw mung beans which can be utilized in swine diets without affecting performance. Liener (1976) reported that heat treating of mung beans failed to improve weight gain in rats. Cannon et al. (1983) reported that performance was improved in chicks fed mung beans heated at 150°C for 45 minutes, when mung beans replaced 100% of the soybean meal on an equal protein basis. However, performance was unaffected when non heat-treated mung beans were used to supply up to 40% of the supplemental lysine. This suggests that raw mung beans may be used to replace a portion of the soybean meal in non-ruminant diets. If mung beans can be fed to meet a portion of the protein requirement, they have the additional advantages of not requiring further processing other than grinding before use in swine rations.

UTILIZATION OF MUNG BEANS

Starter Pigs

Studies to determine the effect of mung beans on the performance of starter pigs are unavailable. However, since the young pig is more susceptible to inhibitors than are older pigs, the use of mung beans in starter diets is not recommended.

Growing-Finishing Pigs

A comprehensive study involving 4 trials with 984 growing (18-55 kg) and three trials with 777 finishing (55-102 kg) swine has been conducted to determine the level of raw mung beans which can be included in growing-finishing swine diets without affecting performance (Maxwell et al., 1983; Maxwell et al., 1984; Maxwell et al., 1986; Maxwell et al., 1989). A control corn-soybean meal diet was fed in each trial, along with levels of mung beans ranging from 10 to 67% of the supplemental lysine (3.73-20.65% mung beans) during the growing phase and ranging from 20 to 75% of the supplemental lysine (5.29-17.88% mung beans) in the finishing period. All diets were formulated on an equivalent lysine basis of 0.75% lysine during the growing phase and 0.62% lysine during the finishing phase. Two sources of mung beans containing 1.86 and 1.63% lysine were used and diet formulations were made based upon the analyzed lysine values.

During the growing phase, daily gain decreased with increasing levels of mung beans in the diet (Table 26.1; Quadratic effect, $P<0.05$). These results indicate that the level of mung beans had little effect on gain at lower levels in the diet but affected gain more dramatically at higher levels. Daily gain was similar in pigs fed up to 10.2% mung beans (30% of the supplemented lysine) but was reduced ($P<0.01$) when mung beans in the diet were increased to more than 30% of the supplemental protein. Pigs fed diets with levels of

30% of the supplemental lysine from mung beans or higher (10.2, 12.83, 15.50 and 20.65% raw mung beans), grew more slowly than those fed the control diet. Feed required per unit of gain followed a pattern similar to that observed for gain, with pigs requiring more feed per unit of gain as mung beans in the diet increased (Table 26.3; Quadratic effect, $P<0.01$). As was observed for gain, level of mung beans in the diet had little effect on feed efficiency at lower levels in the diet but resulted in increased feed required per unit of gain at higher inclusion levels. Feed efficiency was similar in pigs fed up to 30% of the supplemental lysine from mung beans (10.2% mung beans). It should be noted that feed required per unit of gain did not increase greatly until mung beans replaced 67% of the supplemental lysine (20.65% mung beans) where feed required per unit of gain was 10.1% higher than in pigs fed the control diet. Average daily feed intake was similar among all levels of mung beans. This study suggests that mung beans can effectively replace up to 30% of the supplemental lysine (10.2% of the ration) in the diet of growing pigs without significantly affecting performance.

Table 26.3. Performance of Growing Pigs Fed Mung Beans

	Level of Mung Beans (%)						
	0	3.7	7.5	10.2	12.8	15.5	20.7
Daily Gain (kg)	0.68	0.68	0.66	0.62	0.65	0.63	0.62
Daily Intake (kg)	1.85	1.87	1.88	1.78	1.84	1.79	1.89
Feed Efficiency	2.78	2.77	2.82	2.86	2.83	2.87	3.06

Maxwell et al., 1983; 1984; 1986; 1989.

During the finishing period, average daily gain decreased linearly ($P<0.01$) with increasing level of mung beans in the diet (Table 26.4). Although the effect of increasing mung beans resulted in a linear reduction in gain over all levels, it should be noted that the magnitude of the reduction overall was small. Feed required per unit of gain increased (Table 26.4. Quadratic effect, $P<0.01$) with increasing levels of mung beans. As was observed with the growing pig, feed efficiency of pigs fed the lower levels of mung beans was similar and feed required per unit of gain was increased only in pigs fed the highest level of mung bean supplementation (17.88% raw mung beans). Average daily feed intake and adjusted backfat thickness was similar among pigs fed all levels of mung beans. This study indicates that mung beans can be used to replace up to 60% of the supplemental lysine (16.2% mung beans) for finishing pigs with minimal effects on performance. Results of these studies are consistent with the observation of Thompson and Hillier (1942) who reported that ground mung beans were a satisfactory substitute for cottonseed meal when a protein supplement composed of two parts meat scraps, one part cottonseed meal and one part alfalfa meal was used.

Table 26.4. Performance of Finishing Pigs Fed Mung Beans[1]

	Level of Mung Beans (%)						
	0	5.6	8.6	10.7	11.7	16.2	17.9
Daily Gain (kg)	0.91	0.95	0.86	0.90	0.91	0.94	0.86
Daily Intake (kg)	3.16	3.23	3.25	3.06	3.01	3.24	3.17
Feed Efficiency	3.58	3.39	3.44	3.53	3.37	3.64	3.81
Backfat (cm)[2]	3.09	3.21	3.09	3.07	3.27	3.14	3.12

[1]Maxwell et al., 1983; 1984; 1986; 1989.
[2]Average of 1st rib, last rib and last lumbar measurement.

Breeding Stock

A study to determine the effect on weight gain and subsequent reproductive performance of replacing a portion of the soybean meal in the diets of gilts during gestation with raw mung beans has recently been completed (Luce et al., 1989). Treatments were 100% of the supplemental protein from soybean meal; 61% of the supplemented protein from raw mung beans (11.5% mung beans) and 39% from soybean meal; and 89% of the supplemental protein from raw mung beans (19.8% mung beans) and 11% from soybean meal (Table 26.5). Feeding the high level of mung beans decreased weight gain during gestation and reduced weight loss during lactation compared with gilts fed the control diet or the moderate level of mung beans. Little difference was noted for litter size at birth, but litter size at 21 days for gilts fed the moderate level of mung beans was lower than for gilts fed the control diet or the high level of mung beans. Survival rate to 21 or 42 days or individual and litter weights at birth and 21 days were not affected by mung beans. However, pig and litter weights at 42 days were reduced in gilts fed the high level of mung beans. These results suggest that mung beans fed during gestation may affect subsequent performance during lactation although the effects were small.

The effect of feeding mung beans during lactation has not been tested. Extrapolation of data based upon feeding mung beans during gestation may not be advisable since feed intake and therefore daily intake of potential inhibitors would be much greater during lactation. For this reason, feeding mung beans during lactation is not recommended until studies have been completed with lactating sows.

Table 26.5. The Effects of Feeding Raw Mung Beans in Gestation on Gilt Weight Change and Litter Performance

	Control	Mung Beans (11.5%)	Mung Beans (19.8%)
Number of Gilts	289	116	141
Gestation Gain (kg)	55.9^{a}	57.3^{a}	52.5^{b}
Lactation Loss (kg)	15.6^{a}	15.2^{ab}	9.9^{b}
Litter Size at Birth	9.60	9.24	9.56
Litter Size at 42 days	7.72^{a}	7.15^{b}	7.72^{ab}
Piglet Birth Weight (kg)	1.52	1.51	1.48
Piglet 42 Day Weight (kg)	10.88^{a}	11.19^{a}	10.33^{b}
Preweaning Survival (%)	81.4	78.9	82.0

^{ab}Means with different superscripts differ significantly ($P<0.05$).

SUMMARY

Mung beans offer an attractive alternative to soybean meal and may be effectively utilized to replace up to 30% of the supplemental protein (10.2% raw mung beans) in the growing pig and up to 60% of the supplemental protein (16.2% raw mung beans) in the diet of the finishing pig. Results of studies to date would suggest caution in the use of mung beans in the diet of gestating swine.

A general rule of thumb for estimating the value of mung beans (after one determines the level at which mung beans should be included in swine diets without affecting performance) is to estimate the lysine value of mung beans to be 60% of the value of soybean meal (1.75 vs. 2.88% lysine in mung beans and soybean meal, respectively). This means that it would require 1.66 kg of mung beans to replace one kg of 44% crude protein soybean meal. The extra mung beans are added to the ration at the expense of grain. The value of mung beans in swine rations would therefore be worth approximately 60% of the value of soybean meal plus 40% of the value of the grain used in the ration.

REFERENCES

Asian Vegetable Research and Development Center, 1976. Mung bean report for 1975. Shanhua, Taiwan, Republic of China.

Asian Vegetable Research and Development Center. 1978. Mung bean report for 1976. Shanhua, Taiwan, Republic of China.

Borchers, R. and Ackerson, C.W., 1947. Trypsin inhibitor. IV. Occurrence in seeds of the leguminosae and other seeds. Arch. Biochem. Biophys. 13: 291-293.

Cannon, W.N., Maxwell, C.V., Sun, J., Teeter, R.G. and Hintz, R.L., 1983. Mung beans as a protein supplement in poultry. Poult. Sci. 62: 1396 (Abstract).

Chatterjee, S.R. and Abrol, Y.P., 1975. Amino acid composition of new varieties of cereals and pulses and nutritional potential of cereal-pulse combinations. J. Food Sci. Technol. 12: 221-227.

Commonwealth Scientific and Industrial Research Organization, 1987. Feeding Standards for Australian Livestock: IV. Pigs. Australian Agricultural Council. Pig Subcommittee.

De Muelenaere, H.J.H., 1965. Toxicity and haemagglutinating activity of legumes. Nature 206: 827-828.

Gupta, K. and Wagle, D.S., 1978. Antinutritional factors of *Phaseolus mungoreous (Phaseolus mungo* var. M_{1-1} x *Phaseolus aureus* var. T.). J. Food Sci. Technol. 15: 133-136.

Hankins, C.N. and Shannon, L.M., 1978. The physical and enzymatic properties of phytohemagglutin from mung beans. J. Biol. Chem. 253: 7791-7797.

Hymowitz, T. and Collins, F.I., 1975. Relationship between the content of oil, protein and sugar in mung bean seed. Trop. Agric. 52: 47-51.

Kylen, A.M. and McCready, R.M., 1975. Nutrients in seeds and sprouts of alfalfa, lentils, mung beans and soybeans. J. Food Sci. 40: 1008-1009.

Liener, I.E., 1976. Legume toxins in relation to protein digestibility - A review. J. Food Sci. 41: 1076-1081.

Luce, W.G., Maxwell, C.V., Buchanan, D.S., Bates, R.O., Woltmann, M.D., Norton, S.A. and Dietz, G.N., 1989. Raw mung beans as a protein source for bred gilts. J. Anim. Sci. 67: 329-333.

Marquardt, R.R., Ward, T., Campbell, L.D. and Cansfield, P., 1976. Purification of the growth inhibitor in faba beans (*Vicia faba*) Proc. Amer. Soc. Anim. Sci. 27: 138 (Abstract).

Maxwell, C.V., Buchanan, D.S., Walker, W.N. and Cannon, W.N., 1983. Mung beans as a source of protein for swine. Proc. 25th Annual Oklahoma Pork Congress, Oklahoma State University, pp. 61-67.

Maxwell, C.V., Buchanan, D.S., Walker, W.R., Cannon, W.N. and Vencl, R., 1984. Mung beans as a protein source for swine. Oklahoma Agricultural Experimental Station MP-116: 305-309.

Maxwell, C.V., Buchanan, D.S., Luce, W.G., Norton, S.A. and Vencl, R., 1986. Mung beans as a protein source for growing-finishing swine. Oklahoma Agricultural Experimental Station MP-118: 291-294.

Maxwell. C.V., Buchanan, D.S., Luce, W.G., Woltman, M.D., Walker, W.R., Cannon, W.N., Norton, S.A. and Vencl, R., 1989. Raw mung beans as a protein source for growing-finishing swine. Oklahoma Agricultural Experimental Station, Stillwater, MP-127, pp. 230-237.

National Research Council, 1988. Nutrient Requirements of Domestic Animals, No. 2. Nutrient Requirements of Swine. 9th ed. National Academy of Sciences, Washington, D.C.

Sood, D.R. and Wagle, D.S., 1982. Studies on the nutritional quality of some varieties of mung beans. J. Food Sci. Technol. 19: 123-125.

Staten, H.W. 1942. Mung beans for Oklahoma. Experiment Station Circular No. C-104. Oklahoma Agricultural Experiment Station, Stillwater.

Thompson, C.P. and Hillier, J.C. 1942. Mung beans as a protein supplement for growing and fattening swine. 16th Annual Livestock Feeders Day, Oklahoma A & M College. Circular No. 81.

CHAPTER 27

Mustard Meal

J.M. Bell

INTRODUCTION

Mustard was one of the first crops domesticated by man. It is used mainly as a condiment but some is extracted to yield mustard oil and a residue (meal) that is rich in protein. Mustard oil is used either as a food oil or in certain industrial applications while the meal has potential for use as a protein supplement in animal feeding or with further refinement, as human food.

GROWING MUSTARD

Mustards are of several genetic types. Brown mustard, sometimes known as oriental mustard, is a *Brassica juncea* type. Yellow mustard of North America, *Sinapsis alba* (or *Brassica hirta*) is known as white mustard in Europe. Yellow Sarson is a subspecies of *Brassica campestris* rapeseed but in some respects resembles mustards and therefore is included in this discussion.

Mustard is an oilseed crop that grows well in temperate and in high altitude, subtropical areas and is moderately drought tolerant. Mustard is well adapted to brown and dark brown soils but production on sandy soils is not recommended. Most varieties of mustard mature in about 95 days. Typical yields for oriental and brown mustards range from 760 to 1630 kg/ha while yellow mustards have slightly less yield and average 760 to 1300 kg/ha.

NUTRIENT CONTENT OF MUSTARD MEAL

The chemical composition of mustard meals is based on relatively few samples but there is considerable similarity with the composition of rapeseed (Table 27.1). Crude protein level may exceed 40%. Mustard meal contains more crude fiber than soybean meal (National Academy of Sciences, 1971) which contains about 5.2%. However, it may contain slightly less crude fiber than the typical 11 or 12% found in rapeseed or canola meal. As a consequence, digestible energy levels appear to be higher in mustard meal than canola meal.

Table 27.1. Chemical Composition of Mustard Meals (90% DM)

	Brown[1]	Brown[2]	Brown[3]	Yellow[2]	Canola Meal[1]
Crude Protein (%)	46.0	42.2	40.3	41.0	37.1
Ether Extract (%)	1.6	2.0	1.3	2.2	2.9
Ash (%)	6.0	5.7	7.9	5.7	6.4
Crude Fiber (%)	----	6.9	11.0	9.4	11.4
D.E. (kcal/kg)	3340	3360	----	3480	2900

[1]Bell et al., 1984.
[2]Bell et al., 1981.
[3]Musonda, 1988.

The essential amino acid content of brown mustard meal (Table 27.2) and yellow mustard meal (Table 27.3) closely resemble that of canola meal although the isoleucine, leucine and threonine levels may be higher. Some of the sulfur amino acid levels reported appear to be low and perhaps were determined before improved assay procedures were adopted. The lysine content of ammoniated mustard meal was depressed by this treatment (Bell et al., 1984) in contrast to the apparent elevation of its crude protein content, which indicates that certain procedures designed to reduce or eliminate glucosinolates may result in damage to protein quality. The availability of lysine is often most seriously affected. It can be concluded tentatively that the quality of protein (essential amino acid profile), based on amino acid analysis of mustard meals, is comparable to that of rapeseed or canola meal.

Table 27.2. Essential Amino Acid Concentrations (%) in Brown Mustard Meal (90% DM)

	MacKenzie[1]	Bell[2]	Bell[3]	Bell[4]	Musonda[5]
Arginine	2.21	3.04	2.64	2.58	2.50
Histidine	1.21	1.48	1.06	1.11	1.28
Isoleucine	1.46	1.25	1.28	1.55	1.42
Leucine	2.55	2.61	2.69	2.84	2.59
Lysine	2.06	2.00	1.62	2.02	1.97
Methionine	0.25	0.86	0.79	1.04	0.76
Cystine	----	1.13	1.00	1.18	1.17
Phenylalanine	1.62	2.02	1.64	1.66	1.50
Threonine	1.39	1.55	1.67	1.74	1.49
Tryptophan	----	0.40	0.41	----	0.45
Valine	1.81	1.40	1.37	1.88	1.78

[1]MacKenzie, 1973; Average of Five Cultivars.
[2]Bell et al., 1984.
[3]Bell et al., 1984; Ammonia Treated Meal.
[4]Bell et al., 1981.
[5]Musonda, 1988.

Table 27.3. Essential Amino Acid Concentrations (%) in Yellow Mustard Meals in Comparison with Canola Meal (90% DM)

	Mustard A[1]	Mustard B[2]	Canola Meal[2]
Arginine	2.65	2.13	2.45
Histidine	1.13	1.05	0.98
Isoleucine	2.87	2.66	2.44
Lysine	2.13	1.96	2.04
Methionine	0.74	0.91	0.82
Cystine	0.39	0.84	0.93
Phenylalanine	1.69	1.55	1.96
Threonine	1.78	1.75	1.49
Tryptophan	0.39	----	0.43
Valine	1.74	1.83	1.47

[1]Sarwar et al., 1981.
[2]Bell et al., 1984.

UNDESIRABLE CONSTITUENTS IN MUSTARD MEAL

A common characteristic of mustards, rape and other *Brassica* species, including several kinds of common garden vegetables, is their high glucosinolate content. It is the presence of these glucosinolates which give the strong or "hot" flavor of table or condiment mustard. The glucosinolate content of mustard meals varies with the species of mustard, the kinds and amounts of contaminant admixtures present, treatments applied to reduce glucosinolates and perhaps environmental factors during crop growth. However, in most instances, the total glucosinolate content of mustard exceeds that of rapeseed.

Glucosinolates are potentially toxic to man and animals because, depending on amount and kind consumed, they can affect the production of thyroxine in the thyroid gland. Thyroxine performs an important role in regulating metabolic rate and interference with thyroxine production impairs animal performance. In addition, animals show reduced palatability of diets containing high levels of glucosinolates. Brassica meals also contain sinapine which may reduce palatability further.

Several cases of acute illness and death in cattle have been reported following accidental ingestion of large quantities of whole brown mustard seed. Acute inflammation of the gastrointestinal tract was observed. Poulsen (1958) reported mortality in cattle fed myrosinase-free mustard meal containing about 90 micromoles of allyl glucosinolate per gram.

Raw mustard seed also contains an enzyme myrosinase which, with appropriate temperature and moisture conditions, hydrolyzes or breaks down glucosinolates into three major components, sugar, sulphate and either isothiocyanate or thiocyanate, the latter components being the source of goitrogenicity. Intestinal microorganisms may also achieve some glucosinolate breakdown but this seldom appears to be extensive. Cooking the seed, as may be done just prior to oil extraction, inactivates myrosinase thereby reducing the potential risk of thyroid inhibition from feeding the meal.

Several related forms of glucosinolates exist which differ greatly in flavor and toxicity. However, unlike rapeseed, mustards tend to have only one type of glucosinolate per species, with little or none of the other common types of glucosinolates being present. Brown mustard meal may contain over 100 micromoles of allyl glucosinolate per gram of oil free meal (Bell et al., 1984; Keith and Bell, 1985) while yellow mustard may exceed 300 micromoles of p-hydroxybenzyl glucosinolate per gram (Bell et al., 1981; Sarwar et al., 1981). Yellow sarson (*B. campestris*) contains about 50 micromoles of butenyl glucosinolates (Bell et al., 1971; 1972). A sample of mustard meal obtained from India contained about 25 micromoles of allyl and 95 micromoles of butenyl glucosinolate and may have been a mixture of species. For comparison, rapeseed meal contains significant quantities of at least five glucosinolates, with butenyl and hydroxybutenyl types predominating, yielding total glucosinolates of about 150 micromoles per gram. Canola seed contains 30 micromoles of aliphatic glucosinolates per gram of oil-free meal, or less.

Several methods have been proposed for reducing or eliminating the glucosinolates from *Brassica* meals. In 1959, Goering patented an enzyme treatment for removal of allyl glucosinolate from brown mustard (Goering et al., 1960). Youngs et al. (1971) patented a catalytic process involving iron or copper salts for glucosinolate decomposition in rapeseed. In 1980, Coxworth and McGregor patented a method based on ammonia treatment and found it effective with both *B. juncea* and *B. hirta*. Sarwar and Bell (1980) found that moist heat reduced the parahydroxybenzyl glucosinolate in yellow mustard meal (*B. hirta*) and that sodium carbonate (Na_2CO_3) treatment improved palatability with mice, apparently by reducing the sinapine content of the meal. In addition, treatment with 0.5% ferrous sulfate has also been shown to reduce the glucosinolates.

The recent achievement through plant breeding of a low glucosinolate, low erucic acid mustard (Love et al., 1990) may lead to extensive production of this new crop and to production of a new source of high quality food oil and a high protein meal of high acceptability in animal feeds.

FEEDING MUSTARD MEAL TO SWINE

Growing Pigs

Meals prepared by prepress-solvent extraction were obtained from three *Brassica* seed types: *B. hirta* (yellow mustard), *B. campestris* (Yellow Sarson) and *B. napus* (Canola cv. Tower) and fed to growing pigs (22-66 kg) as the only supplementary source of protein at 15% of barley-wheat (2:1) diets (Table 27.4). Daily gains and feed intakes were similar for the three meals and equal to the performance indicated for pigs of this weight range by the National Research Council (1988).

Feeding trials conducted with pigs tested over the 25 to 55 kg weight range, which were limit-fed diets containing 10% mustard meal (*B. juncea*), indicated that pigs fed mustard grew as well as those fed rapeseed meal and nearly as well as those fed soybean meal, although the efficiency of feed utilization was slightly inferior (Marangos and Hill, 1977; Table 27.4). However, the authors observed that limit feeding of the pigs may have masked any adverse flavor factor that may have been present.

Pigs (25-54 kg) fed diets containing ammoniated mustard meal grew and utilized feed as well as pigs fed canola meal as the only protein supplement in barley-wheat (2:1) diets (Table 27.4). Lysine supplementation improved rate of gain and efficiency of feed utilization. A combination of mustard meal and soybean meal was also beneficial. In a subsequent experiment (Keith and Bell, 1985), growing pigs (24-34 kg) fed similar diets responded to a combination of supplementary isoleucine and lysine such that pig performance on diets containing ammoniated mustard meal equalled that of pigs fed the soybean meal control diet.

Treatment of mustard seed meal (*B. juncea*) with aqueous sodium carbonate and heat was ineffective and probably detrimental (Table 27.4) and such meal was not improved by dietary supplementation with amino acids. Pig performance was much inferior to that of pigs fed soybean meal diets (Bell et al., 1981). Similar treatment of *B. hirta* meal depressed both daily feed consumption and daily gain. Apparently sodium carbonate treatment was detrimental for both species of mustard meal.

Table 27.4. Performance of Growing Pigs Fed Diets with Mustard Meal as the Supplementary Protein, Compared with Soybean Meal and Canola Meal[5]

	Weight Range (kg)	Daily Gain (g)	Daily Intake (kg)	Feed Efficiency
Mustard Meal (10%)[1]	25-55	596	1.70	2.85
Rapeseed Meal (10%)		607	1.70	2.80
Soybean Meal (13.5%)		621	1.70	2.74
Ammoniated Mustard Meal (15%)[2]	25-54	633	1.67	2.65
Canola Meal (15%)		699	1.77	2.55
Mustard Meal + Soybean Meal		705	1.73	2.46
Mustard Meal[2]		660[b]	1.74	2.65[b]
Mustard Meal + Lysine		757[a]	1.78	2.37[a]
Ammoniated Mustard Meal[3]	24-34	414[c]	1.23[bc]	3.02[a]
Mustard Meal + Lysine		471[bc]	1.23[bc]	2.63[b]
Mustard Meal + Lysine + Isoleucine		534[ab]	1.33[ab]	2.50[bc]
Canola Meal		444[c]	1.11[c]	2.53[bc]
Soybean Meal		593[a]	1.38[a]	2.35[c]
Mustard Meal (15%)[4]	22-66	622	1.77	2.84
Yellow Sarson Meal (15%)		624	1.87	2.98
Mustard Meal + Lysine + Methionine		594	1.44	2.43
Yellow Sarson + Lysine + Methionine		620	1.75	2.82
Canola Meal (15%)		642	1.73	2.69

[1]Marangos et al., 1976.
[2]Bell et al., 1984.
[3]Keith and Bell, 1985.
[4]Sarwar and Bell, 1980.
[5]Within an experiment, means with different supercripts are significantly different ($P<0.05$)

Finishing Pigs

During the finisher period (54-100 kg), the performance of pigs fed diets containing 10% mustard meal has been shown to be similar to that of pigs fed rapeseed meal or soybean meal (Marangos and Hill, 1976; Table 27.5). In a subsequent experiment, the performance of pigs fed ammoniated mustard meal was compared with canola meal or a combination of soybean meal and mustard meal. All three meals resulted in excellent pig performance (Bell et al., 1984).

It is of interest that pigs performed well when fed diets containing about 15% of various kinds of mustard meal with relatively high levels of glucosinolates. This indicates that some types of glucosinolates, such as those in certain mustards, can be tolerated by animals to a greater extent than others. However, there is also evidence that various kinds of glucosinolates are capable of affecting the thyroid (Lo and Bell, 1972; Bell et al., 1972). It is possible that undetected thyroid changes may have occurred in pigs since the thyroids were not examined in the trials mentioned above.

Table 27.5. Performance of Pigs Fed Diets with Mustard Meal as the Supplementary Protein, Compared with Soybean Meal and Canola Meal

	Weight Range (kg)	Daily Gain (g)	Daily Intake (kg)	Feed Efficiency
Mustard Meal (10%)[1]	55-90	774	2.84	3.66
Rapeseed Meal (10%)		804	2.84	3.52
Soybean Meal (11.75%)		782	2.84	3.71
Ammoniated Mustard Meal (3.9%)[2]	54-100	893	3.04	3.41
Canola Meal (5.2%)		914	3.05	3.35
Mustard Meal + Soybean Meal		877	2.94	3.38
Mustard Meal (13.0%) + Na_2CO_3 + Heat + Lys. + Meth.[3]	27-88	410b	1.46	3.63
Soybean Meal (12%)		720a	2.12	2.95
Mustard Meal (14.4%) + Na_2CO_3 + Heat + Lys. + Meth.		582c	1.97	3.41

[1]Marangos et al., 1976.
[2]Bell et al., 1984.
[3]Bell et al., 1981.

Breeding Stock

In studies involving pregnant and lactating gilts (Marangos and Hill, 1977), mustard meal (*B. juncea*) at 10% of the diet was compared with rapeseed meal and soybean meal (Table 27.6). The gilts had been reared from 22 kg initial weights on diets containing the same test meals. The mustard meal was reported to contain 3.9 mg of potential allyl isothiocyanate per gram (39 micromoles/g), a level lower than reported by others (Table 27.3).

It was concluded that mustard meal impaired reproductive efficiency. Gilts required more services per conception and consequently were two months older than gilts fed soybean meal when they conceived although they were only 7 kg heavier at this time. However, the gilts fed mustard meal produced more pigs per litter. The piglets at birth and weaning weighed less than those farrowed by soybean-fed gilts. Thyroid glands were recovered from gilts that failed to conceive and it was found that thyroids from gilts fed mustard meal were about one-half the weight of thyroids from gilts fed rapeseed meal diets (9.5 vs. 23.6 g, respectively). Unfortunately, no thyroid weights were obtained from gilts fed soybean meal so the goitrogenicity of allyl glucosinolate for gilts could not be assessed from this experiment. Similarly the relatively low glucosinolate content of the mustard meal used could mean that the results obtained may not be typical of mustard meal of *B. juncea* type. The results provide no assistance in determining the potential role of meals containing high levels of benzyl or butenyl glucosinolates.

Table 27.6. Reproductive and Litter Performance of Gilts Fed Soybean Meal, Mustard Seed Meal (*B. Juncea*) and Rapeseed Meals from 22 kg Liveweight[1]

	Soybean[2] Meal	Mustard[2] Seed Meal	Rapeseed[2] Meal
Services per Conception (no.)	1.14	1.50	1.00
Breeding Age (days)	243	300	270
Weight at Breeding (kg)	129	135	139
Lactation Weight Loss (kg)	6.9	18.6	16.1
Days to Service After Weaning	5.5	4.5	39.8
Pigs Born Alive	7.43	11.25	9.5
Birth Weight (kg)	1.38	1.13	1.37
Weaning Weight (kg)	5.42	4.92	5.06

[1]Marangos and Hill, 1977.
[2]Included at dietary levels of 7.85 and 11.75% for soybean meal during pregnancy and lactation and at 10% levels for mustard and rapeseed meals.

These results indicate that mustard meal may cause reduced conception and irregular estrus at the time of first breeding but the short weaning-to-service interval following weaning of the first litter suggests that this problem does not persist into the second breeding cycle. While more research is needed to resolve this matter, prospects for the ultimate use of mustard meal in pregnancy and lactation diets for gilts and sows seem favorable.

SUMMARY

Mustard meal may be derived from any one of several species of mustard, each of which contains a different kind of glucosinolate. These compounds have strong flavor which may be objectionable and they may be somewhat toxic when fed. However, good results have been obtained in swine feeding trials indicating that mustard meal may be used to replace part of the usual protein supplement for market pigs as well as for breeding stock. Special processing such as ammoniation may not be needed if the amount of meal used in the diet is limited to about 5% in sow diets and 10% in diets fed to market hogs. When new low glucosinolate meal becomes available through plant breeding, such meal could be used in the same manner as low glucosinolate rapeseed meal. At such time its competitive position will be determined largely by its digestible energy, crude protein and available lysine contents.

REFERENCES

Bell, J.M., Youngs, C.G. and Downey, R.K., 1971. A nutritional comparison of various rapeseed and mustard seed solvent extracted meals of different glucosinolate composition. Can. J. Anim. Sci. 51: 259-269.

Bell, J.M., Benjamin, B.R. and Giovannetti, P.M., 1972. Histopathy of thyroids and livers of rats and mice fed diets containing Brassica glucosinolates. Can. J. Anim. Sci 52: 395-406.

Bell, J.M., Keith, M.O., Blake, J.A. and McGregor, D.I., 1984. Nutritional evaluation of ammoniated mustard meal for use in swine feeds. Can. J. Anim. Sci. 64: 1023-1033.

Bell, J.M., Shires, A., Blake, J.A., Campbell, S. and McGregor, D.I., 1981. Effect of alkali treatment and amino acid supplementation on the nutritive value of yellow and Oriental mustard meal for swine. Can. J. Anim. Sci. 61: 783-792.

Coxworth, E.C. and McGregor, D.I., 1980. Effect of ammonia treatment on the glucosinolates and related isothiocyanates of yellow and Oriental mustard seed. In: J.A. Kernan, E.C. Coxworth and M.J. Moody, eds. A Survey of the Feed Value of Various Specialty Crop Residues and Forages Before and After Chemical Processing. Saskatchewan Research Council, Saskatoon, Saskatchewan.

Goering, K.J., Thomas, O.O., Beardsley, D.R. and Curran, W.A., Jr., 1960. Nutritional value of mustard and rape seed meals as protein sources for rats. J. Nutr. 72: 210-216.

Keith, M.O. and Bell, J.M., 1985. Amino acid supplementation of ammoniated mustard meal for use in swine feeds. Can. J. Anim. Sci. 65: 937-944.

Lo, M. and Bell, J.M., 1972. Effects of various dietary glucosinolates on growth, feed intake, and thyroid function of rats. Can. J. Anim. Sci. 52: 295-302.

Love, H.K., Rakow, G., Raney, J.P. and Downey, R.K., 1990. Glucosinolates and Brassica. I. Development of low glucosinolate mustard. Can. J. Plant Sci. (in press).

MacKenzie, S.L., 1973. Cultivar differences in proteins of Oriental mustard (*Brassica juncea*). J. Amer. Oil Chem. Soc. 50: 411-414.

Marangos, A.G., Done, S.H. and Hill, R., 1976. Rapeseed meal as a protein supplement in the diet of growing and fattening pigs. Brit. Vet. J. 132: 380-388.

Marangos, A.G. and Hill, R., 1977. Influence of rapeseed and mustard seed meals on reproductive efficiency in gilts. Brit. Vet. J. 133: 46-55.

Musonda, N.M., 1988. Protein quality evaluation of Indian cottonseed, groundnut and mustard seed meals using mice. M.Sc. Thesis, University of Saskatchewan, Saskatoon, Saskatchewan.

National Academy of Sciences, 1971. Atlas of Nutritional Data on United States and Canadian Feeds, National Academy of Sciences, 2101 Constitution Ave., Washington, D.C.

National Research Council, 1988. Nutrient Requirements of Domestic Animals, No. 2. Nutrient Requirements of Swine. 9th ed. National Academy of Sciences, Washington, D.C.

Poulsen, A.E., 1958. Poisoning with myrosinase-free mustard seed cakes in cattle. Nordisk Veterinaermedicin 10: 487-497.

Sarwar, G. and Bell, J.M., 1980. Effects of sodium and ferrous salt treatments on the nutritional value of yellow mustard seed meal (*B. hirta*) for swine and mice. Can. J. Anim. Sci. 60: 447-459.

Sarwar, G., Bell, J.M., Sharby, T.F. and Jones, J.D., 1981. Nutritional evaluation of meals and meal fractions derived from rape and mustard seed. Can. J. Anim. Sci. 61: 719-733.

Youngs, C.G., Sallans, H.R. and Bell, J.M., 1971. Iron or Copper Compound Catalytic Decomposition of Thioglucosides in Rapeseed, U.S. Patent Office No. 3,560,217.

CHAPTER 28

Oats: Naked

J.R. Morris

INTRODUCTION

The high fiber content of domestic oats (*Avena sativa*) is a major factor detracting from their use as a livestock feed. Therefore, a considerable amount of effort has been expended to try and reduce the fiber content of oats. Mechanical removal of the highly fibrous hull from the oat seed to produce oat groats is one option available but the selling price demanded by manufacturers of oat groats is usually higher than can be justified on the basis of their nutrient content. Therefore, other methods of hull removal are being investigated.

Naked seeded cultivars (*Avena nuda*) of oats have been grown in China for centuries. However, it is only recently that plant breeders in other countries have been able to develop varieties of naked oats with sufficient yield potential to be attractive to commercial growers. These new varieties, which have also been referred to by names such as hull-less (Coffman, 1961) and huskless, would appear to have considerable potential for use in swine production.

GROWING NAKED OATS

In oats, each spikelet consists of two thin outer glumes at the common point of attachment of the spikelet parts (Smid and Morris, 1986). Each kernel is tightly surrounded by a large hull (lemma) and a smaller hull (palea). The hull-less genes affect the development of the spikelet so that the lemma and palea of each floret are very thin and membranous (Burrows, 1986a). The kernel or groat in naked oats is held loosely between the lemma and palea. Upon gentle threshing, the groats in each spikelet are separated from the hulls.

Agronomically, naked oat cultivars have not yielded as well as covered seed but the groat yields of the two groups are much closer. Recent variety trials have shown that naked oats produce about 70% of the yield of covered oats (Smid and Morris, 1986). Table 28.1 shows the performance of Tibor and check varieties in 46 trials in Eastern Canada over three years (1982-1984). However, newer and substantially higher yielding varieties are being developed. A new strain, OA826-3, tested at eight locations in Ontario, outyielded Tibor by 24% in 1988 (Burrows, personal communication). These developments will make naked oats even more attractive to pork producers growing their own feed for pigs.

Table 28.1. Field Performance of Tibor Naked Oats and Covered Oats

	Covered			Naked
	Elgin	Woodstock	Oxford	Tibor
Yield (tonnes/ha)	3.7	4.2	4.0	2.8
Groat Yield (tonnes/ha)	2.7	3.0	2.9	2.6
Percentage Hull (%)	28.4	28.0	27.2	5.2
Height (cm)	94	108	104	109
Percentage Lodging (%)	31	29	21	22
Time to Ripen (days)	97	99	100	98
Protein Content (%)	13.0	12.0	12.0	17.0
Oil Content (%)	5.8	6.1	5.7	7.2

Smid and Morris, 1986.

NUTRIENT CONTENT OF NAKED OATS

The nutrient composition of naked oats, common oats, and oat groats are summarized in Table 28.2. It can be seen that the naked oats contain more oil and crude protein than covered oats but are similar in nutrient composition to commercial oat groats. Because of the lack of hulls in naked oats, the crude fiber content is greatly reduced relative to covered oats.

Table 28.2. The Nutrient Analysis of Terra and Tibor Naked Oats, Common Oats and Oat Groats (% as fed)

	Dry Matter	Crude Protein	Crude Fat	Crude Fiber	Ash
Naked Tibor	88.6	17.0	7.2	2.3	2.2
Naked Terra	88.0	13.4	6.6	3.1	2.3
Common Oats	90.0	11.4	5.6	10.0	2.4
Oat Groats	88.5	15.6	6.7	2.6	2.2

Morris (unpublished).

Amino acid profiles for Terra (Christison and Bell, 1980), Coker (Myer et al., 1985; Maurice et al., 1985) and Tibor oats (Friend et al., 1988) are compared with oats and oat groats (National Research Council, 1988) in Table 28.3. Terra naked oats contain lower levels of amino acids than the higher protein cultivars, Tibor and Coker 82-30. Total lysine levels in Tibor and Coker 82-30 naked oats are similar to the requirements for growing-finishing pigs weighing 50-110 kg (National Research Council, 1988). Lysine levels in these two cultivars appear slightly higher than in oat groats.

Table 28.3. Amino Acid Pofiles of 3 Cultivars of Naked Oats, Covered Oats, and Oat Groats (% as fed)

	Naked Oats				
	Terra	Coker 82-30	Tibor	Oats	Oat Groats
Arginine	0.76	1.40	1.17	0.71	0.89
Histidine	0.31	0.39	0.37	0.17	0.27
Isoleucine	0.42	0.67	0.57	0.48	0.51
Leucine	1.00	1.38	1.24	0.87	1.00
Lysine	0.59	0.70	0.65	0.40	0.53
Methionine	0.21	0.35	0.21	0.18	0.21
Cystine	0.32	0.57	0.57	0.19	0.20
Phenylalanine	0.82	0.96	0.82	0.57	0.68
Threonine	0.52	0.59	0.50	0.38	0.44
Valine	0.62	0.98	0.78	0.62	0.71
Crude Protein	13.40	18.12	17.12	11.80	15.80

Morris (unpublished).

The fatty acid composition of a naked oat cultivar (Nuprine) and a husked variety (Condor) grown in Europe are shown in Table 28.4. This information agrees with Saharabudhe (1979) and Maurice et al. (1985), indicating that the three major fatty acids, palmitic (16:0), oleic (18:1) and linoleic (18.2), account for 85 to 95% of the total fatty acids. Friend et al. (1988) found the fatty acids palmitic, oleic (18:1) and linoleic represented 12, 23 and 59% of the total fatty acids in a control corn-soybean based grower-finisher diet and 18, 40 and 37% in a naked oat (Tibor) diet. These data show that naked oat cultivars possess a high proportion of unsaturated fatty acids.

Table 28.4. Fatty Acid Composition (%) of a Husked (H) and a Naked Oat (N) Cultivar Over Two Years

	Condor (H)		Nuprine (N)	
	1973	1974	1973	1974
Palmitic 16:0	17.0	16.4	15.6	15.7
Stearic 18:0	1.7	1.6	1.5	1.6
Oleic 18:1	38.0	39.4	34.0	33.5
Linoleic 18:2	41.0	40.6	46.0	46.5
Linolenic 18:3	2.5	2.3	2.9	2.8

Welch, 1977.

UNDESIRABLE CONSTITUENTS IN NAKED OATS

Concern has been expressed that the high levels of unsaturated fatty acids in naked oats may lead to hydrolytic rancidity as a result of the action of a lipase enzyme (located in the pericarp) on the oil in the kernel (Welch, 1977). Since the pericarp of naked oats is not protected by the husk, the grain is likely to be more susceptible to damage during threshing and handling. Such damage may cause the release of lipase, thus developing higher levels of hydrolytic rancidity. The enzyme acts very slowly at low moisture (13%) and temperature (18°C). Hydrolytic rancidity increases at higher moisture levels but the level in naked oats will only exceed that of husked oats if the grain is severely bruised. Possibly, elevated levels of vitamin E may be needed in swine diets where naked oats are used, to minimize the risk of a deficiency.

A further consideration is that beta glucans may affect the performance of young pigs fed diets containing naked oats. Beta glucans are carbohydrate gums which are believed to limit the rate and extent of nutrient absorption. Enzymatic treatment (pentosanase and glucanase) of corn and Tibor oat diets for young pigs was studied by Morris (1989). The addition of 1 kg of enzyme per tonne of diet did not significantly improve the growth rate of the pigs. However, the enzyme treatment in the Tibor oat diet appeared to benefit pigs more than in the corn diet in terms of feed efficiency. These data suggest that further work needs to be done to assess the value of enzymatic treatment to improve the feeding value of naked oats to very young pigs.

In spite of the lack of hulls, Myer et al. (1985) reported that naked oats were quite bulky and created problems with bridging in self-feeders. Possibly this problem is due to the presence of the hairs (trichomes) on the surface of each groat. Bridging is not a problem when the hairs are removed.

FEEDING NAKED OATS

Starter Pigs

Naked oats have been used with varying degrees of success in diets fed to starter pigs. Christison and Bell (1980) conducted two studies to compare naked oats (cv. Terra) with wheat and corn in diets fed to young pigs from 7 to 14 kg bodyweight. In each study, four diets were formulated with Terra oats substituting for the grain at 0, 33, 67, and 100%. Protein contents were kept constant to evaluate the feed as an energy source only. Substitution of Terra oats had no effect on performance suggesting that Terra oats are equal to wheat or corn as an energy source for young pigs.

Dry matter and crude protein digestibility data indicated that the dry matter of Terra was less digestible than that of oat groats and no improvement in digestibility occurred with age (82.8 and 87.1%, respectively, at 7 kg; 83.4 and 86.7%, respectively, at 17 kg; Christison and Bell, 1980). The protein was less digestible for Terra than for oat groats in pigs weighing 7 kg (70.2 and 79.7%, respectively) but this difference disappeared at 17 kg (78.8 and 80.8%, respectively; Christison and Bell, 1980).

Myer et al. (1985) examined another naked oat cultivar, Coker 82-30, in diets for young pigs. Five dietary treatments containing 0, 18, 37, 58 and 79% naked oats, substituted for corn and soybean meal to maintain a constant 1.1% lysine level across diets, were evaluated. A total of 244 pigs were fed for 30 days beginning at a liveweight of 7.8 kg. No significant differences were observed for daily gain, feed intake or feed efficiency. The naked oats successfully replaced all of the corn and a proportion of the soybean meal (100 kg/tonne of diet) in the diets of young pigs.

Anderson and Tang (1988) showed no difference in growth rate or feed efficiency in a study where 275 three-week old weaned Yorkshire pigs were fed diets containing either corn or hulless oats as a single cereal or 50/50 blend. All diets were formulated at 20% crude protein and 3500 kcal/kg digestible energy.

Unfortunately, the performance of starter pigs fed diets containing naked oats is not always equal to that of pigs fed corn-soybean meal diets. Sixty weaned pigs were placed on weaner decks at 8 kg in an environmentally controlled room for 32 days on diets containing Tibor oats (Morris, 1986). Three experimental diets, including a corn-soybean based starter control diet, a naked oat diet, and one containing naked oats plus supplemental protein (48% soybean meal) to maintain a similar protein level as the control diet, were fed. All diets were supplemented with 4.0% mineral-vitamin premix. Final weights and liveweight gains were significantly less for pigs fed the naked oat diets compared to those receiving the control diet. The pigs fed the protein supplemented diet ate more feed than those fed the unsupplemented naked oat diet. Fowler (1985) also found poorer performance in starter pigs with naked oats.

Table 28.5. Performance of Pigs (8-18 kg) Fed Diets Containing Naked Oats[1,2]

	Level of Naked Oats (%)		
	0	90	96
Daily Gain (kg)	0.34[a]	0.23[b]	0.21[b]
Daily Intake (kg)	0.66[a]	0.52[b]	0.47[b]
Feed Efficiency	1.96	2.31	2.26

[1]Morris, 1986.
[2]Means within columns followed by the same letters are not different at P<0.05.

Roasting naked oats to 130°C appeared to improve feed conversion (2.1 vs. 2.3; P<0.03) compared to non-roasted oats (Morris, 1989). On the other hand, Anderson and Tang (1988) reported no effect on growth rate or feed consumption when naked oats were cooked and fed to three-week old weaned pigs through to 10 kg in weight.

Growing-Finishing Pigs

When pigs were fed from 39 kg to 89 kg liveweight, Froseth (1982) found that those fed naked oats (cv. Terra) grew as quickly and exhibited at least as good, if not better, feed conversion as those fed a corn diet (3.03 vs. 3.21). Average daily gains and average daily feed consumptions were higher for pigs fed Terra oats when the diets contained 0.67% lysine whereas the optimal lysine level for pigs fed the corn diet was 0.59%. It was concluded that the lysine value of Terra oats appeared to be less than that of corn. However, correctly supplemented Terra oat diets produced superior results. The Terra oat diet required only 42% as much supplemented soybean meal as the corn for balancing protein requirements.

Tibor oats were fed to growing-finishing pigs weighing from 32 to 97 kg (Morris and Burrows, 1986). In this study, 128 market pigs were placed into pen groups and fed one of the experimental diets; a corn-soybean meal based control, and three diets containing Tibor oats at 30, 65 and 96.7%. Liveweight gains, carcass backfat, and grading index were not significantly different among dietary treatment means. However, feed conversion and dressing percentage progressively improved as the level of Tibor oats in the diet increased (Table 28.6). The Tibor cultivar of naked oats was developed by Agriculture Canada (Burrows, 1986b).

Table 28.6. Performance of Growing Pigs (32-97 kg) Fed Diets Containing Naked Oats

	Level of Naked Oats (%)			
	0	30	65	97
Daily Gain (kg)	0.82	0.84	0.83	0.84
Daily Intake (kg)	2.61	2.59	2.40	2.36
Feed Efficiency	3.18	3.09	2.89	2.81
Dressing Percentage (%)	77.4	76.9	78.5	78.4
Backfat Thickness (mm)	32.6	32.6	32.2	33.4
Grading Index	104	104	103	104

Morris and Burrows, 1986.

An extensive study with Tibor naked oats by Friend et al. (1988) used three diets, a corn-soybean meal (Diet 0) as a control, 47.6% naked oats (Diet 50), and 95.4% naked oats (Diet 100). Forty Yorkshire pigs (barrows and gilts) were penned individually and fed ad libitum the pelleted experimental diets from approximately 24 kg to 90 kg liveweight. At slaughter, the left side of each carcass was used for carcass quality determinations, meat quality assessment, and backfat samples were taken for fatty acid analysis. On Diet 50, pigs exhibited faster gains and higher daily feed intake than on the other two diets. Lower yields of lean meat and higher levels of ether extract were observed for pigs fed Diet 100. Fatty acid composition in the carcass was a reflection of the dietary fatty acid composition. Diet 0 contained 12, 23 and 59% and Diet 100 contained 18, 40 and 37% palmitic, oleic, and linoleic fatty acids, respectively. The linoleic acid levels in the backfat increased from 12% for pigs fed Diet 0 to 13% for Diet 100.

Pork loin roasts were cooked and evaluated by 10 experienced meat panelists in the Food Research Centre, Agriculture Canada, Ottawa, for flavor, tenderness and juiciness. In addition, instrumental measurements were recorded for meat tenderness. The panel evaluation scores for flavor, tenderness, and juiciness of the pork roasts were higher for pigs fed Diet 100 (Table 28.7). The shear test for meat tenderness showed no significant differences among the diets (Friend et al., 1988).

Friend et al. (1988) used 18 Yorkshire pigs (30-44 kg) in a metabolism trial to determine the digestible and metabolizable energy for three experimental diets containing 0, 47.6 (50), and 95.4 (100) % Tibor oats which were substituted for corn and soybean meal in the control diet. The apparent digestible energy for the control, Diet 50, and Diet 100 were 4000, 4048 and 4120 kcal/kg, respectively and the corresponding digestibility coefficients were 91.9, 91.7 and 91.3%, respectively. Apparent metabolizable energy (AME) was 3905, 3976 and 4048 kcal/kg, respectively. Naked oats are certainly an excellent source of energy for swine, presumably due to their high lipid content.

Table 28.7. Sensory Scores for Pork Loin Roasts From Pigs Fed Diets Containing Naked Oats[1]

Diet	Flavor[2]	Tenderness	Juiciness
Corn-SBM Control	8.0	7.4	6.2
47.6% Naked Oats	8.2	7.8	6.7
94.5% Naked Oats	9.2	9.7	7.9

[1]Friend et al., 1988.
[2]A higher value indicates an improvement in flavor.

Henry and Bourdon (1971) determined the energy value of naked oats (cv. Nuprine) in a digestibility trial with 6 castrate male, Large White pigs weighing 59.5 kg. Each kg of naked oats supplied 4,081 kcal of digestible energy and 3,946 kcal metabolizable energy which are higher than current values for corn (National Research Council, 1988). Digestibility coefficients for energy and protein of the naked oats were 89.1 and 88.7%, respectively.

Breeding Stock

Diets formulated for breeding pigs are commonly formulated with domestic oats. The higher energy value of naked oats (Friend et al., 1988; Morris and Burrows, 1986; Van Lunen et al., 1987; Christison and Bell, 1980; Froseth, 1982) make it a valuable feedstuff for breeding swine. Naked oats would be particularly valuable in herds where supplemental oils and fats are fed to sows during the later stages of gestation and through the nursing period to maintain sow condition. Naked oats in the diet would reduce the need for such supplementation.

SUMMARY

Feeding trials have clearly demonstrated that naked oats can replace the corn and a considerable amount of the soybean meal in the diets of young pigs. For older pigs, naked oats can substitute for most of the corn and soybean meal. Naked oats fed as the only source of protein and energy to growing-finishing pigs appeared to reduce carcass lean yield when compared to soybean-based diets formulated to the same protein or lysine levels. In the latter stages of growth, it might be advisable to limit the amount of substitution to 50% of the diet or to feed higher levels of protein supplementation in naked oat diets. Feeding naked oats produced a significant and favourable effect on sensory evaluation of pork loin roasts. Flavor, tenderness and juiciness scores in one study were higher for the roasts from pigs fed naked oat diets than from those fed the control corn-soybean meal diet.

Breeding swine could benefit substantially from the incorporation of naked oats into gestation and lactation diets. The high energy value of naked oats would substantially reduce the need for supplementation of these diets with oils or fats in cases where dietary energy is a limiting factor in production.

Economically, naked oats are worth a great deal to farmers, especially in cooler season areas where they supply a good energy level combined with good quality protein. The high nutritional value of naked oats make them extremely attractive as an alternate crop for pork producers, despite the fact that they yield 70 to 75% of covered oats because the hulls are left in the field at harvest time.

REFERENCES

Anderson, D.M. and Tang, W., 1988. Comparison of the performance of pigs from weaning to 10 kg liveweight fed diets containing hull-less oats. Can. J. Anim. Sci. 68: 1328 (Abstract).

Burrows, V.D. 1986a. Breeding oats for food and feed: Conventional and new techniques and materials. In: F.H. Hunter, ed. Oats: Chemistry and Technology. Amer. Assoc. Cereal Chem., St. Paul, Minnesota, pp. 13-46.

Burrows, V.D., 1986b. Tibor oats. Can. J. Plant Sci. 66: 403-405.

Christison, G.I. and Bell, J.M., 1980. Evaluation of Terra, a new cultivar of naked oats (*Avena nuda*). Can. J. Anim. Sci. 60: 465-471.

Coffman, F.A., 1961. Oats and Oat Improvement. Publ. Amer. Soc. Agron., Madison, Wisconsin, pp. 1-40.

Fowler, V., 1985. Special Feeding Supplement, Pig Farming, Farming Press Ltd., Ipswich, (Nov): 45-52.

Friend, D.W., Fortin, A., Poste, L.M., Butler, G., Kramer, J.K.G. and Burrows, V.D., 1988. Feeding and metabolism trials and assessment of carcass and meat quality for growing-finishing pigs fed naked oats (*Avena nuda*). Can. J. Anim. Sci. 68: 511-521.

Froseth, J.A., 1982. Comparative protein value of corn, Otana white oats and Terra hull-less oats for finishing pigs. J. Anim. Sci. 55 (Supp.1): 270.

Henry, Y. and Bourdon, D., 1971. Energy value of naked oats for the pig. Annals Zootech. 20: 577-579.

Maurice, D.V., Jones, J.E., Hall, M.A., Castaldo, D.J., Whisenhunt, J.E. and McConnell, J.C., 1985. Chemical composition and nutritive value of naked oats (*Avena nuda* L.) in broiler diets. Poult. Sci. 64: 529-535.

Morris, J.R., 1986. Tibor naked oats in weaner pig diets. Can. J. Anim. Sci. 66: 1181 (Abstract).

Morris, J.R., 1989. The effect of heat and enzymatic treatment of corn and Tibor oats in weaner pig diets. Can. J. Anim. Sci. 69: 281 (Abstract).

Morris, J.R. and Burrows, V.D., 1986. Naked oats in grower-finisher pig diets. Can. J. Anim. Sci. 66: 833-836.

Myer, R.O., Barnett, R.B. and Walker, W.R., 1985. Evaluation of hull-less oats (*Avena nuda* L.) in diets for young pigs. Nutr. Rep. Int. 32: 1273-1277.

National Research Council, 1988. Nutrient Requirements of Domestic Animals, No. 2. Nutrient Requirements of Swine. 9th ed. National Academy of Sciences, Washington, D.C.

Saharabudhe, M.R., 1979. Lipid composition of oats (*Avena nuda* L.). J. Amer. Oil Chem. Soc. 56: 80-84.

Smid, A.E. and Morris, J.R., 1986. Naked or hull-less oats for pigs. Highlights of Agricultural Research in Ontario 9: 22-24.

Van Lunen, T.A. and Anderson, D.M., 1987. Hull-less oats in diets for grower-finishing pigs. Can. J. Anim. Sci. 67: 1199 (Abstract).

Welch, R.W., 1977. The development of rancidity in husked and naked oats after storage under various conditions. J. Sci. Food Agric. 28: 269-274.

Acknowledgements - I wish to thank Dr. Vern D. Burrows, Dr. Doug Friend, and Dr. Arend Smid for taking the time to review and make corrections to this paper.

CHAPTER 29

Peanut Kernels

G.L. Newton, O.M. Hale and K.D. Haydon

INTRODUCTION

Peanuts (*Arachis hypogaea* L.) have been known to man for many centuries but achieved economic importance only about 1850 when France began importing them from West Africa to obtain the oil they contained. Since then, peanut production has grown tremendously throughout the world. China and India are the largest producers, accounting for over 59% of the world's total production while the U.S.A. is a distant third with about 9% (USDA, 1988). It is estimated that worldwide production of peanuts exceeded 20,500,000 metric tonnes in 1988.

There has been renewed interest on the part of some swine producers, principally those in the midwestern United States, in feeding oil-stock peanuts as well as peanuts which are deemed unsuitable by candy and peanut butter manufacturers. Much of this interest is due to the high oil content of peanuts. Also, new standards have been proposed, and are being tested, which will require that peanuts go through an additional cleaning step at the purchase point. If adopted, the peanuts which are not eligible for human consumption may create an additional supply available for use in swine production.

NUTRIENT COMPOSITION OF PEANUT KERNELS

A chemical analysis of peanut kernels and soybean seed is shown in Table 29.1. Peanut kernels contain less water, protein and crude fiber than soybean seed. However, the fat content of peanuts is more than twice that of soybeans. The protein and fat content of peanut kernels varies widely depending upon variety, cultural practices and the environment where they are grown (Woodroof, 1973). For example, the fat content of peanut kernels may vary from 36 to 54%.

Table 29.1. Proximate Analysis of Peanuts and Soybeans (%)

	Peanut Kernels[1]	Soybean Seed[2]	Peanut Meal[2]	Soybean Meal[2]
Dry Matter	95.0	90.0	90.0	89.0
Crude Protein	28.5	36.7	47.0	44.0
Ether Extract	47.5	18.8	1.2	0.8
Crude Fiber	2.9	5.2	13.1	7.3

[1]Woodroof, 1973.
[2]National Research Council, 1988.

The lipids present in peanut kernels are listed in Table 29.2. Of the eight saturated fatty acids shown to be present in peanuts by Ahmed and Young (1982), palmitic, stearic, arachidic and behenic account for about 95% of the total amount. Peanuts contain large amounts of both monounsaturated and polyunsaturated fatty acids. These high levels of unsaturated fatty acids (83% of total fatty acids) are chiefly responsible for the soft, oily carcasses obtained from pigs fed large amounts of peanuts (Ellis, 1933; Myer et al., 1985; West and Myer, 1987) and such carcasses are discounted at the market because of difficulties in processing and marketing.

Another limitation to using peanuts as the sole source of supplemental protein for swine diets is the low level of certain essential amino acids. A comparison of the amino acid contents of peanuts and soybeans is shown in Table 29.3. Peanut protein is deficient in methionine, lysine and tryptophan, with methionine being the first limiting amino acid. If peanuts are to be used as the sole source of supplemental protein in swine diets, it will be necessary to add crystalline lysine, methionine and tryptophan to satisfy the pig's amino acid requirement (National Research Council, 1988). Anurag and Geervani (1987) reported that dry heat processing of peanuts reduced the availability of lysine and methionine.

Table 29.2. Lipid Content of Peanut Kernels (per 100 grams)

Saturated	
Palmitic 16:0	4.3 g
Stearic 18:0	1.4 g
Arachidic 20:0	0.7 g
Behenic 22:0	1.3 g
Total Saturated	8.0 g
Monounsaturated	
Oleic 18:1	23.9 g
Eicosenic 20:1	0.5 g
Nervonic 24:1	0.6 g
Total Monounsaturated	25.3 g
Polyunsaturated	
Linoleic 18:2	12.6 g
Total Polyunsaturated	12.7 g
Phytosterols	
Capesterol	26.0 mg
Beta-sitosterol	185.0 mg
Stigmasterol	7.0 mg
Cholesterol	0.0 mg

Peanut Advisory Board, 1985.

Table 29.3. Amino Acid Composition of Peanuts and Soybeans (% of Meal)

	Peanut Kernels[1]	Full-Fat Soybeans[2]	Peanut Meal[1]	Soybean Meal[2]
Arginine	3.11	2.54	5.82	3.20
Histidine	0.67	0.87	1.46	1.12
Isoleucine	0.90	1.60	1.84	2.00
Leucine	1.73	2.64	3.27	3.37
Lysine	0.89	2.25	1.45	2.90
Methionine	0.24	0.46	0.44	0.52
Cystine	0.31	0.55	0.73	0.66
Phenylalanine	1.32	1.80	2.12	2.10
Tyrosine	1.07	1.26	----	1.50
Threonine	0.67	1.42	1.37	1.70
Tryptophan	0.28	0.54	0.48	0.64
Valine	1.07	1.62	2.16	2.02

[1]Peanut Advisory Board, 1985.
[2]National Research Council, 1988.

Although peanut kernels contain appreciable quantities of minerals (Table 29.4), calcium, copper, iron, manganese and potassium are only about a third of the level found in soybean seeds while magnesium and phosphorous are about one-half that of soybean seeds. Sodium levels are drastically lower, with peanut kernels having only 5 to 10% as much sodium as soybean seeds. The zinc content of peanut kernels is slightly higher than that of soybeans. If mineral content is expressed on a per unit of protein basis, peanut kernels (with the exception of sodium) contain 40 to 160% of the levels of essential minerals found in soybean meal.

Table 29.4. Vitamin and Mineral Content of Peanut Kernels (mg/100 grams)

Minerals	Amount	Vitamins	Amount
Calcium	80	Niacin	15.4
Copper	1.1	Pantothenic Acid	2.7
Iron	2.9	Thiamin	0.83
Magnesium	190	Riboflavin	0.15
Manganese	1.2	Vitamin B_{12}	0.13
Phosphorus	370	Folacin	100
Potassium	660	Ascorbic Acid	0
Sodium	14	Vitamin E	10
Zinc	2.9		

Peanut Advisory Board, 1985.

With the exception of riboflavin (Table 29.4), peanut kernels are substantially higher in B vitamins than are soybean seeds. On a per unit of protein basis, riboflavin may be about 80% of the level found in soybean meal while other B vitamins are 3 to 7 times greater in peanut kernels than in soybean meal. Although peanut kernels contain appreciable vitamin E, additional vitamin E and antioxidants should be included in diets containing peanut kernels due to their high level of unsaturated fatty acids.

UNDESIRABLE CONSTITUENTS IN PEANUT KERNELS

The testa or skins of peanuts contain 16 to 19% tannins. The detrimental consequences of feeding high levels of tannins are well known (Mangan, 1988). Hale and McCormick (1981) found that the addition of 10 or 20% peanut skins to a swine diet significantly reduced dry matter and protein digestibility as well as feed efficiency. Peanut skins react strongly with copper (Randall et al. 1975). Therefore, Newton (1986) found that the addition of 250 mg/kg copper to diets containing 5 or 10% peanut skins restored pig performance to near that of those fed a diet containing neither peanut skins nor added copper.

Testa account for only 4.1% of the peanut kernel (Woodroof, 1973), giving a moderately low tannin level for the whole kernel. If the phenolic acid content of peanut flour (Dabrowski and Sosulski, 1984) is added to the estimated tannin contribution from the skins, the total tannin content of whole peanut kernels is about 10% of the level found in bird resistant sorghum grain (Mitaru et al., 1984) and about four times that found in barley (Soerensen and Truelsen, 1985). The tannin content of peanut kernels appears to be of little practical consequence considering the results of trials in which peanuts have been included in excess of 90% of the diet (McCormick 1950, unpublished).

Like other legume seeds, peanut kernels contain trypsin inhibitor. Although not extensively studied, one sample of raw peanut kernels contained 2800 TIU/g. When sampled after roasting to 138°C they contained 2400 TIU/g (Newton and Haydon, 1988). This may indicate some resistance to heat treatment, but these levels are only about 5% of the levels often found in raw soybean seeds and are within the range for roasted soybean seeds (Woodson-Tenent Laboratories, personal communication). The trypsin inhibitor in peanut kernels is probably of limited concern.

Mycotoxins, toxic metabolites of fungi found in grain, seed and feeds, have been a problem for man and animals for many years. Aflatoxin, a mycotoxin produced by the fungi *Aspergillus flavus* (Diener et al., 1982), is of major importance in peanuts since it can be produced by the invasion of the peanut kernel by these fungi before, during and after harvest. Aflatoxin, when present in large amounts, is acutely toxic to man and animals as was first reported by Lancaster et al. (1961). The long term effects of ingesting small amounts of aflatoxin have not been determined. However, the aflatoxin content of peanuts can be minimized by preventing drought stress with irrigation and by reducing kernel moisture to about 7% during storage.

PROCESSING PEANUT KERNELS

Peanut kernels must be ground before including them in swine feeds. However, grinding presents some difficulty. Presumably due to their high oil content, the grinding of peanut kernels often produces a sticky mass resembling peanut butter and this may present severe handling and mixing problems in conventional feed equipment. The problem is especially prevalent in high speed hammer mills. Therefore, peanut kernels are most easily processed through a roller mill. However, peanut kernels can be ground using a hammer mill if they are mixed with part of the grain portion of the diet prior to grinding. In-shell, raw peanuts are ground most easily and efficiently through a hammer mill.

Peanuts are sometimes roasted in a small rotating drum roaster to an internal temperature of 139 to 141°C over a 1.25 to 1.5 hour period. Roasted peanut kernels which are still hot from the roaster are easily ground through a roller mill. However, roasted peanut kernels which have cooled to ambient temperature are difficult to grind without producing at least some peanut butter. When grinding cooled peanut kernels, the feed rate to the mill and roller spacing are critical.

UTILIZATION OF PEANUT KERNELS

Starter Pigs

Weanling pigs (4 wk old) were fed 0, 2.5, 5.0 or 10.0% roasted peanut kernels in either simple corn-soy or complex diets containing 10% whey and 10% oats for 28 days (Haydon and Newton, 1987). For pigs fed the simple diet, additions of 2.5 or 5% tended to increase gain while the same was true for 5 and 10% peanut kernels in the complex diet (Table 29.5). Feed intake followed essentially the same pattern, being greater for the simple diets containing 2.5 and 5% peanuts and the complex diets containing 5 and 10% peanuts. Feed efficiency tended to be improved with the addition of 5 and 10% peanuts to the simple diet while there was no improvement as a result of the addition of peanuts to the complex diet. Five percent peanut kernels appeared to be about optimum for weaner pigs as pigs fed 5% peanuts in the simple diet gained 22% faster and were 8% more efficient in feed conversion than pigs not fed peanuts. Roasting also appears to be beneficial.

Table 29.5. Effect of Diet Type and Level of Roasted Peanut Kernels on Young Pig Performance

Diet Type	Simple				Complex			
Percent Peanut	0	2.5	5	10	0	2.5	5	10
Daily Gain (kg)	0.26	0.30	0.32	0.27	0.25	0.23	0.31	0.30
Daily Intake (kg)	0.45	0.56	0.49	0.44	0.46	0.41	0.52	0.55
Feed Efficiency	1.70	1.86	1.55	1.60	1.77	1.68	1.72	1.82

Haydon and Newton, 1987.

Growing-Finishing Pigs

In the early part of the nineteenth century, peanuts were grown extensively in the southern region of the United States as a crop for hog feed. Hogs were turned into the peanut fields after the nuts reached maturity which was usually late October or early November. The hogs remained in the field until they harvested the nuts and were then removed and taken to market. This practice was commonly referred to as "hogging-off".

Southwell and Treanor (1949) conducted a seven year study on the feasibility of letting pigs harvest mature peanuts (hogging-off). A crude mineral mixture composed of equal parts (W/W) of steamed bone meal, ground limestone and salt was supplied ad libitum. Pigs were placed in the peanut fields in late November and removed in late February. Average daily gain and feed efficiency were 0.52 kg and 5.66, respectively. McCormick (1950, unpublished) fed pigs in drylot peanut kernels and the mineral mixture of Southwell and Treanor (1949) ad libitum. Pigs weighed 25 kg initially and about 101 kg when removed from test. Average daily gain and feed efficiency were 0.73 kg and 2.14, respectively. More recently, Myer et al. (1985) studied the performance and carcass characteristics of swine that consumed peanuts remaining in the field after harvest. The average daily gain and feed efficiency of pigs consuming peanut kernels and a mineral mixture designed to satisfy the pig's mineral needs ad libitum were 0.60 kg and 4.6, respectively.

In all three of the above mentioned studies, the performance of pigs consuming large quantities of peanuts was acceptable. Unfortunately, peanuts contain large amounts of unsaturated fatty acids and consumption of high levels of unsaturated fatty acids result in soft, oily carcasses which are undesirable. As a consequence, pigs should not make more than one-third of their weight gain during the growing-finishing period from the consumption of peanuts in order to prevent such soft, oily carcasses (Myer et al., 1985).

At lower levels of inclusion, peanuts can be used successfully as a means of increasing the caloric density of the diet. In a recent trial, growing pigs (18 to 100 kg) were fed a control diet, a 2.5% added fat diet, a 5% roasted peanut diet, or a 5% raw peanut diet. Peanuts were substituted for corn and soybean meal on a lysine content basis. Compared to the control diet, 100 kg of peanuts were substituted for 79 kg of corn and 21 kg of soybean meal. Compared to the 2.5% added fat diet, 100 kg of peanuts were substituted for 25 kg of corn, 25 kg of soybean meal and 50 kg of fat. Raw peanuts were included at 5.25% compared to 5% for roasted peanuts to compensate for the loss of moisture during roasting. Pigs fed the diets containing raw or roasted peanuts gained the same or slightly better than those fed the control diet or the diet containing 2.5% added fat (Table 29.6). Peanuts and added fat produced similar improvements in feed efficiency. However, pigs fed the added fat diet tended to have more backfat than pigs fed the control diet, and to a greater degree than pigs fed peanut diets.

Table 29.6. Performance and Backfat of Pigs (18-100 kg) Fed 5% Peanuts

	Control	Fat	Roasted	Raw
Daily Gain (kg)	0.85	0.86	0.89	0.86
Daily Intake (kg)	2.95	2.75	2.87	2.77
Feed Efficiency	3.48	3.18	3.21	3.22
Backfat (cm)	2.77	3.07	2.95	2.87

Newton and Haydon, 1988.

In the above experiment, pigs fed raw peanuts seemed to require some time to adapt to the inclusion of peanuts in their diet as their performance during the first four weeks of the trial was poor. However, by the time they were 12 to 13 weeks old, they seemed to be able to produce enough digestive enzymes, or other factors, to overcome any inhibitory factors present in raw peanuts.

Two digestion-metabolism trials (with 8 and 16 week old pigs) were conducted to investigate the effect of pig age on the utilization of roasted and raw peanuts and also to compare edible peanuts with out-graded (belt grader) peanuts (Newton et al., 1989b). The diets fed were an added fat control diet and four diets containing 5% peanuts. These peanut diets contained roasted or raw edible peanuts or roasted or raw culled peanuts. The pigs were offered an amount of feed equal to 3% of their body weight, and the two trials were similar except for age of the pigs. Apparent digestion coefficients for dry matter, fat, crude fiber, nitrogen and NFE are shown in Table 29.7. All five diets were highly digestible and roasting did not improve the digestibility of either type of peanuts. In general, digestibilities were slightly greater for diets containing peanuts than for the control diet which contained animal fat. However, there did not appear to be any difference between culled and edible peanuts.

Table 29.7. Apparent Digestibility (%) of Diets Containing 5% Raw or Roasted Edible or Culled Peanuts

		Edible Peanut		Culled Peanut	
	Control	Raw	Roasted	Raw	Roasted
Trial 1					
Dry Matter	88.2	88.4	88.5	88.8	88.2
Fat	84.7	86.4	86.5	86.1	85.2
Crude Fiber	52.9	56.6	61.7	58.8	54.7
Nitrogen	85.3	86.8	86.2	86.1	85.1
NFE	92.7	92.5	92.6	92.9	92.6
Trial 2					
Dry Matter	88.1	89.3	87.4	88.8	89.4
Fat	85.3	87.8	85.1	87.3	89.9
Crude Fiber	53.3	57.5	50.4	59.5	58.0
Nitrogen	84.6	87.4	85.1	86.5	87.5
NFE	92.8	93.4	91.8	92.7	93.0

Newton et al., 1989b.

One interesting finding from this experiment was that from the 1,656 kg of belt grader fall-through that was shelled and cleaned, only 134 kg of sound kernels and 163 kg of splits (mechanically damaged) and shrivels (immature kernels) were recovered. Therefore, recovery of material suitable for swine feeding (roughly equivalent to oil-stock peanuts) was only 17.9% of the total. If the expense of handling, shelling and cleaning is taken into account, assuming that the material used in this trial is typical of all such material, belt grader fall-through will probably not be an economical swine feed.

Very little information is available on the processing of peanuts for swine feeding. In order to investigate the effects of grinding and of feeding in-shell raw peanuts, 48 pigs (average weight 51.6 kg) were housed in eight pens of six pigs each and allotted to one of four diets balanced for lysine content (Newton et al., 1989a). The diets were a control diet containing 4.5% of a mixed fat, a whole peanut diet containing 10% peanut kernels, a ground peanut diet containing 10% peanut kernels ground through a roller mill, and a peanut in shell diet containing 13.9% in-shell peanuts ground through a hammer mill (1.27 cm screen).

Mean performance and backfat of pigs fed the four diets are shown in Table 29.8. Pigs fed whole peanut kernels gained slower than pigs fed all other diets. More feed also disappeared from the feeders (feed intake) in pens fed whole peanut kernels. As a result, pigs fed whole kernels tended to be less efficient in feed conversion than pigs fed other diets. Average backfat for these pigs was much greater than for pigs on other treatments, but the variation was great and excessive backfat was primarily limited to gilts. The reason for the undesirable results from feeding whole peanut kernels was readily apparent. Most of the pigs fed this diet apparently developed an appetite for peanuts to the exclusion of the other dietary ingredients. Pigs learned to search the feed for peanuts, and pushed the remaining feed out of the feeder onto the slotted floor. As a result, pigs fed unground peanuts did not consume a balanced diet, therefore they grew more slowly and became fatter.

Table 29.8. Effect of Processing on the Performance of Finishing Pigs Fed Peanuts

	Control	Whole Kernels	Ground Kernels	Ground Peanuts in Shell
Daily Gain (kg)	0.93	0.78	0.87	0.90
Daily Intake (kg)	2.33	2.97	2.26	2.42
Feed Efficiency	2.49	3.84	2.61	2.70
Backfat (cm)	1.70	3.02	1.55	1.70

Newton et al., 1989a.

Possibly the most interesting result of this trial was the unexpected good performance of pigs fed the ground in-shell peanuts. Peanut shells and foreign material were estimated to be 3.9% of this diet and pigs fed this diet consumed 3.9% more feed than pigs fed the control diet and they were 3.8% less effficient in feed conversion than pigs fed ground peanut kernels. Therefore, the pigs were able to compensate for the dilution effect of including in-shell peanuts in the diet by increasing intake. Feeding ground in-shell, raw peanuts may be the most economical practice for including peanuts in diets for finishing swine.

The overall results of these experiments indicate that peanuts are as effective as added fat in promoting feed efficiency and when substituted for other protein feeds on a lysine basis, peanut protein is as efficiently utilized. The use of peanuts in swine feeds should be based on economics and availability. When substituted into diets containing added fat for growing-finishing swine, peanuts fed at 5% have a value of approximately 25% of the price of corn + 25% of the price of soybean meal + 50% of the price of fat. Exact substitution values for a given lot of peanuts should be based on an analysis for fat and lysine.

Breeding Stock

Lactating sows, especially primiparous sows farrowing during summer, may derive benefit from the inclusion of fat in their diet (Britt, 1986). Fat additions in lactation diets have been reported to increase milk yield (Boyd et al., 1982; Shurson et al., 1986). Haydon et al. (1990) conducted a trial to determine if the addition of raw or roasted peanut kernels was equivalent to the addition of fat to sow lactation diets. One hundred and five crossbred sows (Yorkshire x Landrace x Duroc) were allotted on day 110 of gestation, to one of three treatment groups to determine the effect of dietary addition (12%) of roasted or raw peanut kernels during two farrowing seasons, (July to September for summer and December to February for winter). The experimental diets were corn-soybean meal based and were formulated to contain 0.70% lysine, 0.87% Ca, and 0.70% P. Trace minerals and vitamins were provided to meet or exceed National Research Council (1988) requirements for the lactating sow. Dietary treatments were a control diet, containing 5% animal fat; or the 12% dietary substitution of roasted or raw peanut kernels (ground through a roller mill) for corn, animal fat and soybean meal. Based on lysine and fat content, 100 kg of peanut kernels replaced 27 kg corn, 31 kg soybean meal and 42 kg fat.

The effect of lactation diet on sow performance is shown in Table 29.9. No differences were observed for sow weight loss during lactation for sows fed diets with either 12% raw or 12% roasted peanuts as compared to sows fed the 5% added fat control diet. Source of dietary fat had no effect on daily feed intake during lactation. Sows fed roasted peanuts had higher milk fat and protein percentages (9.77 and 7.90%, respectively) at 3 days postfarrowing than sows fed animal fat (7.67 and 6.60%, respectively) or raw peanuts (7.92 and 6.05%, respectively). At 7 days postfarrowing, roasted peanut-fed sows had a higher percentage protein (5.72%) in milk than control-fed sows (4.85%) with the raw peanut-fed sows (5.13%) being intermediate. Sows receiving roasted or raw peanuts during the winter tended to have a numerically higher percentage of fat and protein in milk at 14 and 21 days postfarrowing than sows fed 5% animal fat. The addition of roasted or raw peanuts had no effect on litter size or piglet survival postfarrowing. Also, varying dietary fat source during lactation had no effect on piglet or litter weight gains.

Table 29.9. Effect of Peanuts on Performance of Sows Farrowing in Summer and Winter

	Summer			Winter		
	Control	Roasted	Raw	Control	Roasted	Raw
Number of Sows	18	20	20	15	16	16
Weight Loss (kg)	20.6	19.2	18.1	14.7	8.9	15.1
Feed Intake (kg/d)	4.8	4.7	4.7	5.6	5.4	5.5
Return to Estrus (d)	6.2	7.0	6.9	5.2	5.3	5.2
Day 3 Milk Fat (%)	6.76	7.81	7.32	7.67	9.77	7.92
Day 3 Milk Protein (%)	5.78	5.98	5.92	6.60	7.90	6.05
Day 21 Milk Fat (%)	5.94	5.73	6.91	6.28	6.91	7.82
Day 21 Milk Protein (%)	4.39	4.24	5.24	4.64	5.78	6.19
Pigs Born Alive	10.60	10.08	10.10	9.59	9.18	9.31
Birth Weight (kg)	1.52	1.46	1.55	1.51	1.47	1.42
Pigs Weaned	8.64	8.54	8.52	8.63	8.51	8.64
Weaning Weight (kg)	6.49	6.62	6.30	5.95	5.09	6.03
Survival (%)	81.51	84.72	84.35	90.00	92.70	92.80

Haydon et al., 1990.

The results of this trial indicate that 12% peanuts may be substituted for 5% animal fat in sow lactation diets with no effect on lactation performance. Additionally, the sow appears to utilize raw peanut kernels as well as roasted peanut kernels, which would decrease the cost of peanuts in lactation diet formulations since the expense of roasting is eliminated. When substituted into fat-containing lactation diets, peanuts fed at 12% have a value of approximately 27% of the price of corn plus 31% of the price of soybean meal plus 42% of the price of fat.

SUMMARY

Peanut kernels are a effective source of energy and protein for swine diets. They contain only low levels of anti-nutritional factors, but their level of use is limited by their amino acid composition and high levels of unsaturated fatty acids. When included in diets for pigs of less than 35 kg, peanut kernels should be roasted. There appears to be little or no advantage to roasting peanuts for finishing pigs or sows. Including roasted peanut kernels at approximately 5% appears to be optimum for performance in weanling diets. The economic optimum depends upon the cost of peanuts compared to other feedstuffs. Although the maximum level of peanut kernels that may be fed during the grower phase is limited only by their protein and amino acid content, finishing swine should be fed peanut kernels at less than 10% of their diet due to high unsaturated fatty acid content. Peanut kernels are an excellent source of supplemental dietary fat for sows and breeding swine.

REFERENCES

Ahmed, E.M. and Young, C.T., 1982. Composition, quality and flavor of peanuts. In: H.E. Pattee and C.T. Young, eds. Peanut Science and Technology. American Peanut Research and Education Society, Yoakum, Texas, pp. 655-688.

Anurag, C. and Geervani, P., 1987. Effect of heat processing on availability of lysine, methionine and tryptophan in selected varieties of groundnuts. Nutr. Rep. Int. 36: 175-181.

Boyd, R.D., Moser, B.D., Peo, E.R. Jr., Lewis, A.J. and Johnson, R.K., 1982. Effect of tallow and choline chloride addition to the diet of sows on milk composition, milk yield and preweaning pig performance. J. Anim. Sci. 54: 1-7.

Britt, J.H., 1986. Improving sow productivity through management during gestation, lactation and after weaning. J. Anim. Sci. 63: 1288-1296.

Dabrowski, K. J. and Sosulski, F.W., 1984. Composition of free and hydrolyzable phenolic acids in defatted flours of ten oilseeds. J. Agric. Food Chem. 32: 128-130.

Diener, U.L., Pettit, R.E. and Cole, R.J., 1982. Aflatoxins and other mycotoxins in peanuts. In: H.E. Pattee and C.T. Young, eds. Peanut Science and Technology. American Peanut Research and Education Society, Yoakum, Texas, pp. 486-519.

Ellis, N.R., 1933. Changes in quality and composition of fat in hogs fed a peanut ration followed by a corn ration. USDA Technical Bulletin No. 368, pp. 1-14.

Hale, O.M. and McCormick, W.C., 1981. Value of peanut skins (testa) as a feed ingredient for growing-finishing swine. J. Anim. Sci. 53: 1006-1010.

Haydon, K.D. and Newton, G.L., 1987. Effect of whole-roasted peanuts fed in either simple or complex diets on starter pig performance. University of Georgia, Athens, Swine Report Spec. Publ. No. 44: 25-28.

Haydon, K.D., Newton, G.L., Dove, C.R. and Hobbs, S.E., 1990. Effect of roasted or raw peanut kernels on lactation performance and milk composition of swine. J. Anim. Sci. (In Press).

Lancaster, M.C., Jenkins, F.P. and Philip, J.M., 1961. Toxicity associated with certain samples of groundnuts. Nature 192: 1095-1096.

Mangan, J.L., 1988. Nutritional effects of tannins in animals feeds. Nutr. Res. Rev. 1: 209-231.

Mitaru, B.N., Reichert, R.D. and Blair, R., 1984. The binding of dietary protein by sorghum tannins in the digestive tract of pigs. J. Nutr. 114: 1787-1796.

Myer, R.O., West, R.L., Gorbet, D.W. and Brasher, C.L., 1985. Performance and carcass characteristics of swine as affected by the consumption of peanuts remaining in the field after harvest. J. Anim. Sci. 61: 1378-1386.

Newton, G.L., 1986. Added copper in high tannin diets for growing-finishing swine. J. Anim. Sci. 63: (Suppl. 1) 37.

Newton, G.L. and Haydon, K.D., 1988. Raw or roasted peanuts in growing-finishing diets. University of Georgia, Athens, Swine Report Spec. Publ. No. 56: 41.

Newton, G.L., Haydon, K.D. and Dove, C.R., 1989a. Use of whole or ground raw shelled peanuts or ground raw peanuts with shells in finishing pig diets. University of Georgia, Athens, Swine Report Spec. Publ. No. 67.

Newton, G.L., Haydon, K.D. and Willis, S.R., 1989b. Digestibility of two types of raw or roasted peanuts by pigs at two ages. University of Georgia, Athens, Swine Report Spec. Publ. No. 67.

National Research Council, 1988. Nutrient Requirements of Domestic Animals. No. 2. Nutrient Requirements of Swine. 9th ed. National Academy of Sciences, Washington, D.C.

Peanut Advisory Board, 1985. A Food Technologist's Guide to Peanuts and Peanut Products. Ketchum Public Relations, New York, New York.

Randall, J.M., Reuter, F.W. and Waiss, A.C., 1975. Removal of cupric ion from solution by contact with peanut skins. J. Appl. Polymer Sci. 19: 1563-1571.

Shurson, G.C., Hogberg, M.G., DeFever, N., Radecki, S.V. and Miller, E.R., 1986. Effects of adding fat to the sow lactation diet on lactation and rebreeding performance. J. Anim. Sci. 62: 672-680.

Soerensen, C. and Truelsen, E., 1985. Chemical composition of barley varieties with different nutrient supplies. I. Concentration of nitrogen, tannins, phytate, B-glucans and minerals. Tidsskr. Planteavl. 89: 253-261.

Southwell, B.L. and Treanor, K., 1949. Hogging-off crops in the Coastal Plain. University of Georgia Coastal Plain Exp. Station, Coastal Plain, Bull. No. 41, 63 pp.

USDA, 1988. World Indices of Agricultural and Food Production 1977-1986. USDA Economic Research Service, Washington, Statistical Bull. No. 759: 1-16.

West, R.L. and Myer, R.O., 1987. Carcass and meat quality characteristics and backfat fatty acid composition of swine as affected by the consumption of peanuts remaining in the field after harvest. J. Anim. Sci. 65: 475-480.

Woodroof, J.G., 1973. Peanuts: Production, Processing Products. The AVI Publishing Company, Westpoint, Connecticut. 330 pp.

CHAPTER 30

Popcorn: Unpopped

G.C. Shurson

INTRODUCTION

Popcorn (*zea mays everta*) is one of the oldest, truely North American foods currently grown in the United States. The origin of popcorn dates back to the days of the early American Indians who cultivated popcorn and developed over 700 varieties before introducing it to the first English colonists arriving in North America. The Indians used popcorn as a dietary staple as well as for decorative dress.

Recent estimates suggest that 12.9 billion quarts of popped popcorn are consumed annually in the United States (The Popcorn Institute, 1989). Processors of popcorn use a comprehensive screening and quality assurance system to remove kernals that are cracked, broken, bird damaged, over or under-sized as well as those of non-uniform color. Depending on annual growing conditions, between 2 and 6% of the popcorn harvested fails to meet quality and uniformity standards at popcorn processing plants. This unpopped popcorn then becomes a by-product of the popcorn processing industry which has potential as a livestock feed. Although quantities are limited, opportunities exist for some swine producers, under certain price conditions, to reduce diet cost by using popcorn in their swine feeds.

POPCORN PRODUCTION

Almost all of the popcorn consumed throughout the world is grown in the United States (The Popcorn Institute, 1989). The major popcorn producing states include Illinois, Indiana, Iowa, Kansas, Kentucky, Missouri, Michigan, Nebraska and Ohio. In 1988, a total of 206,700 acres were used for popcorn production which produced about 230 million kilograms of shelled popcorn (The Popcorn Institute, 1989). Popcorn yields are typically 50% lower than that obtained for yellow dent corn and generally range between 450 to 900 kg/ha with an average yield of about 570 kg/ha.

Agronomic and management practices used to produce popcorn are very similar to those used in yellow dent corn production. The growing season is generally 110 to 150 days in length from planting to harvest. This requires early planting, particularly in the northern part of the Midwest. Popcorn is not as vigorous as yellow dent corn and grows better in light, well-drained soils when planted at a shallow depth to encourage faster germination and emergence. If poor germination and emergence occurs, popcorn can compensate for lower plant density by producing two ears.

Popcorn is ideally harvested when the moisture content of the grain is about 17%. Prolonged harvest seasons are avoided because of the potential grain loss due to the reduced standability of popcorn. Once the grain is harvested, it is slowly dried using aeration and low heat. This slow drying, low heat method is used to avoid kernal breakage and damage while reducing the moisture content to about 13.5 to 14% which is ideal for popping.

NUTRIENT CONTENT OF POPCORN

Information related to the nutrient content and availability of popcorn is limited. The nutrient content of popcorn may be quite variable due to potential differences in varieties, soil types, growing conditions and the amount of weed seeds and foreign material in cull popcorn. The metabolizable energy content of popcorn appears to be 5 to 6% lower than yellow dent corn. However, popcorn appears to be slightly lower in crude fiber and higher in crude fat than yellow dent corn. Published nutrient values for popcorn (Fonnesbeck et al., 1984; Carlson and Peo, 1985; Shurson, 1989a) relative to yellow dent corn are shown in Table 30.1.

The crude protein content of popcorn is considerably higher than found in yellow dent corn but the level of lysine and other essential amino acids are comparable to yellow dent corn (Table 30.2). These relationships suggest that some of the additional crude protein present in popcorn is in the form of non-essential amino acids or non-protein nitrogen and is not of much nutritional or economic significance when fed to swine.

Table 30.1. Nutrient Composition of Yellow Dent Corn and Popcorn (as fed)

	Yellow Dent Corn	Popcorn
Dry Matter (%)	88.0[1]	88.0[2]
Crude Protein (%)	8.5[1]	10.6[2]
Crude Fiber (%)	2.3[1]	1.8[3]
Crude Fat (%)	3.6[1]	4.9[3]
Calcium (%)	0.03[1]	0.02[4]
Phosphorus (%)	0.28[1]	0.29[3]
M.E. (kcal/kg)	3412[1]	3247[2]

[1]National Research Council, 1988.
[2]Shurson, 1989a.
[3]Fonnesbeck et al., 1984.
[4]Carlson and Peo, 1985.

Table 30.2. Amino Acid Composition of Yellow Dent Corn and Popcorn (% as fed)

	Corn[1]	Popcorn[2]
Arginine	0.43	0.39
Histidine	0.27	----
Isoleucine	0.35	0.33
Leucine	1.19	1.46
Lysine	0.25	0.22
Methionine	0.18	0.23
Phenylalanine	0.46	----
Threonine	0.36	0.36
Tryptophan	0.09	----
Valine	0.48	0.44

[1]National Research Council, 1988.
[2]Shurson, 1989a.

Unpublished data (Shurson, 1989c) comparing the apparent digestibility of nitrogen, net protein utilization and biological value of protein when increasing levels of popcorn were added to diets of weanling pigs indicate that as dietary nitrogen from popcorn increased, nitrogen or protein utilization was not improved (Table 30.3).

Popcorn appears to be comparable to yellow dent corn regarding calcium and phosphorus content. Although some values for a few trace minerals and vitamins have been determined (Fonnesbeck et al., 1984) they are of little significance when considering the use of popcorn in practical swine diets.

Table 30.3. Effect of Dietary Dent Corn (DC) and Popcorn (PC) Level on Nitrogen (Protein) Digestibility and Utilization in Weanling Pigs

	100% DC 0% PC	67% DC 33% PC	33% DC 67% PC	0% DC 100% PC
Apparent N Digestibility (%)	84.0	83.4	79.4	82.6
Net Protein Utilization (%)	66.6	59.7	60.2	60.6
Biological Value of Protein	79.3	71.6	75.8	73.4

Shurson, 1989c.

UTILIZATION OF POPCORN

Carlson and Peo (1985) conducted a growth performance trial utilizing a total of 48 starter pigs weaned at 28 days of age and averaging 5.9 kg in initial body weight. Pigs fed the control diet containing yellow dent corn grew 20% faster and were about 9% more efficient in converting feed to gain than pigs fed diets containing popcorn at a 33, 67 or 100% replacement rate for yellow dent corn. Pigs consuming the popcorn diets also consumed about 12% less feed than those fed the control diet. These unexpected results were difficult to interpret since the popcorn diets exceeded the control diet in crude protein (19.6%) and lysine (1.1%) content. Perhaps fineness of grind or the presence of some foreign material in the popcorn diets reduced palatability and consumption which consequently resulted in reduced gains. Popcorn should be ground to a medium degree of fineness to provide optimum nutrient utilization.

Further studies have shown that substituting popcorn for dent corn at a replacement rate of 33, 67 or 100% had no adverse effect on growth performance of starter, grower and finisher pigs (Table 30.4). Comparable gain, feed consumption and feed conversion were obtained for starter, grower and finisher pigs regardless of the substitution rate of popcorn. The higher crude protein content of the popcorn diets compared to the yellow dent corn control diet was of no apparent benefit due to the constant lysine level in the experimental diets used. Furthermore, if the metabolizable energy level of popcorn is 5% lower than dent corn as suggested previously, it had no apparent influence on growth performance of pigs in these studies. Therefore, although information is limited, popcorn has been shown to satisfactorily replace all of the yellow dent corn in diets for growing pigs without affecting performance.

There is also no reason to expect any detrimental effects from utilizing popcorn in diets for gestating or lactating sows. However, popcorn quality may be quite variable and could influence performance analogous to any other ingredient of lowered quality, and therefore should be evaluated before utilizing it in practical swine diets.

Table 30.4. Effect of Level of Yellow Dent Corn (DC) and Unpopped Popcorn (PC) on Pig Performance

	100% DC 0% PC	67% DC 33% PC	33% DC 67% PC	0% DC 100% PC
Starter Period (8-18 kg)				
Daily Gain (kg)	0.35	0.36	0.37	0.35
Daily Intake (kg)	0.67	0.69	0.70	0.66
Feed Efficiency	1.95	1.91	1.88	1.88
Growing Period (24-55 kg)				
Daily Gain (kg)	0.73	0.71	0.74	0.73
Daily Intake (kg)	2.00	2.01	2.08	2.02
Feed Efficiency	2.75	2.82	2.81	2.78
Finishing Period (55-102 kg)				
Daily Gain (kg)	0.84	0.81	0.84	0.83
Daily Intake (kg)	3.09	2.95	3.19	3.08
Feed Efficiency	3.66	3.63	3.78	3.69

Shurson, 1989a,b.

SUMMARY

The nutrient composition of cull popcorn may vary because of differences between varieties, growing conditions, and the quality (condition of grain and amount of foreign material) of the popcorn used. The metabolizable energy value of popcorn appears to be slightly lower and the crude protein content higher than in yellow dent corn, but these differences have not significantly affected growth performance of growing pigs. The higher crude protein level of popcorn appears to be of no nutritional advantage since the lysine content is equal to that than found in yellow dent corn.

Popcorn can be satisfactorily utilized as a partial or complete substitute for yellow dent corn or other cereal grains in all swine diets. Therefore, if cull popcorn is available at a price equal to or lower than the value of yellow dent corn, it can be effectively used as an economical alternative energy source in swine diets.

REFERENCES

Carlson, R. and Peo, E.R. Jr., 1985. Alternate feed grains for weanling pigs. Nebraska Swine Report, University of Nebraska, Lincoln, pp. 9-10.

Fonnesbeck, P.V., Lloyd, H., Obray, R. and Romesburg, S., 1984. IFI Tables of Feed Composition. International Feedstuffs Institute, Utah State University, Logan.

National Research Council, 1988. Nutrient Requirements of Domestic Animals, No. 2. Nutrient Requirements of Swine. 9th ed. National Academy of Sciences, Washington, D.C.

Shurson, G.C., 1989a. Performance of growing pigs fed diets containing various levels of cull unpopped popcorn. Ohio Swine Research and Industry Report, Ohio State University, Columbus, pp. 15-19.

Shurson, G.C., 1989b. Substitution of cull, unpopped popcorn in diets for starter, grower and finisher pigs. J. Anim. Sci. 67 (Suppl. 2): 129.

Shurson, G.C., 1989c. Energy and Nitrogen Utilization of Diets Containing Various Levels of Unpopped Popcorn by Starter Pigs. Unpublished.

The Popcorn Institute, 1989. Popcorn Industry Factsheet. Chicago, Ilinois.

CHAPTER 31

Potato and Potato Products

S.A. Edwards and R.M. Livingstone

INTRODUCTION

Potatoes (*Solanum tuberosum*) are grown primarily for human consumption. However, surplus potatoes or those unfit for human consumption are often available for use as a swine feed. The fleshy potato tuber provides an excellent source of carbohydrates, protein, essential vitamins and minerals. As with most root crops, the major drawback of potatoes is their relatively low dry matter content and consequent low nutrient density.

In addition to raw potatoes, the processing of potatoes prior to sale as crisps (chips), chips (french fries) or dehydrated potato products has become increasingly common in western Europe and North America. Potatoes are also used in the industrial production of starch and alcohol. Therefore, the materials available for swine feeding include not only the potato itself, but also a wide array of by-products resulting from these processing industries. The nutritive value of these potato by-products depends on the industry from which they are derived. Potato protein concentrate provides a high quality protein source, whereas potato pulp, the total residue from the starch extraction industry, or steamed peelings from the human food processing industry, provide lower quality products for swine feeding because of their higher crude fiber content.

GROWING POTATOES

The potato is grown in more countries than any other crop except maize while its global volume of production ranks fourth behind rice, wheat and maize (Horton and Sawyer, 1985). On a worldwide basis the crop is superior to any of the major cereal crops in its yield of dry matter and protein per hectare. World production of potatoes has remained relatively unchanged in recent years at around 300 million tonnes per year. The major areas of production are in Northern Europe, while Asia and North America also make significant contributions to world production. The estimated regional production of potatoes is shown in Table 31.1.

Table 31.1. Estimated Production and Yield of Potato in Different Regions

	Annual Production (million tonnes)	Average Yield (tonnes/ha)
World	299	14.8
Europe	111	21.2
USSR	73	11.3
Asia	75	12.6
North America	22	29.2
South America	10	10.6
Africa	6	8.4
Oceania	1	25.4

FAO, 1985.

NUTRIENT CONTENT OF POTATO AND POTATO PRODUCTS

The nutritional composition of the potato tuber varies according to many factors including variety, season, tuber size and yield level. As mentioned previously, the major drawback of potatoes is their relatively low dry matter content. The dry matter concentration of raw potatoes varies from 18 to 25%. Consequently, when fed fresh, its low dry matter content results in a very low concentration of nutrients per unit of bulk. However, when cooked, potatoes have a nutrient value which is very similar to that of cereals and, in a dehydrated powdered or flaked form, can directly substitute for them. Expressed on a dry matter basis, whole potatoes contain 6 to 12% crude protein, 0.2 to 0.6% fat, 2 to 5% crude fiber and 4 to 7% ash. The protein content tends to be inversely related to dry matter content.

On a dry matter basis, raw potato has a digestible energy (D.E.) value slightly less than that of cereal grain. The starch in raw potatoes is resistant to attack by amylase in the foregut of the pig and a large proportion passes undigested into the cecum and large intestine where it is degraded by bacterial action (Whittemore, 1977; Livingstone, 1985). However, this bacterial fermentation is a less efficient process than enzymic digestion for two reasons; first because of the associated losses of heat and combustible gases, and second because of the lower efficiency of utilization of the volatile fatty acids produced from fermentation in comparison with the simple sugars resulting from digestion. The apparent digestibility of energy over the whole gastrointestinal tract therefore needs to be corrected to take into account this reduced efficiency (Livingstone, 1985). Average values for the composition of the potato tuber are given in Table 31.2.

Table 31.2. Chemical Composition of Potato and Potato Products (% as fed)

	Dry Matter	Crude Protein	Crude Fat	Crude Fiber	Ash	D.E. (kcal/kg)
Raw Potato	20.0	2.2	0.1	0.5	0.9	516
Boiled Potato	22.0	2.4	0.1	0.6	1.0	741
Cooked Flake	92.5	10.2	0.5	2.3	4.2	3537
Dried Pulp	88.0	6.2	0.3	8.8	2.9	3197
Steamed Peelings	11.3	1.7	0.1	0.7	1.0	327
Potato Silage	16.6	1.2	0.1	1.1	0.5	476
Potato Chips	90.0	6.5	30.0	1.2	3.7	5250
Protein Concentrate	94.0	80.0	1.9	5.6	1.4	3928

Burton, 1966; Whittemore, 1977; Duke and Atchley, 1986; Boucque and Fiems, 1988.

Despite the low absolute content of protein in the fresh potato tuber, potatoes provide a potentially useful source of protein because the biological value of the protein is very good. Of the total nitrogen in the potato tuber, 30 to 50% is in the form of soluble protein, 10% is insoluble protein (located mainly in the skin) and the remainder is non-protein nitrogen (NPN). This soluble NPN is approximately 50% amides, glutamine and asparagine, with the rest being amino nitrogen.

The concentrations of individual amino acids in the protein of potato can vary considerably between different cultivars (FAO, 1970). Values on a fresh weight basis can vary widely according to the dry matter and protein content of individual samples, and it is better to consider the balance of amino acids within each unit of protein present. Potato tuber typically contains 5.3 g lysine, 2.7 g cystine + methionine, 3.2 g threonine and 1.1 g tryptophan in each 100 g of crude protein. In potato protein concentrate, these values are higher (6.8, 3.6, 5.5 and 1.2 g, respectively) and compare very favorably with the composition of the "ideal protein" for pig growth (Agricultural Research Council, 1981), which contains 7.0 g lysine, 3.5 g cystine + methionine, 4.2 g threonine and 1.0 g of tryptophan per 100 g crude protein. The first limiting amino acid in potato is generally methionine or isoleucine, and the high lysine content makes it a good complement to cereal protein. Table 31.3 shows the typical amino acid concentrations in potato tubers and in potato by-products.

Table 31.3. Amino Acid Composition (g/kg) of Potato and of Potato By-products (as fed)[1]

	Potato Tuber	PPC[2]	Cooked Flake	Diced Pulp	Steamed Peelings	Potato Silage	Potato Crisps
Lysine	1.2	5.6	4.7	2.2	0.7	0.9	3.1
SCAA[3]	0.6	2.7	1.8	2.8	0.4	0.5	2.6
Threonine	0.7	4.2	3.4	2.6	0.6	0.6	2.0
Tryptophan	0.2	0.9	0.9	0.7	0.1	0.2	1.0
Isoleucine	0.6	4.1	2.8	2.3	0.4	0.6	1.2
Leucine	1.0	7.6	4.2	4.2	0.9	0.9	1.4
Histidine	0.4	1.6	1.9	0.8	0.2	0.3	1.3
Tyrosine + Phenylalanine	1.5	8.6	6.2	5.0	0.7	0.8	3.7
Valine	1.1	5.1	3.8	3.0	0.6	0.7	2.7
Arginine	1.1	3.8	4.9	2.0	---	---	---

[1]Burton, 1966; Whittemore, 1977; Duke and Atchley, 1986; Boucque and Fiems, 1988.
[2]Potato protein concentrate.
[3]Sulfur-containing amino acids.

The typical mineral and vitamin composition of the potato tuber is shown in Table 31.4. The potato is a good source of vitamin C and provides smaller amounts of the B vitamins. Heat processing of the potato will substantially reduce the vitamin content.

Table 31.4. The Typical Mineral and Vitamin Composition of Potato Tubers

Mineral Content (g/kg as fed)		Vitamin Content (mg/kg as fed)	
Calcium	0.2	Ascorbic acid	140
Phosphorus	0.5	Niacin	15
Potassium	3.9	Nicotinic acid	1
Sodium	0.02	Thiamine	1
Magnesium	0.3	Riboflavin	0.4
Iron (mg/kg)	6.0		

Burton, 1966; Whittemore, 1977; Duke and Atchley, 1986; Boucque and Fiems, 1988.

Although the data for potato and its by-products shown in Tables 31.2 and 31.3 are typical, it must be emphasized that these products are very variable in their nutritive value depending on the processing they have undergone. This is particularly true of potato pulp, whose protein and fiber content depends on the proportion of potato solubles added back into the material. Therefore, it is a wise precaution to have such materials chemically analyzed before using them for feeding to swine. Steamed peelings, produced under heat and pressure, contain partly denatured starch and little proteolic enzyme inhibitor. However, their higher crude fiber content results in a dry matter energy value slightly less than for raw potato and they havc a low and variable dry matter content. Potato protein concentrate is a high quality product, widely used in the human food industry because of its good digestibility and high biological value.

The values given in Table 31.2 are also indicative of other similar products. Uncooked peelings will have a dry matter of about 22% but a similar composition in the dry matter to steamed peelings. Ensiled potatoes will have a dry matter of about 16.5% and, as a result of some loss of fermentable carbohydrate, crude fiber in the dry matter is increased to between 6.0 and 7.0%. Potato silage made from raw material has a slightly improved D.E. value in comparison with the starting material and better digestible protein content. Ensiled cooked potato will have a similar dry matter nutrient value to the starting material.

UNDESIRABLE CONSTITUENTS IN POTATO AND POTATO BY-PRODUCTS

Green potato tubers and sprouts contain the toxin solanine (Burton, 1966). This is an alkaloid glycoside which causes "potato poisoning" which has symptoms similar to gastroenteritis. Therfore, green potatoes should not be fed to pigs.

Raw potato also contains a powerful chymotrypsin inhibitor in its tubers (Livingstone et al., 1979; 1980). This protease inhibitor has been shown to cause a reduction in nitrogen digestibility (Whittemore et al., 1975). Digestibility of nitrogen is particularly poor for the potato itself, and may be impaired in other feedstuffs given simultaneously. However, the inhibitor is inactivated by heat.

PROCESSING POTATOES FOR SWINE FEEDING

It has been known for many years that pigs utilise raw potato much less well than cooked material, showing reduced intake and poor liveweight gain. Cooking the potato ruptures the starch grains, rendering them more digestible, so that the real energy value of the material is increased. Cooking also destroys the anti-nutritive factors which reduce protein digestibility. The advantage obtained by cooking is much greater for nitrogen than for energy utilization (Whittemore et al., 1973). Since cooking of potatoes requires expensive energy inputs, it is important to determine the minimum processing necessary to improve nutritional quality. Table 31.5 shows the results of experiments on this subject.

Heating potatoes to 100°C completely denatured the chymotrypsin inhibitor and gelatinized the starch grains, whereas heating to 70°C only partially did so and resulted in lower digestibility of energy and protein in the foregut (Livingstone et al., 1979). Prolonged heating, or slow cooling after heating, resulted in damage to the protein and a reduction in its digestibility. For best results potatoes should therefore be boiled for 30 to 40 minutes, steamed at 100°C for 20 to 30 minutes or simmered for one hour, to ensure thorough cooking through to the center, and then rapidly cooled.

Table 31.5. The Effects of Processing Method on the Digestibility of Nutrients in the Foregut of the Pig as Measured at the Terminal Ileum

Processing Method	Gross Energy	Starch	Nitrogen
Raw Minced	0.32		0.15
Raw Ensiled (4 months)	0.43		0.38
Steamed Whole	0.71		0.60
100°C for 60 min, Slowly Cooled		0.92	0.46
100°C for 20 min, Rapidly Cooled		0.90	0.60
70°C for 20 min, Rapidly Cooled		0.85	0.31

Livingstone et al., 1979.

The possibility of ensiling raw potato as an alternative means of improving nutritional quality was investigated by Livingstone (1980). Minced raw potato was ensiled with 1% molasses in a sealed silo. After four months, the chymotrypsin inhibitor activity was abolished but there was little disruption of the starch granules and little improvement in palatibility. The process also resulted in a loss of 5% in gross energy of the ensiled material.

FEEDING RECOMMENDATIONS FOR SWINE

Raw Potato

Raw potato is unpalatable to swine and should only be fed to dry sows and finishing swine. Approximately 6 kg of potato is needed to replace 1 kg of barley, with additional protein supplementation required. Dry sows can easily meet their daily energy requirements from raw potato fed ad libitum with a small amount of concentrate supplement (Sissins, 1983). To avoid overfatness, it will generally be necessary to restrict the potato to 6-8 kg per day fed in conjunction with about 1 kg of a concentrate supplement providing the necessary additional protein, vitamins and minerals. For finishing swine about 25% of the dietary dry matter can be provided by raw potato, but poorer performance should be expected compared with cereal diets. Higher inclusion rates may be possible but are likely to result in significant feed refusal and greatly reduced performance. Inclusion of raw potato in the diet will reduce the killing out percentage of finishing swine. Feeding the potato and concentrate separately at different daily meals may reduce adverse effects of the raw potato on protein digestion in the supplement (Whittemore, 1977).

Boiled or Steamed Potato

Cooked potato is readily acceptable to swine and problems with reduced nutrient intake are only likely to occur in young animals unable to consume the necessary daily bulk. Approximately 4 kg of cooked potato can totally replace 1 kg of barley in both energy and protein terms. To avoid overfatness, dry sows will need to be restricted to 4 to 6 kg per day of cooked potato fed in conjunction with 1 kg of a supplementary concentrate. A similar daily amount (5-7 kg depending on liveweight and litter size) of cooked potato can be given to lactating sows in conjunction with higher supplement levels. Growing/finishing swine can receive up to 50% of their dietary dry matter as cooked potato (equivalent to 4-6 kg/day of fresh material).

Cooked Potato Flake or Flour

Dehydrated cooked potato flakes or flours are directly interchangeable with the cereal component of the diet. They are very palatable and can be included at 50 to 60% in the diet of any class of swine without adversely affecting performance (Whittemore, 1977). However, because of the high energy costs involved in producing such materials, they will be most economically employed in diets for sucklers and early weaned swine, where highly digestible nutrients are required.

Ensiled Potato

Ensiling provides a means of enhancing the feeding value of raw potato and preserving either raw or cooked potato over long periods. Raw potatoes should be partially sliced or minced before ensiling, and molasses (1% fresh weight) or formic acid (3%) should be added to enhance fermentation. About 1.2 tonnes of raw potato produces 1 tonne of silage which should keep for 6 to 8 months. Ensiling does little to improve the palatability of raw potato and guidelines for feeding to swine are similar to those for the fresh tubers. Cooked potato can be very successfully ensiled for long term storage, provided that the silo or clamp is made airtight. The silage should be fed in the same way as fresh cooked potato.

Potato Pulp

This by-product of the potato starch industry comprises the residue remaining after starch has been extracted. It is generally pressed and dried but the composition of the dried product can be quite variable (Friend et al., 1963; Boeve and Smits, 1973; Roth and Kirchgessner, 1975; Boucque and Fiems, 1988), depending on the proportion of potato solubles contained. The product has similar characteristics to those of raw potato and should not be used in diets for young swine. Inclusion rates of up to 15% can be used in the diet for growing/finishing swine with little detriment to performance, provided that adequate protein supplementation exists, but will result in a reduction of killing-out percentage. Higher levels can cause marked deterioration in growth rate and food conversion efficiency.

Steamed Peelings

This product, sometimes called "liquid potato feed" contains only partially cooked starch and has a low and variable dry matter content (Edwards et al., 1986; Nicholson et al., 1988). For this reason, it is best treated in the same way as raw potato and restricted to use in diets fed to dry sows and finishing swine. Dry sows can be given 6 to 8 kg/day with 1.0 to 1.5 kg of a concentrate supplement. Steamed peelings can be included in finishing swine rations at up to 25% of the dry matter but a 5 to 10% reduction in performance in comparison with cereal diets should be expected. Higher inclusion levels result in decreased feed intake and poor performance. Because of the high potassium level in the product, swine should have a plentiful supply of water available. The product is best fed as soon as possible after production since it undergoes fermentation during storage with consequent loss of nutrient value. Separation of solids, which float towards the top of the container, and moulding of the crust can also occur over time.

Waste Potato Chips

Potato crisps or chips, which have been cooked in oil for human consumption, are very palatable and high in energy. They can be included in diets for sows and growing swine at up to 25% of the diet. They are also suitable for inclusion in diets for younger swine at up to 15%. They frequently contain a high salt content and a plentiful supply of fresh water should be made available if they are fed.

Potato Protein Concentrate

This is a high quality protein source, suitable for use in rations for any class of swine. However, its high cost makes it most appropriate for use in weaner diets. Potato protein concentrate can replace milk and fish protein in diets for early weaned swine at inclusion levels of up to 15%, with no detrimental effects on growth or food conversion ratio (Seve, 1977).

SUMMARY

Potato tubers provide a succulent, starchy feed with a high quality protein but low nutrient density. Despite their good chemical composition, raw potatoes are poorly utilized by swine because they contain starch resistant to enzyme attack and chymotrypsin inhibitors which depress protein digestion. Cooking raises the nutritive value of potato dry matter to a level similar to that of cereals. A wide variety of by-products from potato processing exist for use in swine rations but their composition can be very variable.

Recommended dietary inclusion rates for a range of potato products are shown in Table 31.6.

Table 31.6. The Recommended Maximum Inclusion Rates of Potato Products in Diets for Swine (% of diet DM)

	Starters	Growers	Finishers	Dry Sows	Lactating Sows
Raw Potato	0	0	25	50	0
Boiled Potato	0	30	50	50	25
Cooked Flake	60	60	--	--	--
Ensiled Raw	0	0	25	50	0
Ensiled Cooked	0	30	50	50	25
Dried Pulp	0	0	15	50	0
Steamed Peelings	0	0	25	50	0
Waste Crisps	15	25	10	25	25
Protein Concentrate	15	15	--	--	15

Where no maximum is indicated levels similar to those in diets for starters can be fed but are unlikely to be economic.

REFERENCES

Agricultural Research Council, 1981. The Nutrient Requirements of Pigs. Commonwealth Agricultural Bureau, Slough, England.

Boeve, J. and Smits, B., 1973. Verteerbaarheid bij varkens van enkele produkten van de aardappelzetmeelbereiding. Verslagen van Landbouwkundige Onderzoekingen No. 802: 14 pp.

Boucque, Ch. V. and Fiems, L.O., 1988. Vegetable by products of agro-industrial origin. In: F. De Boer and H. Bickel, eds. Livestock Feed Resources and Feed Evaluation in Europe. Elsevier, Amsterdam, pp. 97-135.

Burton, W.G., 1966. The Potato. Veenman and Zonen NV, Wageningen.

Duke, J.A. and Atchley, A.A., 1986. Handbook of Proximate Analysis Tables of Higher Plants. CRC Press, Boca Raton, Florida.

Edwards, S.A., Fairbairn, C.B. and Capper, A.L., 1986. Liquid potato feed for finishing pigs: Feeding value, inclusion rate and storage properties. Anim. Feed Sci. Technol. 15: 129-139.

FAO, 1970. Amino acid content of foods and biological data on proteins. Nutritional Studies No. 24. Food and Agriculture Organization, Rome.

FAO, 1985. FAO Production Yearbook, Food and Agriculture Organization, Rome, 39: 126-127.

Friend, D.W., Cunningham, H.M. and Nicholson, J.W.G., 1963. The feeding value of dried potato pulp for pigs. Can. J. Anim. Sci. 43: 241-251.

Horton, D. and Sawyer, R.L., 1985. The potato as a world food crop, with special reference to developing areas. In: P.H. Li, ed. Potato Physiology. Academic Press, London, pp. 1-34.

Livingstone, R.M., 1980. Potatoes as feed for pigs. Annual Report of Studies in Animal Nutrition and Allied Sciences, Rowett Research Institute, Aberdeen, 30: 63-64

Livingstone, M., 1985. Alternative feeds are good for sows. In: J. Hardcastle, ed. Science, Pigs and Profits. Agricultural and Food Research Council, London, pp. 8-9.

Livingstone, R.M., Baird, B.A., Atkinson, T. and Crofts, R.M.J., 1979. The effect of different patterns of thermal processing of potatoes on their digestibility by growing pigs. Anim. Feed Sci. Technol. 4: 295-306.

Livingstone, R.M., Baird. B.A., Atkinson, T. and Crofts, R.M.J., 1980. The effect of either raw or boiled liquid extract from potato (*Solanum tuberosum*) on the digestibility of a diet based on barley in pigs. J. Sci. Food Agric. 31: 695-700.

Nicholson, J.W.G., Snoddon, P.M. and Dean, P.R., 1988. Digestibility and acceptability of potato steam peel by pigs. Can. J. Anim. Sci. 68: 233-239.

Roth, F.X. and Kirchgessner, M., 1975. Zum futterwert getrockneter kartoffel- und maisschlempe beim schweln. Wirtschaftseigene Futter 21: 225-232.

Seve, B., 1977. Utilisation d'un concentre de proteine de pommes de terre dans l'aliment de sevrage du porcelet a 10 jours et a 21 jours. Journees de la Recherche Porcine en France, pp. 205-210.

Sissins, M.J., 1983. The value of raw potato (*Solanum tuberosum*) as a feed for pregnant sows. M.Sc. Thesis, University of Aberdeen.

Whittemore, C.T., 1977. The potato (*Solanum tuberosum*) as a source of nutrients for pigs, calves and fowl: A review. Anim. Feed Sci. Technol. 2: 171-190.

Whittemore, C.T., Taylor, A.G. and Elsley, F.W.H., 1973. The influence of processing upon the nutritive value of the potato: Digestibility studies with pigs. J. Sci. Food Agric. 24: 539-545.

Whittemore, C.T., Taylor, A.G., Moffat, I.W. and Scott, A., 1975. Nutritive value of raw potato for pigs. J. Sci. Food Agric. 26: 255-260.

CHAPTER 32

Probiotics

T.P. Lyons and J.D. Chapman

INTRODUCTION

Both beneficial and potentially harmful bacteria can normally be found in the digestive tract of swine. Some examples of harmful bacteria are *Salmonella, E. coli* and *Clostridium perfringens*. Not only can these bacteria produce specific diseases known to be detrimental to the host but through competition for essential nutrients, they can also reduce animal performance. In contrast to the effects of these disease causing micro-organisms, bacteria such as *Lactobacillus* and the vitamin B-complex producing bacteria can actually be beneficial to the host. By encouraging the proliferation of these bacteria in the intestinal tract, it may be possible to improve animal performance.

The term probiotic was first coined by Parker (1974) to describe "organisms and substances which contribute to intestinal microbial balance." The origin of the word comes from two Greek words meaning "for life." Yeast culture, on the other hand, is a term used to describe the use of living yeast and the medium upon which they were grown in animal production.

Over the last several years, considerable attention has been given to the use of probiotics, yeast cultures and other natural feed additives in pig feeds (Pollmann, 1985). Recent research efforts have been directed towards the use of probiotics as a means of reducing the symptoms of stress, acting as a natural growth promoter, enhancing production and improving the general health of the animal.

MODE OF ACTION OF PROBIOTICS

A most important characteristic of a well-functioning intestinal tract is the balance of its bacterial microflora. A healthy intestinal tract has a preponderance of lactic acid-producing bacteria, such as *lactobacilli* and *streptococci*. However, this equilibrium within the intestinal tract may be upset when the pig is put under stress, such as at castration or weaning. At these times, the balance can swing in favor of pathogens such as *E. coli*. Smith (1971) showed that while *lactobacilli* counts were similar in healthy and diseased pigs, large differences existed in the *E. coli* population of the upper regions of the digestive tract. In fact, there was a 10,000 fold increase in the *E. coli* population in the case of a diseased pig (Table 32.1).

Table 32.1. Bacterial Content of the Alimentary Tract in Eight Pairs of Healthy and Diseased Piglets (Log of Number of Viable Bacteria)[1]

	Small Intestine[2]				
	1	3	5	7	Large Intestine
E. coli					
Diseased	7.3	8.5	9.4	9.5	9.6
Healthy	3.6	4.8	7.3	8.3	9.0
Lactobacillus					
Diseased	8.1	8.4	8.6	7.4	8.8
Healthy	7.9	8.0	8.0	8.3	9.0

[1]Smith, 1971.
[2]Level 1 was distal to the stomach and level 7 was proximal to the large intestine.

There is an increasing body of evidence which suggests that certain micro-organisms, such as *Lactobacillus acidophilus* and *Streptococcus faecium*, can help maintain a favorable microbial population in the gut, either as part of the natural population or as a microbial supplement to the diet. As part of the natural flora, they can exert beneficial effects through competitive exclusion. This means that lactic acid bacteria actually compete for receptor sites or space along the intestinal wall with certain types of pathogenic organisms such as coliforms, particularly *E. coli*. Work by Nemeskery (1983), with a *streptococcus* species, demonstrated that this micro-organism was effective in inhibiting the colonization of various types of pathogens common to livestock and poultry (Table 32.2).

Table 32.2. Effect of *Streptococcus* on the Inhibition of Pathogenic Bacteria

Bacteria	Inhibition (%)
Enteropathogenic *E. coli*	80
Enterotoxin-producing *E. coli*	75
Salmonella strains	45
Shigella strains	50
Pseudomonas	50
Clostridium perfringens strains	100

Nemeskery, 1983.

In addition to the competitive exclusion principle, many beneficial species of *lactobacilli* and *streptococci* are acid-producing. Lactic acid, secreted by these species, exerts many positive effects towards the maintenance of a healthy intestinal environment. One such positive effect is that lactic acid appears to be a stimulus for the development of *L. acidophilus* and *S. faecium*. Secondly, as the pH is reduced through the production of acid, the intestinal environment becomes unfavorable for certain pathogenic bacteria. Finally, an acidic environment is conducive to increased enzymatic activity within the digestive system.

This balance of beneficial versus pathogenic organisms, referred to as "eubiosis" by Gedek (1987), has been demonstrated by Robinson et al. (1984), who examined and determined the bacterial numbers and species found on normal and dysenteric intestinal epithelium of swine. Their results showed that the predominant species in a normal intestine were *L. acidophilus* (11.9%) and *S. faecium* (54.4%), with *E. coli* levels of less than 1%. However, in dysenteric pigs, *E. coli* levels were elevated to 14% with *L. acidophilus* disappearing and *S. faecium* decreasing to 6%. Manfredi (1986) suggested that replacement of the *lactobacilli* would be a reasonable course of action since these organisms are more directly associated with animal health and are more severely affected by adverse conditions or stresses.

In addition to the production of lactic acid, the inhibitory effects of *lactobacillus* on other bacteria species have also been attributed to hydrogen peroxide formation (Dahiya and Speck, 1968) and production of an inhibitory substance termed "acidolin" (Hamden and Mikolajcik, 1974). However, controversy still exists as to the exact mode of action of probiotics. Some additional mechanisms are listed in Table 32.3.

Table 32.3. Possible Modes of Action of Probiotics

a) Neutralization of Toxins
b) Suppression of Viable Numbers of Specific Bacteria
 -Antibacterial Agents
 -Competition for Adhesion Sites
c) Alteration of Microbial Metabolism
d) Stimulation of Immunity

Fuller and Cole, 1988.

FACTORS DETERMINING THE EFFECTIVENESS OF PROBIOTICS

Many reports have shown no response to *L. acidophilus* or *S. faecium* inclusion in the diet during stressful periods, where a response might have been expected. Gilliland (1981) suggested that poor viability of the bacteria, if indeed any were present, was probably the reason for the lack of response. The response to *L. acidophilus* is dependent upon having sufficient quantities of viable bacteria in the diet which have the potential to successfully colonize the animal's intestinal tract. Gilliland (1981) examined 15 commercially available probiotic supplements and found that only two contained more than one million viable *lactobacilli* per gram.

Lyons (1987), when discussing factors affecting the potential of probiotics to colonize in the gut, stressed the importance of the ability to attach to the gut epithelium and the ability to grow in the gut environment using available substrates. Barrow et al. (1980), in an in vitro study, assayed a selection of bacteria isolated from the pig's gut for their ability to adhere to stomach squamous cells and found that the number of bacteria adhering per cell ranged from zero for *E. coli* to 42 for *S. salivarius*. Fuller (1977) contended that an organism could establish itself in the stomach either by attaching to the epithelium or by rapid growth. When a *lactobacillus* strain, selected on the basis of its high adhesion index and good growth in vitro, was fed to pigs, there was a significant reduction in the coliform count in the stomach and duodenum. However, many bacteria are host specific, so a probiotic should ideally contain a microorganism isolated from that type of animal.

The effectiveness of probiotic bacteria is also dependent on their resistance to hydrochloric acid and to bile salts. It is well established that gastric acidity is an important barrier to gut colonization. It is essential, therefore, that the probiotic bacteria have the ability to survive acidic conditions in the stomach. Gilliland (1981) suggested that bile resistance was also an important criterion in the selection of an effective probiotic and found that only one of 15 commercially available products examined contained more than one million bile resistant *lactobacilli* per gram. Many bacteria are known to metabolize bile salts and to have their growth stimulated by this substance.

In summary, for a probiotic to be successful it must meet certain criteria:

a) The bacteria must be capable of reaching and colonizing the intestinal tract. The bacteria are normally of a species isolated from the particular animal and are first grown on a plate containing bile acids. At the end of the harvesting process the bacteria are screened again for bile resistance.
b) The bacterium must be a rapid acid producer. Each bacterium has its own specific rate of acid production and only those capable of producing above a specific rate are acceptable. The importance of acid lies not only in its anti-*E. coli* effect but also in the fact that acidified conditions in the gut are conducive for good digestion.
c) The bacteria should be present in sufficient numbers to be significant. As both *E. coli* and *lactobacillus* are facultative anaerobes, both can survive under similar conditions, but the specific growth rate of *E. coli* is so much greater that higher levels of lactic acid bacteria must be present to ensure their survival. For newborn calves, a rate of one gram at an activity of 20 billion colony forming units is adequate. For baby pigs, the rate should be one-tenth of this.
d) The bacteria must be quickly activated and have a high specific growth rate.
e) The bacteria must be durable enough to withstand the duress of commercial manufacturing, processing and distribution so the product can be delivered alive to the intestine.

MODE OF ACTION OF YEAST CULTURES

Yeast cultures are believed to act in a different manner to bacterial probiotics. The work of Williams (1989) indicates the ability of certain strains of yeast to remove simple sugars and oligosaccharides from rumen fluid thereby overcoming the post concentrate feeding lactate production which often leads to acidosis. In young piglets, the same action would avoid digestive upsets when they are weaned to high energy diets. In sows, stimulation of fiber digestion may be occurring in the hindgut in a similar way to that observed in the rumen. The ability of the yeast strain "1026" to survive to the ileum (Williams, 1989) would indicate the possibility of such a role. Some additional mechanisms are listed in Table 32.4.

Table 32.4. Mechanisms of Action for Yeast Cultures

a) Removal of simple sugars from the diet.
b) Reduction in post feeding lactate levels.
c) Overcoming carbohydrate overload in the gastrointestinal tract.
d) Stimulation of cellulytic bacteria.
e) Absorption of toxins produced by pathogens.
f) Source of nutrients for other bacteria.
g) Stimulation of hindgut bacteria.

PROBIOTICS AND THE NURSING PIG

Probiotics have shown their greatest potential in very young and rapidly growing pigs. Bacteria normally inhabit the intestinal tract of young animals, as demonstrated by Robinson et al. (1984). However, during periods of stress or mismanagement, the balance between beneficial species and pathogens can shift in favor of *E. coli*. The producer normally observes the result as poor performance, increased incidence of disease and higher mortality.

The baby pig first encounters bacteria in the birth canal at farrowing. It is further exposed to bacteria, many of the coliform type, through continued association with the sow, its own litter mates and from the environment. If the intestinal tract becomes overwhelmed with enteropathic *E. coli* bacteria, piglets may begin to scour with symptoms developing before they are seven days old. In many cases, *E. coli* have been detected in the feces of young pigs within two hours of birth, while *lactobacillus* are not detected for at least 18 hours. However, by introducing the beneficial bacteria early in the life of the young pig, the possibility of saving more pigs from enteric complications could be increased.

In a recent experiment conducted at the University of the Philippines, Lim (1988) compared the efficacy of an orally administered probiotic containing a mixed bacteria culture of *L. acidophilus* and *S. faecium* with a routinely used antibiotic treatment program for the prevention of enteric problems in newborn pigs. Incidence and type of diarrhea, mortality rates and weight gain were used to evaluate the two programs (Table 32.5).

Table 32.5. Performance of Piglets (0-15 days) Receiving Either an Oral Probiotic at Birth or Antibiotic Treatment as Needed.[1]

	Antibiotic[2]	Probiotic[3]
Birth Weight (kg)	1.28	1.30
Average Weight Day 15 (kg)	2.93	3.25
Average Daily Gain (kg)	0.11	0.13
Normal Feces (%)	75.00	79.62
Pasty Feces (%)	14.28	14.81
Yellow Diarrhea (%)	10.71	3.70

[1]Lim, 1988.
[2]Trimethoprim and colistine sulfate.
[3]All-Lac (0.25 g/piglet).

Mortality rate in the probiotic group was 13.1% with neonatal diarrhea-associated deaths of 8.2%. Mortality rate in the antibiotic group was 15.2% with 12.1% attributed to neonatal diarrhea. Use of a probiotic was found to be effective in preventing and controlling the incidence of diarrhea

(3.7% vs. 10.7% yellow diarrhea for the probiotic and antibiotic groups, respectively). It was also noted that, since the incidence of scouring was reduced, the piglets that received the probiotic were superior in average daily gain and were 320 grams heavier at day 15 than piglets treated for scours with antibiotics as needed.

The beneficial effects of using a probiotic in creep feeds for piglets for combating stress was also demonstrated by Mordenti (1986). He clearly showed the value of feeding lactic acid-producing bacteria in conjunction with peptides for improving performance (Table 32.6). In this study, while the peptides and lactic acid bacteria alone improved piglet productivity, the combined treatment resulted in a synergy which significantly reduced mortality following diarrhea and improved average daily gain over both the control and individual treatments. Mordenti concluded that the addition of lactic acid bacteria to creep feeds can be an effective management practice for the prevention of pathological disorders of the digestive tract in piglets. The peptides used were not identified. However, it was infered that the lactic acid producing bacteria may have produced a peptide metabolite which had anti-bacterial effects on the pathogenic strains.

Table 32.6. Effects of Probiotics and Peptides on Piglet Performance[1,2]

	Control	Peptides	Probiotic	Combination[3]
Number of Piglets	471	481	484	499
Daily Gain (g)	186[a]	195[b]	207[c]	221[d]
Diarrhea Rate (%)				
1st Week	20.4[a]	16.4[ab]	9.1[b]	6.0[b]
2nd Week	28.8	28.8	21.4	14.7
3rd Week	11.7[ab]	23.5[a]	7.8[b]	5.8[bc]
Mortality (%)				
Total	11.7[a]	12.7[a]	11.5[a]	8.4[b]
Diarrhea	6.7[ab]	7.3[a]	5.9[b]	4.2[c]
Other	5.0	5.4	5.6	5.2

[1]Mordenti, 1986.
[2]Means of two trials carried out for 34 days in autumn and winter.
[3]Means with different superscripts differ ($P<0.05$).

PROBIOTICS AND THE WEANED PIG

Nature intended that weaning should be a gradual process. However, intensive swine management practices make weaning an abrupt event. The young pig, possibly only three to four weeks of age with an immature digestive function, is suddenly subjected to the stress of weaning and to dramatic changes in diet. On many occasions, this stress results in a post-weaning growth check, associated with the onset of scouring.

Rosell (1987) compared post-weaning performance of pigs fed a starter feed containing *L. acidophilus* and *S. faecium* with those fed a feed grade antibiotic. It was observed that the pigs receiving the probiotic in their feed were equal to or superior in daily gain, intake, and feed efficiency compared to pigs fed the antibiotic (Table 32.7).

Table 32.7. Effects of Probiotic and Antibiotic Addition to a Pig Starter Diet on Weight Gain and Feed Efficiency (8-40 kg)[1]

	Antibiotic[2]	Probiotic[3]
Number of Pigs	36	41
Daily Gain (kg)	0.47	0.52
Daily Intake (kg)	1.10	1.14
Feed Efficiency	2.32	2.17

[1]Rosell, 1987.
[2]Olaquinox (50 mg/kg).
[3]Lacto-Sacc (1 kg/ton).

USE OF PROBIOTICS IN GROWER DIETS

In a recent report from the College of Livestock and Agriculture in Korea, Won (1988) observed that growing pigs fed a lactic acid bacteria plus yeast culture combination were 4.8 kg heavier after 56 days and converted feed 9.6% more efficiently than pigs in the control group (Table 32.8).

Table 32.8. Effect of a Combination of Yeast Culture and Lactic Bacteria on the Performance of Growing Pigs (19-60 kg)[1]

	Control	Yeast Culture and Bacteria Product[2]
Daily Gain (kg)	0.69	0.77
Daily Intake (kg)	1.88	1.89
Feed Efficiency	2.72	2.46

[1]Won, 1988.
[2]Lacto-Sacc (1 kg/ton).

A summary of two recent trials conducted in the United Kingdom (Harker, 1988) examined the response of growing pigs to a yeast culture/bacterial complex (Table 32.9). The probiotic was incorporated in the diet in addition to the standard antibiotic growth promoter. The results showed marked responses in terms of feed intake and growth rate for the probiotic group.

Table 32.9. The Effect of a Probiotic on Performance of Growing Pigs (18-35 kg)[1]

	Control	Probiotic[2]
Number of Pigs	86	87
Daily Gain (kg)	0.48	0.61
Daily Intake (kg)	0.85	1.06
Feed Efficiency	1.78	1.75

[1]Harker, 1988.
[2]Lacto-Sacc (1 kg/ton).

APPLICATION OF PROBIOTICS IN SOW DIETS

The use of microbial cultures has also been examined in regard to the sow. The probiotic species *Bacillus subtilis* has been thought to exert its effect by reducing the number of *E. coli* shed from the intestinal tract. This action could be valuable since *E. coli* is a common bacteria associated with piglet scours and the sow is a major contributor of these organisms. Use of a probiotic could reduce the incidence of *E. coli* scours and result in more pigs weaned. Pollmann (1987) reviewed the results from three studies that were conducted with sows where a microbial culture was added to the diet (Table 32.10). Results showed that, in general, sows consuming diets supplemented with probiotic bacteria weaned larger, heavier litters.

Table 32.10. Effect of Probiotic *(Bacillus Subtilis)* on Sow Performance

	Item	Control	Probiotic
Trial 1	Number of Sows	26	22
	Piglets Weaned	7.65	8.32
	Weaning Weight (kg)	6.63	6.26
Trial 2	Number of Sows	20	20
	Piglets Weaned	8.55	9.05
	Weaning Weight (kg)	5.30	5.48
Trial 3	Number of Sows	39	41
	Piglets Weaned	9.74	10.07
	Weaning Weight (kg)	5.40	7.42

Pollmann, 1987.

The mode of action for probiotics in sows has yet to be established. However, it is possible that the *lactobacillus* component modifies an imbalance in the sow's gut microflora, thereby reducing the pathogenic challenge to the newborn pig via the sow's feces. Such imbalances may be created by the stresses associated with movement into farrowing facilities, parturition and lactation. Williams (1988) suggested that hindgut or cecal fermentation is stimulated through the inclusion of probiotics. This process, which results in increased production of volatile fatty acids, could contribute as much as 30% of the sow's energy needs. The result could be improved nutrient availability and milk production in the sow. Results from growing pig trials, field observations and responses obtained in other species (Fallon and Harte, 1987; Hughes, 1988) indicate that yeast culture could play a significant role in stimulating the sow's appetite, which is often limiting during lactation.

SUMMARY

As intensive management practices place additional pressures upon newborn and weanling pigs, the importance of "natural biologicals" for improving health and performance becomes even greater. Stress-induced scours are a major contributing factor in piglet mortality. Research now suggests and supports the theory that the addition of live *lactobacillus* and *streptococcus* given either orally shortly after birth and (or) in the creep feed, would be an important deterrent to colibacilliosis. Probiotics are now currently available which are easily administered to the newborn as an oral drench or paste.

The probiotic concept is concerned with promoting the growth of *lactobacilli* and in reducing the numbers of enteropathogenic bacteria. Since the early weaned pig is not physiologically equipped to withstand the stresses of enteropathic bacteria, the addition of live *lactobacillus* and *streptococcus* bacteria are necessary for the young pig to maintain a functional and balanced intestinal microflora for the effective utilization of nutrients and the inhibition of coliform bacteria. In practice, the authors are of the opinion that probiotics have a role in a number of specific areas. These include the early days of an animal's life and times of stress. Ideally, the products should be orally dosed or added to feed. Antibiotics should be used before probiotics and not in conjunction with them.

REFERENCES

Barrow, P.S., Brooker, B.E., Fuler, R. and Newport, M.J., 1980. The attachment of bacteria to the gastric epithelium of the pig and its importance in the microecology of the intestine. J. Appl. Bacteriol. 48: 147-154.

Dahiya, C.P. and Speck, M.I., 1968. Hydrogen peroxide formation by *lactobacillus* and its effects on *Staphylococcus aurius*. J. Dairy Sci. 51: 1568-1572.

Fallon, R.J. and Harte, F.J., 1987. The effect of yeast culture inclusion in the concentrate diet on calf performance. J. Dairy Sci. 70 (Suppl. 1): 143.

Fuller, R., 1977. The importance of *lactobacilli* in maintaining normal microbial balance in the crop. Brit. Poult. Sci. 18: 85-94.

Fuller, R. and Cole, C.B., 1988. The scientific basis of the probiotic concept. In: B.A. Stark and J.M. Wilkinson, eds. Probiotics - Theory and Applications. Chalcombe Publications, United Kingdom, pp. 1-14.

Gedek, B., 1987. Probiotics in Animal Feeding: Effects on Performance and Animal Health. Feed Management International. Watt Publishing Company, Mount Morris, Illinois.

Gilliland, S.E., 1981. Enumerations and identification of *lactobacilli* in feed supplements marketed as sources of *Lactobacillus acidopohilus*. Animal Science Research Report. Oklahoma Agricultural Experimental Station, Stillwater, pp. 61-63.

Hamden, E.Y. and Mikolajcik, E.M., 1974. Acidolin, an antibiotic produced by *Lactobacillus acidophilus*. J. Antibiotics 27: 632.

Harker, A.J., 1988. Probiotics and acidification as part of a natural pig production programme. The Feed Compounder (June/July) pp. 12-14.

Hughes, J., 1988. The effect of high strength yeast culture in the diet of early weaned calves. Anim. Prod. 46: 483 (Abstract).

Lim, A.M., 1988. Effects of probiotics on the incidence of yellowish diarrhea in piglets. M.Sc. Thesis, University of the Philippines.

Lyons, T.P., 1987. The role of biological tools in the feed industry. In: T.P. Lyons, ed. Biotechnology in the Feed Industry. Alltech, Inc. Nicholasville, Kentucky, pp. 1-50.

Manfredi, E.T., 1986. The application of microbial ecology and biotechnology to animal feeding. World Biotech. Rep. 2: 81.

Mordenti, A., 1986. Probiotics and new aspects of growth promoters in pig production. Information Zootechnology 32(5): 69.

Nemeskery, T., 1983. Probiotics for young animals. Feed Int. (Dec) :46.

Parker, R.B., 1974. Probiotics, the other half of the antibiotic story. Animal Nutrition and Health 29: 4-8.

Pollmann, D.S., 1985. Feed modifiers - what are they? Proceedings of the Guelph Pork Symposium, University of Guelph, Ontario, pp. 59-74.

Pollmann, D.S., 1987. Probiotics in pig diets. In: W. Haresign and D.J.A. Cole, eds. Recent Advances in Animal Nutrition. Butterworths, London, pp. 193-205.

Robinson, I.M., Whipp, S.C., Bucklin, J.A. and Allison, M.J., 1984. Characterization of predominant bacteria from the colons of normal and dysenteric pigs. Appl. Environ. Microbiol. 48: 964.

Rosell, V., 1987. Acidification and probiotics in Spanish pig and calf rearing. In: T.P. Lyons, ed. Biotechnology in the Feed Industry. Alltech, Inc. Nicholasville, Kentucky, pp. 177-180.

Smith, W.H., 1971. The bacteriology of the alimentary tract of domestic animals suffering from *E. coli* infection. Annals New York Academy of Science 176: 110-125.

Williams, P.E.V., 1988. Understanding the biochemical mode of action of yeast culture. In: T.P. Lyons, ed. Biotechnology in the Feed Industry. Alltech, Inc., Nicholasvile, Kentucky, pp. 79-100.

Williams, P.E.V., 1989. The mode of action of yeast culture in ruminant diets: A review of the effects on rumen fermentation patterns. In: T.P. Lyons, ed. Biotechnology in the Feed Industry. Volume 3, Alltech Technical Publications, Nicholasvile, Kentucky.

Won, S.D., 1988. Effect of Lacto-Sacc on performance of early-weaned pigs. Yonam Jr. College of Livestock and Horticulture, Korea.

CHAPTER 33

Pumpkin, Melon and Other Gourd Seeds

E. Nwokolo

INTRODUCTION

Melons, pumpkins, squashes and gourds belong to the *Cucurbitaceae* family, which is native to Africa but grows well in most tropical and subtropical climates and in temperate climates with long, warm and humid summers. Important genera in this family are *Cucurbita, Citrullus, Lageneria, Telfaria* and *Luffa*, all of which produce seeds that have potential for use in the nutrition of swine and poultry, particularly in the tropics.

Melon seeds are highly nutritious, containing higher levels of protein and oil than other edible oilseeds (Nwokolo and Sim, 1987). While the watermelon (*Citrullus lunatus*) is cultivated for its red, sweet and juicy pulp, a less improved variety referred to as *C. vulgaris, C. colocynthis* or *Colocynthis citrullus*, is cultivated for its seed. This variety is small-fruited (10-15 cm in diameter), bitter-tasting and filled with seeds. Seed production is about 2000 kg/ha, with a kernel yield of 70 to 80% (Nwokolo and Sim, 1987).

The calabash gourd (*Lageneria siceraria*), grown purely for its hard shell calabash, also has considerable seed yield which recent analyses show to be a good potential source of protein and oil (Okoli and Mgbeogu, 1983). Although seeds of the calabash gourd are yet to be used in animal nutrition, they have a tremendous potential for use, once the nature of their toxic components have been determined and ways devised to detoxify them.

The fluted pumpkin (*Telfaria occidentalis*, Hook) produces large, fluted fruits, 60 to 80 cm in length, and 5 to 10 kg in weight. The fruits contain a considerable number of large seeds (50-70) embedded in a fleshy, bitter pulp. Yields of pumpkin seeds are between 2500 and 3000 kg/ha, with a kernel yield of 60 to 70% (Nwokolo and Sim, 1987). Dried pumpkin seeds have been shown to be very good sources of fat and protein, the defatted meal being extremely rich in protein (Longe et al., 1983; Nwokolo and Sim, 1987).

In the past decade, considerable attention has been focussed on the potential value of the wild Buffalo gourd (*Cucurbita foetidissima*) as a source of dietary oil and protein for human and animal use (Bemis et al., 1978; Berry et al., 1976; Tu et al., 1978; Vasconcellos et al., 1980). The Buffalo gourd produces many fruits between 5 and 7 cm in diameter. Each fruit contains 200 to 300 seeds (Bemis et al., 1978) and 100 seeds weigh about 3.7 grams (Berry et al., 1976). A conservative estimate of seed yield in the Buffalo gourd has been put at 2000 kg/ha (Hogan and Bemis, 1983), the seed being reported as a potentially excellent source of protein and fat in human and animal diets (Vasconcellos et al., 1980; Tu et al., 1978).

NUTRIENT COMPOSITION

Melon Seeds and Meals

The most distinctive feature of melon seed is the very high content of edible oil in the seed and the extremely high content of protein in the defatted seed meal. Oyenuga and Fetuga (1975) observed a protein content that ranged from 30.6 to 35.7% in undefatted seed, and from 68.8 to 76.2% in defatted melon seed. In toasted whole melon seed, protein content ranged from 30.8 to 36.9%, while in toasted defatted melon seed, protein content varied from 69.4 to 77.7%. This is much higher than the protein content of soybean meal, groundnut (peanut) meal and sunflower seed meal. The nutritional value of melon seed is demonstrated in Table 33.1.

Melon seed meal has an excellent amino acid configuration, being almost as good as soybean meal in its content of all essential amino acids with the exception of lysine (Table 33.2). The high content of essential amino acids in the defatted protein meal will enable it to adequately supplement the diets of all types and ages of pigs. Amino acid availability of toasted melon seed meal, determined with broiler chicks, was uniformly high for all amino acids (average 96.1%) and was not significantly different from soybean meal (94.9%). This indicates the potential for utilization of these amino acids by broiler chicks, and possibly by growing pigs.

Table 33.1. Proximate Analysis of Melon, Pumpkin and Gourd Seed (% as fed)

	DM	Fat	CP	Fiber	Ash	NFE
Melon Seed Products[1]						
Raw Melon Seed	86.6	45.6	27.4	5.9	7.7	13.4
Pressed Melon Cake	90.1	20.3	45.3	6.8	8.9	18.7
Defatted Melon Meal	94.7	0.9	66.3	7.2	9.1	16.6
Pumpkin Seed Products[1]						
Raw Pumpkin Seed	86.8	53.4	26.6	3.1	2.3	14.6
Pressed Pumpkin Cake	90.3	19.1	46.5	6.9	8.6	9.0
Defatted Pumpkin Meal	95.0	0.9	71.9	8.0	9.4	10.7
Gourd Seed Products[2]						
Raw Buffalo Gourd Seed	95.1	33.0	32.9	---	3.1	---
Buffalo Gourd Flour	92.4	0.3	70.2	4.0	8.8	---

[1]Nwokolo and Sim, 1987.
[2]Tu et al., 1978; Berry et al., 1976.

Table 33.2. Amino Acids (% of sample) in Melon, Pumpkin and Gourd Seed Meals Compared With Amino Acid Requirements of a 25 kg Pig

	Melon[1]	Pumpkin[1]	Buffalo[2] Gourd	Luffa[3] Gourd	Requirement[4] (25 kg pig)
Arginine	2.03	5.36	6.25	9.82	0.25
Histidine	1.80	1.98	1.64	1.12	0.22
Isoleucine	2.23	2.74	2.26	1.83	0.46
Leucine	4.00	5.01	4.30	3.20	0.60
Lysine	0.92	2.46	2.70	2.25	0.75
Methionine + Cystine	2.01	1.54	2.76	1.09	0.41
Phenylalanine + Tyrosine	4.77	4.87	4.89	4.22	0.66
Threonine	3.14	2.31	1.55	1.38	0.48
Valine	2.50	2.88	2.65	1.55	0.48

[1]Nwokolo and Sim, 1987.
[2]Tu et al., 1978.
[3]Kamel and Blackman, 1982.
[4]National Research Council, 1988.

The fat content of melon seed ranges from 45.6 to 56.8% in raw undefatted samples. Melon seed oil is a semi-drying oil, consisting mainly of glycerides of oleic and linoleic acids. In fatty acid composition, melon seed oil is similar to soybean oil, except that soybean oil has a higher percentage of linolenic acid than melon seed oil (Table 33.3).

Table 33.3. Fatty Acid Profile (% of total) and Physico-chemical Characteristics of Melon, Pumpkin and Buffalo Gourd Seed Oil

	Melon Seed Oil[1]	Pumpkin Seed Oil[1]	Buffalo Gourd Seed Oil[3]
Fatty Acids			
12:0	1.6	1.6	---
14:0	1.2	0.6	---
16:0	11.8	17.1	11.8
18:0	10.7	15.0	3.5
18:1	14.5	35.4	22.2
18:2	57.6	27.1	60.6
18:3	2.1	1.2	---
22:1	---	1.7	---
Characteristics[2]			
Free Fatty Acid	1.1-2.9	---	0.5-1.7
Saponifation Value	189-191	190	190-195
Hydroxyl Value	2.6	---	7.3
Iodine Number	115-121	120	123-138
Refractive Index (25°C)	1.47	1.47	1.47
Specific Gravity (25°C)	0.92	0.92	0.92

[1]Nwokolo and Sim, 1987.
[2]Nwokolo, unpublished data.
[3]Vasconcellos et al., 1980.

Melon seed meal has a significantly higher phosphorus, magnesium and copper content than soybean meal, while soybean meal has significantly more calcium, potassium, manganese and iron. Content of trace minerals is generally low (Table 33.4), as is usual with most oilseed meals. Average availability of minerals in melon seed (54%), determined with broiler chicks, is in the same range as soybean meal and other oilseeds (55-65%). Availability of potassium (41%) is as low as in soybean meal (44%). Availabilities of iron, phosphorus, magnesium and copper are moderate (50-55%), while those of calcium, zinc and manganese are fairly high (58-61%).

Pumpkin Seeds and Meals

The proximate composition of various pumpkin seed products (raw pumpkin seed, pressed pumpkin seed cake and defatted pumpkin seed meal) are shown in Table 33.1. Whole pumpkin seed has a very high lipid content

(53.4%) and a moderate protein content (26.6%). Mechanically-pressed pumpkin cake has a fair amount of residual fat (19.1%) and a protein content (46.5%) similar to that of solvent-extracted soybean meal. With such high levels of protein and residual fat, pressed pumpkin seed meal will most likely be as good as soybean meal as a source of dietary amino acids and energy for swine. Solvent-extracted pumpkin meal has a variable residual fat content (1-10%), depending on the extent of extraction. Pumpkin meal that has been repeatedly extracted has a very low fat content (1%) and an extremely high protein content (71.9%), almost as high as in protein isolates from various oilseed meals.

The amino acid content of defatted pumpkin meal is excellent and is comparable to soybean meal. Pumpkin seed meal has a low content of methionine but a moderate content of lysine (Table 33.2). Longe and coworkers (1983) concluded, based on whole egg reference protein standard, that the sulfur amino acids are the most limiting amino acids, while lysine and threonine are the second and third limiting amino acids. Amino acid availability in defatted, toasted pumpkin seed meal, determined with broiler chicks, is uniformly high (93.8%), as high as in soybean meal (94.9%) and in melon seed meal (96.1%).

Dry seeds of fluted pumpkin contain over 51% fat, the recovered oil having a bland taste and odor, and a dark reddish-brown color. Physicochemical properties of the oil (Table 33.3) indicate that pumpkin seed oil closely resembles other vegetable oils that have been successfully used in swine nutrition. There is a high level of carotenoids in pumpkin seed oil (105 mg/g), similar to the level in cotton seed oil. Fluted pumpkin lipids contain about 17% palmitic acid, 15% stearic acid, 35% oleic acid and 27% linoleic acid (Table 33.3). This is in comparison with 12% palmitic acid, 11% stearic acid, 14% oleic acid, 58% linoleic acid and 2% linolenic acid in melon seed oil.

In general, defatted pumpkin seed meals have a high content of potassium, magnesium and phosphorus. Calcium content is low, as is usual with many oilseeds. There is also a very low concentration of the micro elements manganese, zinc, iron and copper (Table 33.4). Average availability of minerals in defatted pumpkin seed meal determined with broiler chicks is moderate (58.8%), similar to average availability of minerals in soybean meal (54.9%; Nwokolo and Sim, 1987). Availability of potassium is poor (39%) as is usual with most oilseed meals, while availability of magnesium and iron is moderate (52-59%). Availability of such minerals as copper, manganese, phosphorus and copper is high (63-73%).

Gourd Seeds and Meals

Gourds include the calabash or bottle gourd, the wild buffalo gourd and the luffa gourd. Protein content is high in many gourd seeds (33-39%) and very high in defatted gourd seed flours (70%). In general, there is a high concentration of the indispensible amino acids in flours produced from gourd seeds. Buffalo gourd protein has an amino acid profile that is distinctly better than the luffa gourd profile in respect to most amino acids. Luffa gourd has only 39% of the methionine, 58% of the valine and 83% of the lysine in buffalo gourd meal. Content of indispensible amino acids in melon seed meal, pumpkin seed meal, buffalo gourd seed meal and luffa gourd seed meal have been compared to nutrient requirements of young rapidly growing pigs in Table 35.2. It is likely that incorporation of these protein meals at levels up to 15% of the diet will meet the requirements of young pigs for the essential amino acids, thereby minimizing the need for supplementary amino acids.

The few reported analytical data on gourd seeds indicate that crude fiber is lower in gourd seeds and meals than in melon and pumpkin seeds and meals. Fat content varies from 33% in buffalo gourd seeds to 47.3% in calabash gourd seeds, while luffa gourd seeds have an intermediate fat content. Fatty acid composition of luffa gourd seed oil is also distinctly different from buffalo gourd seed oil. Luffa gourd seed oil is higher in palmitic acid, stearic acid and oleic acid and lower in linoleic acid than buffalo gourd seed oil. The physicochemical characteristics of buffalo gourd seed oil and luffa gourd seed oil are similar to each other and not different from those of melon and pumpkin seed oils (Table 33.3).

Calcium is low, magnesium high, and phosphorus very high in gourd seed meals. There is considerable similarity among the trace minerals in melon seed, pumpkin seed and luffa gourd seed (Table 33.4).

Table 33.4. Mineral Content of Melon, Pumpkin and Gourd Seed Meals (mg/100g)

Minerals	Melon Seed Meal[1]	Pumpkin Seed Meal[1]	Luffa Gourd Seed Meal[2]
Calcium	83	43	59
Phosphorus	803	468	991
Magnesium	338	144	311
Potassium	590	552	---
Manganese	3.0	1.1	---
Iron	7.6	2.5	10.1
Zinc	3.6	3.3	5.5
Copper	2.3	1.8	2.1

[1]Nwokolo and Sim, 1987.
[2]Kamel and Blackman, 1982.

UNDESIRABLE CONSTITUENTS

Melon Seed and Meal

The presence of anti-nutritional factors in melon seed was suspected by Nwokolo and Sim (1987), since broiler chicks consuming raw melon seed showed feed refusal within 24 hours. Weight loss was severe after 5 days and by the 7th day, some mortality had occured and the feeding trial had to be discontinued. Subsequently, when the melon seeds were toasted, growth rate was normal and there was no mortality. Post-mortem examination of birds that died indicated evidence of hepatic and pancreatic hypertrophy. Solvent extraction without heat treatment did not eliminate the growth-depressing factor(s) in melon seed. Trypsin and chymotrypsin inhibitors are usually associated with pancreatic hypertrophy but the content of such inhibitors in melon seed was not determined. Whatever the anti-nutritional factors in melon seed are, they seem to be heat-labile since cooking, toasting and other heat treatments are able to enhance the nutritional quality of melon seed.

Pumpkin Seed and Meal

Longe et al. (1983) reported 100% mortality within 8 days among albino rats fed diets in which raw pumpkin seed meal supplied 15% of the dietary protein. Raw full-fat pumpkin seed meal, fed at levels which supplied 10% of dietary protein, while not causing death, was associated with a drastic weight loss in rats. For pumpkin seed meal, growth is improved with cooking or toasting, suggesting that growth inhibitors are destroyed by heat treatment. The chemical nature of the anti-nutritional factors in fluted pumpkin have not been reported, although heat-labile trypsin and chymotrypsin inhibitors are suspected to be present.

Gourd Seed and Meal

Defatted buffalo gourd seeds have been reported to produce a neuromuscular condition in young chicks, a condition characterized by an abnormal condition of the neck and inability to keep the head upwards (Daghir et al., 1980; Daghir and Sell, 1980). These workers established that the factor(s) responsible were localized in the embryo but not the hulls of the seed. Saponins, which are triterpenoids, have also been found in buffalo gourd seed, although their concentration in the seed is not high enough to depress growth (Daghir and Zaatari, 1983).

Seeds as well as oilcakes from luffa gourd seeds are bitter, the bitterness being reported to be as a result of the presence of cucurbitacin B, which has been reported to be as high as 0.67% of the seed (Watt and Breyer-Brandwijk, 1962). The bitter, water soluble cucurbitacins can be removed from the seeds by soaking or water treatment.

The nutritive value of luffa gourd seeds can be drastically improved by boiling with water prior to milling and solvent extraction. In addition, Varshney and Beg (1977) reported that the saponins present in luffa gourd seeds were responsible for the toxicity of the gourd seeds. Many gourd seeds are reported to contain several terpenoid glycosides, including the purgative colocynthin. Proper heat treatment, especially the use of moist heat, alone or in combination with some fermentation, seems to be enough to detoxify or inactivate these anti-nutritional factors in cucurbit seeds and meals.

Buffalo gourd oil seems to be free of any toxic components, except very low levels of conjugated fatty acids which have no important effect on the possible use of the oil as an edible oil (Vasconcellos et al., 1980). This absence of toxicity was confirmed by Khoury et al. (1982) who showed an absence of toxicity when buffalo gourd oil was fed to chicks.

ANIMAL FEEDING TRIALS

Melon Seed and Meal

Weanling albino rats have been used to evaluate raw defatted and toasted defatted melon seed meal (Oyenuga and Fetuga, 1975; Akpapunam and Markakis, 1981; Achinewhu, 1983). For raw defatted melon seed meal, protein efficiency ratio, net protein utilization and biological value were 0.93, 49.74 and 52.84, respectively. For toasted defatted melon seed meal, these parameters were 1.01, 50.12 and 54.18, respectively. In soybean meal the same parameters were 1.91, 61.33 and 65.85 and in whole egg they were 4.77, 94.86 and 96.94, respectively. Improvement in growth rate of rats following amino acid supplementation of a 10% protein diet containing only toasted defatted melon seed meal confirmed that lysine was the first limiting amino acid, with methionine and threonine as second and third limiting amino acids, for rat growth (Oyenuga and Fetuga, 1975). Akpapunam and Markakis (1981) observed that supplementation of cowpea diets with melon seed resulted in a significant improvement in apparent digestibility of protein, and increased protein efficiency ratio. It is believed that lysine, methionine and threonine are also the three most limiting amino acids in melon seed meal-based diets for swine.

Pumpkin Seed and Meal

Weanling rats, chickens and pigs have been used to assess the nutritional value of pumpkin meals and pumpkin fodder. The earliest evaluation of pumpkin seed protein was by Zucker et al. (1958) who observed that soybean meal was superior to pumpkin seed meal for weight gains in protein-depleted rats and pigs. Bressani (1963) and Bressani and Arroyave (1963) reported that at marginal dietary protein levels for rats, protein efficiency ratio of pumpkin meal was approximately 80% of that of skim milk powder. At higher dietary protein levels, pumpkin seed meal was equivalent to skim milk powder in diets of rats. They also observed that lysine supplementation of pumpkin seed protein drastically improved the protein efficiency ratio of the meal, and while supplementation with either methionine or threonine did not influence protein efficiency ratio, supplementation with both amino acids increased not only growth rate but also feed efficiency in rats.

Bezares and Avila (1981) report experiments in which broiler chicks were fed pumpkin meal-based diets which were supplemented with various amino acids. Diets were unsupplemented or individually supplemented with lysine, methionine or threonine, or combinations of lysine and methionine, lysine and threonine or lysine, threonine and methionine. Interpretation of growth data indicated that in pumpkin seed meal, lysine was the first limiting amino acid and threonine was the second limiting amino acid. Tryptophan supplementation did not increase weight gains. Gain and feed efficiency were significantly greater with soybean meal than with pumpkin seed meal.

The replacement value of soybean meal with pumpkin seed meal in diets of fattening pigs was reported by Manjarrez et al. (1976). These workers fed fattening pigs (59.4 kg) diets in which 0, 50 or 100% of the soybean meal component was replaced by pumpkin seed meal. Daily gains were 970, 950 and 910 g, respectively, and did not significantly differ. Feed conversion ratios were 3.71, 3.65 and 3.95, respectively. For weanling pigs, weight gains were lower in the pumpkin seed diets. In diets where pumpkin seed meal replaced all the soybean meal, supplementation with lysine increased daily weight gain to the same level as in the soybean diet.

Earlier, the use of pumpkin fodder as a source of dietary carotene was reported (Betskov and Kapko, 1972). These workers noted that when isonitrogenous diets for growing pigs were supplemented with 2 kg of pumpkin fodder, daily gains were increased.

Gourd Seed and Meal

To date, almost all feeding experiments conducted with buffalo gourd seed and meal have been carried out using rats, mice or chicks. In spite of an exhaustive literature search, the author is not aware of any feeding trials involving pigs. This is not because buffalo gourd meals cannot be fed to pigs, but because usually available seed or meal is small and researchers would prefer to use animals that consume small amounts of feed in order to maximize data output. Experiments by Daghir and Zaatari (1983) using growing chickens indicated that soybean meal was superior to hexane or acetone-extracted buffalo gourd meal, with respect to growth rate and feed efficiency. Supplementation of buffalo gourd meal diets with methionine, lysine, methionine and lysine or a combination of methionine, lysine and threonine, did not significantly improve performance of experimental chicks. Chicks fed soybean meal diets still performed better than those fed buffalo gourd meal diets.

It is possible to conclude that buffalo gourd oil and meal have a potential for use in diets of swine and other monogastric animals. Further work is needed to determine the most effective procedure to detoxify the meal and enhance its utilization in swine diets. Additional experiments are also needed to determine the optimum levels of incorporation of gourd seed meals and oil in diets of growing-finishing pigs and sows.

SUMMARY

Melon and pumpkin seed meals are good sources of protein and dietary energy. For effective utilization, there is a need to defat and toast these meals prior to their use in swine diets. Because of a relative deficiency of lysine and methionine, melon or pumpkin seed meals should not constitute more than 30% of the dietary protein for starter pigs. However, a level of inclusion providing up to 50% of the dietary protein for grower pigs, finisher pigs and breeding stock should sustain acceptable growth and production rates. When higher levels of melon and pumpkin seed meals are used, care should be taken to ensure that amino acids which may be limiting are provided. There appears to be no data available concerning the feeding of gourd seed meals to pigs. Therefore, even though they have excellent amino acid and fatty acid profiles, defatted, toasted meals should be fed with caution and only to finishing pigs.

REFERENCES

Achinewhu, S.C., 1983. The nutritive qualities of plant foods. 2. Protein quality of bread fruit (*Artocarpus ultilis*), climbing melon (*Colocynthis vulgaris*) and creeping melon (*Citrullus vulgaris*). Nutr. Rep. Int. 27: 541-545.

Akpapunam, M.A. and Markakis, P., 1981. Protein supplementation of cowpeas with sesame and watermelon seeds. J. Food Sci. 26: 960-961.

Bemis, W.P., Berry, J.W. and Weber, C.W., 1978. The buffalo gourd, a potential crop for arid lands. Arid Lands Newsletter 8: 1-7.

Berry, J.W., Weber, C.W., Dreher, M.L. and Bemis, W.P., 1976. Chemical composition of buffalo gourd, a potential food source. J. Food Sci. 41: 465-466.

Betskov, I. and Kapko, P., 1972. Grass meal and fodder pumpkins in feeds for pigs. Nutr. Abstr. Rev. 43: No. 1509 (Abstract).

Bezares, S.A. and Avila, G. E., 1981. Limiting amino acids in pumpkin seed meal (*Cucurbita pepo*, L.) for the chick. Nutr. Abstr. Rev. 52: No. 3024 (Abstract).

Bressani, R., 1963. Nutritional value of various mixtures of vegetable oilseed proteins developed to combat malnutrition in Latin America. Qualitas Plant. 10: 73-108.

Bressani, R. and Arroyave, R., 1963. Nutritive value of pumpkin seed: Essential amino acid content and protein value of pumpkin seed (*Cucurbita farinosa*). J. Agric. Food Chem. 11: 29-33.

Daghir, N.J., Mahmoud, H.K. and El-Zein, A., 1980. Buffalo gourd (*Cucurbita foetidissima*) meal: Nutritive value and detoxification. Nutr. Rep. Int. 21: 837-847.

Daghir, N.J. and Sell, J.L., 1980. Buffalo gourd (*Cucurbita foetidissima*) seed and seed components for growing chickens. Nutr. Rep. Int. 22: 445-452.

Daghir, N.J. and Zaatari, I.M., 1983. Detoxification and protein quality of buffalo gourd meal (*Cucurbita foetidissima*) for growing chickens. Nutr. Rep. Int. 27: 339-346.

Hogan, L. and Bemis, W.P., 1983. Buffalo gourd and jojoba: Potential new crops for arid lands. Advances Agron. 36: 317-349.

Kamel, B.S. and Blackman, B., 1982. Nutritional and oil characteristics of seeds of angled luffa (*Luffa acutangula*). Food Chem. 9: 277-282.

Khoury, N.N., Dagher, S. and Sawaya, W., 1982. Chemical and physical characteristics, fatty acid composition and toxicity of buffalo gourd seed oil, (*Cucurbita foetidissima*). J. Food Technol. 17: 19-26.

Longe, O.G., Farinu, G.O. and Fetuga, B.L., 1983. Nutritional value of the fluted pumpkin (*Telfaria occidentalis*, Hook). J. Agric. Chem. 31: 989-992.

Manjarrez, B., Enriquez, F., Avila, G.E. and Shimada, A.S., 1976. Replacement of soya meal with pumpkin seed meal in rations of fattening pigs. Nutr. Abstr. Rev. 49: No. 3789 (Abstract).

National Research Council, 1988. Nutrient Requirements of Domestic Animals, No. 2. Nutrient Requirements of Swine. 9th ed. National Academy of Sciences, Washington, D.C.

Nwokolo, E.N. and Sim, J.S., 1987. Nutritional assessment of defatted meals of melon (*Colocynthis citrullus* L.) and fluted pumpkin (*Telfaria occidentalis* Hook) by chick assay. J. Sci. Food Agric. 38: 237-246.

Okoli, B.E. and Mgbeogu, C.M., 1983. Fluted pumpkin, *Telfaria occidentalis*: West African vegetable crop. Econ. Bot. 37: 145-149.

Oyenuga, V.A. and Fetuga, B.L., 1975. Some aspects of the biochemistry and nutritive value of the watermelon (*Citrullus vulgaris*, Schrad). J. Sci. Food Agric. 26: 843-854.

Tu, M., Eustace, W.D. and Deyoe, C.W., 1978. Nutritive value of buffalo gourd seed protein. Cereal Chem. 55: 766-772.

Varshney, I.P. and Beg, M.F.A., 1977. Saponins from the seeds of *Luffa aegyptinin*, *L. aegyptinin* A. and *L. aegyptinin* B. Indian J. Chem. 15: 394-397.

Vasconcellos, J.A., Berry, J.W., Weber, C.W., Bemis, W.P. and Scheerens, J.C., 1980. The properties of *Cucurbita foetidissima* seed oil. J. Amer. Oil Chem. Soc. 57: 310-313.

Watt, J.M. and Breyer-Brandwijk, M.G., 1962. The Medicinal and Poisonous Plants of Southern and Eastern Africa. E. & S. Livingston Ltd., Edinburgh and London.

Zucker, H., Hays, V.W., Speer, V.C. and Catron, D.V., 1958. Evaluation of pumpkin seed meal as a source of protein using a depletion-repletion technique. J. Nutr. 65: 327-334.

CHAPTER 34

Rice and Rice Milling By-Products

D.J. Farrell and K. Hutton

INTRODUCTION

Rice (*Oryza sativa* L.) is an ancient cereal grain which was first cultivated in Asia over 6,000 years ago (Luh, 1980). It now forms the staple food for over two-thirds of the world's human population. Cultivation of rice is still confined largely to the Asian region where over 90% of the world's production is grown. However, new cultivars, with a range of environmental tolerances, could allow rice to be cropped at latitudes reaching well into the temperate regions and at high altitudes.

The polished white rice which is consumed in the human diet is obtained by milling paddy rice. In the process, a wide variety of by-products are produced and many have the potential to be used in swine production. In addition, off-grade paddy rice, which results from delayed harvesting or heating and discoloration in storage, may also be available for swine feeding.

PRODUCTION OF POLISHED RICE AND BY-PRODUCTS

There are approximately 400 million tonnes of paddy rice produced annually (FAO, 1980). To produce polished white rice, paddy rice is first dehulled, resulting in a yield of about 80% brown rice and 20% hulls. The brown rice fraction is then further milled to remove the germ and other layers, yielding rice bran (10%) and white rice (70%). After this process, small particles of broken seeds are removed. This process yields about 48% head rice and 22% broken rice. These values will vary with season and with variety. Medium grain cultivars produce more head rice than long grain cultivars with less broken seeds (58% head rice and 12% broken rice).

The primary rice milling by-products available for use as swine feeds are therefore rice bran, broken rice and rice hulls. Another product of minor significance is screenings, which are mainly a mixture of small broken seeds and weed seeds. Screenings represent a minor fraction resulting from the cleaning of paddy rice after the harvest trash has been removed, prior to milling. Rice pollard is a term used in Australia to describe the fraction milled from brown rice to white rice and includes both bran and rice polishings.

NUTRIENT CONTENT OF RICE AND RICE BY-PRODUCTS

Milling techniques and efficiency will determine to a large extent the chemical composition of rice by-products. However, a wide range of milling methods and tolerances make rice milling by-products highly variable in chemical composition from country to country and mill to mill. Precise statements about their nutritional value are therefore difficult to make. It is therefore essential to chemically analyze rice by-products as the first stage in nutritional evaluation. It would also be advantageous to be aware of the milling system used to produce the by-products.

Typical changes that occur as a consequence of adequate milling of rice are given in Table 34.1. The major change is a reduction in fat, fiber and ash with an increase in the carbohydrate fraction, primarily as starch.

Table 34.1. Chemical Composition of Rice and Rice By-Products before and after Milling (% as fed)

	Protein	Fat	Crude Fiber	Ash	Moisture
Paddy Rice	8.4	1.7	9.1	4.7	11.0
Rice Hulls	2.0	0.8	41.0	20.0	10.0
Brown Rice	8.3	1.7	0.8	1.0	12.0
Rice Pollard	13.5	19.7	8.0	10.0	11.0
White Rice	7.3	0.4	0.4	0.5	11.0

Farrell (unpublished).

Paddy Rice

Paddy rice or rough rice is sometimes available at a price that makes it economical to use in swine rations. It contains on average about 11%

moisture, 8.4% crude protein, 1.7% fat, 9.1% crude fiber and 4.7% ash. Paddy rice has an energy value lower than corn or wheat because of the low digestibility of its fibrous hull. Determined digestible energy (D.E.) for swine is 3380 kcal/kg DM (Ensminger and Olentine, 1978).

Rice Hulls

Although rice hulls are the by-product produced in greatest quantity from rice milling, they are of low nutritional value and are not used in large quantities in animal feeds. Rice hulls contain about 66% neutral detergent fiber or 41% crude fiber, 2% crude protein, 0.8% oil and 20% ash, mainly as silica. Determined D.E. for swine is 900 kcal/kg DM (Farrell and Warren, 1982), which is only about a quarter of the energy value of corn or wheat.

Full-Fat Rice Bran

Full-fat rice bran is the most common feed ingredient produced from rice milling and world production is estimated at almost 50 million metric tonnes a year. The bran fraction contains some of the embryo and small amounts of endosperm. It can comprise 20 to 25% of the total protein, 80% of the oil and 70% of the minerals and vitamins contained in paddy rice. Some rice brans obtained from older and less sophisticated rice mills are very variable in composition due to irregular contamination by hull fractions. A simple and inexpensive analytical technique, using phloroglucinol, which allows an estimate of hull contamination through staining of the hulls, has been described by Tagendjaja and Lowry (1985).

Warren and Farrell (1989a) analysed rice bran from rice cultivars commonly grown in Australia from up to four harvests (1980-1983). Means of values for chemical composition are given in Table 34.2, with samples from Southeast Asia and California included for comparison.

Table 34.2. Composition of Rice Bran (% DM Basis) from Different Regions

Region	Dry Matter	Crude Protein	Ether Extract	Ash	Crude Fiber
Australia	91.8	15.4	22.0	10.5	---
Southeast Asia	89.3	12.9	16.0	9.4	8.2
California	91.2	12.2	13.5	9.3	11.3

Warren and Farrell, 1989a; Creswell, 1988.

D.E. values for rice bran have been reported by Robles and Ewan (1982) in the Unites States and by Farrell and Warren (1982) in Australia to be 3880 and 3870 kcal/kg DM, respectively. The values are substantially higher than the 3100 to 3260 kcal/kg found by Brooks and Lumantra (1975) in the USA.

Campadal et al. (1976) also reported lower values of 2540 to 3390 kcal/kg. The differences probably relate to oil level and amount of hull contamination. An indication of the variation in the fibrous fractions of rice bran is shown in Table 34.3.

Table 34.3. Mean and Range of Fibrous Fractions and Acid Insoluble Ash (AIA) of Full-Fat Rice Bran from Four Australian Cultivars (% DM)

	NDF	ADF	Lignin	AIA
Mean	21.5	10.7	3.8	0.77
Range	20.1-22.2	9.4-11.6	2.9-5.3	0.7-1.2

Farrell (unpublished).

Rice bran is a good source of quality protein. The amino acid content of rice brans from Australia and Southeast Asia are shown in Table 34.4. Values were generally higher for the Australian rice bran than those obtained from Southeast Asia.

Table 34.4. Amino Acid Contents (% DM) of Full-Fat Rice Brans from Three Cultivars Grown in Australia; Mean Values for 10 Samples of Australian Full-Fat Brans and 15 Good and 15 Poor Samples from Southeast Asia

	Australia				Southeast Asia	
	Pelde[1]	Star-bonnet[1]	Cal-rose[1]	Australian Average	Good Samples[2]	Poor Samples[2]
Arginine	1.24	1.59	1.23	1.16	1.18	0.94
Isoleucine	0.56	0.66	0.51	0.51	0.43	0.40
Leucine	1.01	1.15	1.17	1.03	1.04	0.85
Lysine	0.85	0.91	0.82	0.82	0.70	0.59
Methionine	0.24	0.43	0.24	0.28	0.30	0.25
Phenylalanine	0.50	0.59	0.76	0.65	0.69	0.56
Threonine	0.54	0.64	0.53	0.58	0.56	0.46
Tyrosine	0.51	0.60	0.59	0.46	0.49	0.39
Valine	0.76	0.88	1.14	0.77	0.77	0.64

[1]Warren and Farrell 1989a.
[2]Creswell, 1978.

Some mineral analyses are shown in Table 34.5. Rice bran is high in phytate with values of up to 4% but more commonly about 3% (Warren and Farrell, 1989a). Nelson et al. (1968) found Californian rice bran to contain

5% phytic acid. This may reduce the availability of some minerals in rice bran and in rice bran-based diets (Warren and Farrell, 1989b; Warren et al., 1989).

Table 34.5. Mean and Range of Mineral Content of Full-Fat Rice Bran from Three Australian Cultivars of Rice[1]

	Calcium	Phosphorus	Magnesium	Zinc	Iron
Mean	0.39[2]	17.1[2]	6.9[2]	49.0[3]	42.8[3]
Range	0.27-0.51	16.2-18.1	6.1-7.7	44.2-53.9	37.9-48.1

[1]Farrell (unpublished).
[2]mg/g DM.
[3]ug/g DM.

Fat Extracted Rice Bran

Rice bran oil is sometimes extracted from rice bran for human consumption. As a consequence, fat-free or extracted rice bran may be available from oil extraction plants. The resulting fat free bran has a longer storage life than full-fat rice bran and is sometimes referred to as "stabilized rice bran". However, this is a confusing misnomer as it is obviously very different in composition to rice bran. Also, the term "stabilized rice bran" is used to describe the fraction milled from parboiled brown rice, which is stabilized during the parboiling process and still contains all of its original oil. Stabilized rice bran from parboiling plants usually has a place in human nutrition either through health food outlets or as an ingredient in multigrain breads.

Removal of the valuable oil from full-fat rice bran by solvent extraction has the direct effect of concentrating other nutrients by about 20%. There is a reduction in D.E. from 3830 to 2800 kcal/kg DM due to the decrease in oil level and an increase in neutral detergent fiber to 27% of dry matter (Warren and Farrell, 1989a). Creswell (1987) has compiled mean data for 11 samples of extracted rice bran (Table 34.6) collected from Thailand and India, where adulteration with hulls occurs.

Table 34.6. Mean Chemical Composition of 11 Samples of Fat-Extracted Rice Bran from Thailand and India (% as fed)

Dry Matter	Crude Protein	Ether Extract	Crude Fiber	Ash
89.6	15.8	1.6	12.5	12.0

Creswell, 1987.

Hussein and Kratzer (1982) showed that free fatty acid content of fresh rice bran was 13.7%; this increased to 43% when stored at 23°C for 3 months. Rancidity was prevented by the addition of 0.5 g of disodium ethylenediaminetetracetate per kg bran. The fatty acid ester composition of oil extracted from three samples of rice bran is shown in Table 34.7 (Warren, 1986). The oil is particularly rich in the essential fatty acid, linoleic.

Table 34.7. Fatty Acid Composition (%) of Oil Extracted from Individual Samples of Three Rice Bran Varieties

	Calrose 1982	Calrose 1983	Starbonnet 1982
Palmitic 16:0	15.8	15.3	16.5
Palmitoleic 12:1	trace	0.1	0.3
Stearic 18:1	1.4	1.4	1.6
Linoleic 18:2	42.2	42.0	42.9
Linolenic 18:3	1.9	1.3	1.2
Arachidonic 20:0	0.3	trace	0.4

Warren, 1986.

Broken Rice

This valuable feed ingredient is made up of the fragments of grain which result during the milling of rice. Some large brokens are used in the production of extruded breakfast cereals, muesli bars and rice flour and also for brewing. Therefore, broken rice is of limited availability for use as a feedstuff for swine.

The chemical composition of a good sample of broken rice is similar to that of whole white rice but feed quality material may contain some contaminants. Chemical composition will vary with rice cultivar, growing conditions and season. Creswell (1987) reported a wide range in chemical composition of broken rice produced in Thailand (Table 34.8).

Table 34.8. Range of Chemical Compositions of Broken Rice Samples from Thailand (% as fed)

Moisture	Protein	Fat	Crude Fiber	Ash
11.6-13.1	6.7-7.5	0.5-1.3	0.0-0.6	0.4-0.6

Creswell, 1987.

Broken rice has a high D.E. of 3480 kcal/kg DM (Farrell and Warren, 1982) which compares with 3700 kcal/kg DM for polished rice. Broken rice is relatively low in protein (6.7-8.6%) but the amino acid balance is good. The amino acid composition of broken rice from Malaysia and Thailand is shown in Table 34.9 (Creswell, 1988). These analyses would suggest that broken rice is at least of similar nutritive value to corn.

Table 34.9. The Amino Acid Composition of 7 Samples of Broken Rice from Malaysia and 9 Samples from Thailand (% as fed)

	Malaysia	Thailand
Moisture	12.8	12.5
Arginine	0.65	0.55
Histidine	0.19	0.16
Isoleucine	0.35	0.30
Leucine	0.66	0.57
Lysine	0.31	0.26
Methionine	0.23	0.18
Phenylalanine	0.44	0.38
Threonine	0.28	0.24
Tyrosine	0.32	0.25
Valine	0.74	0.40

Creswell, 1988.

Rice Screenings

Rice screenings make up a very small portion (<0.25%) of the paddy rice after deadheads, sticks and trash have been removed and comprise a mixture of weed seeds and small amounts of by-products, mainly small broken grains. A survey of samples of Australian rice screenings was carried out in 1982 by the Ricegrowers' Co-operative Limited which produced the data summarized in Table 34.10. These analyses are consistent with an energy value similar to broken rice for Australian rice screenings. However, this is probably not so for screenings from other countries.

Table 34.10. Mean (± S.E.) Chemical Components of Rice Screenings (% DM)

	Crude Protein	Ether Extract	Crude Fiber	Ash	Ca	P
Mean	8.4	4.3	2.6	3.5	0.18	0.39
S.E.	0.37	0.82	0.13	0.44	0.014	0.07

Ricegrowers' Co-operative Limited (unpublished).

FEEDING RICE AND RICE BY-PRODUCTS TO SWINE

Paddy Rice

Paddy rice must be ground for swine feeding and is a suitable replacement for part of the grain portion of the complete ration. The two major factors restricting its use are the level of hulls, which result in its low digestibility, and the possibility of contamination with mycotoxins. Although mycotoxins have not been a problem in our experience, it must be stressed that it is unlikely that good quality rice which could be milled to human feed grade material quality standards, would be available for swine. The offgrade paddy available for feed results from late harvesting or poor storage conditions and as such must always be monitored for toxic fungal contamination. In theory, it could be milled to give a product suitable for livestock feed as the mold is unlikely to penetrate the hull except for really poor quality material. However, in our experience, the quantities available for livestock are so small that it is not worth contamination and subsequent cleaning of the rice mill to accommodate such second grade material. It is therefore a problem and is best fed to mature ruminants. It is certainly not recommended for young swine. However, if prices justify, it can be used for larger swine and breeders if only discolored and free of mycotoxins.

Rice Hulls

In a study of sweet lupinseed meal as a protein source for growing pigs (Taverner, 1975), ground rice hulls were added to one of the experimental diets (8.15% of the diet) to raise the concentration of crude fiber to approximately that of the sweet lupinseed meal diet. No deleterious effects of the rice hulls were observed. In a second study, conducted to predict the D.E. of pig diets from an analyses of its fiber content (King and Taverner, 1975), ground rice hulls (17.04% of the diet) were used to reduce the D.E. of one of the experimental diets. Again, no deleterious effects were observed, even at such a high level of inclusion. It is apparent from these observations that hulls can be used to reduce the nutrient concentration of diets. This may remove the need to restrict the feed intake of swine, a practice often used to reduce backfat thickness.

Roese and Howard (personal communication) replaced barley with ground rice hulls at 15% of the diet fed to swine between 25 and 50 kg, and increased this to 30% from 50 to 85 kg (Table 34.11). The consequence was to reduce dietary D.E. content from 3050 to 2620 and 2095 kcal D.E./kg, respectively. There was also a grower (18% CP) and a finisher (14%) barley/soybean based diet fed either ad libitum or in restricted amounts. In the grower and finisher phases, diet 3 reduced growth by 100 and 160 g/day, respectively, compared to diet 2. In the grower phase, diets 1 and 2 reduced growth by 60 and 100 g/d, respectively, compared to diet 3. Corresponding reductions were 80 and 160 g/d during the finisher phase. Feed efficiency was much poorer on diet 3 due to the low D.E. content of the diet. However, backfat thickness was lower on diets 1 and 3 compared to the ad libitum fed pigs.

Table 34.11. The Effect of Addition of Rice Hulls to Diets for Pigs during the Growing and Finishing Periods

	Restrict Fed	Ad lib Fed	Ad lib + Rice hulls
Diet	1	2	3
Grower (25-50 kg)			
Daily Gain (g)	580	620	520
Daily Intake (kg)	1.7	2.0	2.0
Feed Efficiency	2.9	3.3	3.9
Finisher (50-90 kg)			
Daily Gain (g)	510	590	430
Daily Intake (kg)	2.1	2.6	2.8
Feed Efficiency	4.2	4.5	6.5
Grower and Finisher (25-90 kg)			
Daily Gain (g)	530	600	460
Daily Intake (kg)	2.0	2.4	2.5
Feed Efficiency	3.9	4.0	5.5
Dressing (%)	73.1	73.9	73.5
P2 backfat (mm)	20	29	19

It is clear that some fine-tuning is necessary to determine the optimum dietary inclusion level of rice hulls. On the one hand, it does have the advantage of removing the need to feed swine diets in restricted amounts and hence improve carcass quality. However, some disadvantages are the additional cost associated with increased bulk and transport as well as the possible abrasive action which may increase fecal nitrogen excretion (Farrell, 1973). There is also some evidence that with young pigs and improved strains of grower pigs at high stocking densities, as little as 10% of ground hulls in pelleted diets reduces appetite for energy to such a degree that growth rates are unacceptable.

Full-Fat Rice Bran

The performance of swine fed diets based on rice bran will vary depending largely on the level of adulteration with hulls. Campabadal et al. (1976) concluded that rice bran could be successfully used in weaner diets at up to 15% and at up to 30% in grower-finisher diets. However, in their experiment, weaners grew 54 g/d less (nonsignificant) on diets with 15% rice bran compared with those on diets with no bran.

Brooks and Lumantra (1975), in two trials, grew swine from 16 to 79 kg liveweight on diets based on corn, molasses or sugar with up to 80% rice bran from Californian mills. Up to 40% rice bran additions did not significantly depress liveweight gain nor adversely affect feed efficiency or carcass characteristics.

Roese (1977) replaced up to 30% barley in the diet with Australian rice bran for swine grown from 18 to 58 kg. Weight gain and feed efficiency were highest on the highest rice bran inclusion level but were only marginally better than on the control diet. Backfat thickness was reduced and carcass quality was improved at the highest level of rice bran inclusion. However, there was a linear relationship between bran inclusion and iodine number of fat, indicating an increase in the degree of unsaturation of the carcass fat and therefore in fat softness.

Full-fat rice bran is high in polyunsaturated oil. Therefore, rancidity of the oil must always be considered in hot, humid environments and/or where storage times in inadequate conditions are excessive. Hydrolysis of the oil is very rapid with up to 10% occurring within hours of production. As a consequence, vitamin E responsive conditions in young rapidly growing pigs are still observed under practical conditions in Australia. This is despite levels of supplementation well in excess of National Research Council (1979) requirements, so vitamin E is of special interest. A ratio of vitamin E to linoleic acid of 0.6 mg/g has been recommended.

The other area of potential concern relates to carcass quality and the level of linoleic acid in rice bran. Soft oily pork has been shown to be a problem in pigs fed large quantities of linoleic acid. However, linoleic acid has little effect on fat quality as long as it makes up less than 15% of the fatty acids in carcass fat. Values greater than 15% linoleic acid in body fat will generally not result from diets with less than 1.6% linoleic acid.

A further nutritional factor affecting fat quality in pig carcases is the use of copper at a level sufficient to promote growth. The melting point of carcass fat was reduced by 10°C in one study with 250 mg/kg copper in the feed and in another study 80% of pigs on a copper supplemented diet showed soft fat compared with only 5% of controls. Copper is used to promote growth in many countries in the world so care is needed with rice bran in swine feeds where carcass quality is critical.

Very recent work on processing rice bran noted an improvement in dry matter digestibility and D.E. when fresh rice bran was pelleted. This process also significantly improved feed efficiency as did extrusion (Tangendjaja et al., 1988b). However, addition of cellulose and amylase to rice bran did not increase growth rate of pigs beyond that of diets with untreated rice bran when included at 40% in the diet (Tangendjaja et al., 1988a). In these experiments, pigs were grown from only 6 to 12 kg and this is probably insufficient time to evaluate fully the rice bran treatments.

Fat-Extracted Rice Bran

Warren et al. (1981) substituted defatted rice bran at up to 30% into a commercial grower diet for weaner pigs (19 kg) without any adverse effect on performance to 45 kg liveweight (Table 34.12). At 30% inclusion, rice bran contributed 30% of the dietary CP. Pigs were fed on an individual basis an allowance such that D.E. intake was constant on all treatments. This meant that intake (g/d) increased with increasing rice bran inclusion but without changing feed efficiency. Compared to the basal diet, some pigs took significantly fewer ($P<0.05$) days to reach 45 kg. Wheat bran and rice bran gave similar performance at 20% inclusion, but daily gain on the latter was 53 g above that of pigs on the basal diet. This may have been due to higher ($P<0.05$) daily intake of crude protein on the rice bran-based diet (Table 34.12) but was probably an energy response.

Table 34.12. Production Parameters for Swine on Diets with Fat-Extracted Rice Bran and Wheat Bran

	Fat-Extracted Bran Inclusion (%)[1]				
	0	10	20	30	20[2]
Daily Intake (kg)[3]	1.28[a]	1.32[ab]	1.40[bc]	1.43[c]	1.38[bc]
Daily Gain (kg)	0.53	0.54	0.59	0.57	0.55
Feed Conversion	2.50	2.50	2.50	2.70	2.60
Days to 45 kg	48[a]	47[ab]	43[b]	44[b]	46[ab]

[1]Warren et al., 1981.
[2]Wheat bran.
[3]Values with the same superscript are not significantly different ($P<0.05$).

When offered feed ad libitum, growth rate of pigs to 69 kg was reduced on diets containing 40% fat-extracted rice bran compared with diets containing 40% raw rice bran, but feed efficiency was not depressed (Feedstuffs, 1985). In a second experiment, raw and fat-extracted rice brans were included at 40% in diets of pigs grown from 24 to 100 kg liveweight. Compared to the corn/soybean control diet, pig growth was inferior ($P<0.05$) on the rice bran-based diets, and particularly on the diet with fat-extracted rice bran. Rib eye area was reduced by 3 cm ($P<0.05$) on the experimental diets compared to the control. Tillman et al. (1951) fed swine on fat-extracted rice bran and reported that up to 30% inclusion did not adversely affect performance of swine grown to 90 kg liveweight.

Whole and Broken Rice

It is assumed that data for whole white rice corresponds to that of broken rice. Robles and Ewan (1982) observed improved growth rates over 28 days in pigs initially weighing 6 kg, with increasing supplementation of ground, polished rice to a corn/soybean diet. The authors concluded that the net energy of rice was similar to that of corn. Feeding of broken rice to pigs is restricted only by nutrient constraints and up to 30% of the pelleted weaner, grower and breeder rations have been made up of broken rice in years of good availability in Australia, with no deleterious effects.

Rice Screenings

As previously indicated, rice screenings in Australia are very similar in composition to broken rice. The main contaminant is the seed of hairy millet which is of high nutritional value and toxic weed seeds are not a problem. As screenings are in such short supply in Australia they are usually mixed with small brokens to produce "stockfeed brokens," and these are treated nutritionally as whole or broken milled rice.

SUMMARY

There are six products of rice milling which may be available in different locations. These are offgrade paddy rice, rice hulls, rice pollard (rice bran and polishings), fat-extracted rice bran, broken rice and rice screenings. Of these, the major volumes are in rice hulls, rice brans and broken rice. Rice hulls are available as 20% of the paddy crop but must be ground for swine feed and are of limited value due to their low nutritive value relative to grains. Rice brans and broken rice are high energy by-products. Broken rice presents no problems in balanced feeds for swine and pellets well in commercial feedmill systems. All rice products should be monitored carefully for quality, but rice brans are particularly variable due to the effect of storage and handling conditions on fat quality and the level of contamination by rice hull fractions. Good quality rice brans are equal to wheat in feeding value whereas rancid samples adulterated with rice hulls may be unsuitable for swine feeding. Fat-extracted rice bran should not result in any problems when fed in balanced rations, but it is probably worth noting that high phytate levels in rice brans generally necessitate adequate supplementation with a good source of calcium.

Because of the quality aspects and variability of rice milling by-products worldwide, any recommendations regarding maximum inclusion levels should only be used as guidelines. Table 34.13 summarizes the preferred recommended values used by the commercial stockfeed division of Ricegrowers' Co-operative Limited in Australia. They are presented to indicate the values which have been used successfully in practice for at least ten years. They take into account by-product availability, acceptability, swine breeding performance, feed conversion efficiency, growth rates, pig carcass quality, stockfeed pellet quality and rate of production of pelleted swine feeds in the mill and as such are commercial guidelines with adequate safety margins. Nevertheless, they are unlikely to be of direct application elsewhere without some modifications for reasons previously discussed.

Table 34.13. Maximum Inclusion Levels for Rice By-Products for Swine Used in Practice under Australian Commercial Conditions (% as fed)[1]

Livestock	Liveweight	Paddy[2]	Hulls[3]	Brokens[4]	Pollard[5]
Starter	to 10 kg	--	--	30	10
Starter	to 20 kg	--	--	30	10
Grower	to 30 kg	--	--	30	10
Grower	to 65 kg	50	10	30	10
Finisher	to 90 kg	50	10	30	10
Breeder	-------	62.5	12.5	30	15

[1]Ricegrowers' Co-operative Limited (unpublished).
[2]This refers to discolored grain only. Moldy or stack-burned grain which is unpalatable and may be toxic should be used with extreme caution.
[3]These levels are in balanced rations but may be too high for maximum performance of improved strains with low appetites.
[4]This restriction is based on availability. There is probably no need for any restriction on good broken rice for swine in balanced rations.
[5]This restriction is primarily for carcass quality and pellet quality considerations and refers to good quality material with no hull contamination and no rancidity.

REFERENCES

Brooks, C.C. and Lumantra, I.G., 1975. Rice bran composition and digestibility by the pig. Proc. West. Sec. Amer. Soc. Anim. Sci. 26: 112-114.

Campabadal, C., Creswell, D., Wallace, H.D. and Combs, G.E., 1976. Nutritional value of rice bran for pigs. Tropic. Agric. (Trinidad) 53: 141-149.

Creswell, D., 1987. A survey of rice byproducts from different countries In: D. Creswell, ed. New Developments in Feed and Technology. Monsanto Technical Symposium, Bangkok, pp. 4-35.

Creswell, D., 1988. Amino acid composition of feedgrade rice products from several countries. In: T.H. Applewhite ed. World Congress on Vegetable Protein Utilization in Human Food and Animal Feedstuffs. Singapore, pp. 474-479.

Ensminger, M.E. and Olentine, C.G., 1978. Feeds and Nutrition. Ensminger Publishing Company, Clovis, California.

FAO, 1980. Production Yearbook. Food and Agriculture Organization of the United Nations, Rome.

Farrell, D.J., 1973. Digestibility by pigs of the major chemical components of diets high in plant cell-wall constituents. Anim. Prod. 16: 43-47.

Farrell, D.J. and Warren, B.E., 1982. The energy concentration of rice by-products for sheep, pigs and poultry. Anim. Prod. Aust. 14: 676.

Feedstuffs, 1985. Raw, stabilized rice bran in corn-soy diet examined. Feedstuffs (May 27) p. 12.

Hussein, A.S. and Kratzer, F.H., 1982. Effect of rancidity on the feeding value of rice bran for chickens. Poult. Sci. 61: 2450-2455.

King, R.H. and Taverner, M.R., 1975. Prediction of the digestible energy in pig diets from analyses of fibre contents. Anim. Prod. 27: 275-284.

Luh, B.S., 1980. Rice in its temporal and spatial perspectives. In: B.S. Lutz, ed. Rice: Production and Utilization. AVI Publishing Company, West Point, Conneticut, pp. 1-74.

Nelson, T.D., Ferrara, L.W. and Storer, N.L., 1968. Phytate phosphorous content of feed ingredients derived from plants. Poult. Sci. 47: 1372-1376.

National Research Council, 1979. Nutrient Requirements of Domestic Animals, No. 2. Nutrient Requirements of Swine, 8th ed. National Academy of Sciences, Washington, D.C.

Robles, A. and Ewan, R.C., 1982. Utilization of rice and rice bran by young pigs. J. Anim. Sci. 55: 572-577.

Roese, G.J., 1977. Rice pollard proves its worth in pig growing rations. Rice Mill News (March): 31-32.

Tagendjaja, B. and Lowry, J.B., 1985. Improved utilization of rice hulls: A rapid method of estimating hull content. Ilmu Dan Peternakan 1: 323-327.

Tangendjaja., B., Johnson, Z.B. and Noland, P.R., 1988a. Effect of cooking and addition of enzymes on feeding value of rice bran for swine. Nutr. Rep. Int. 37: 449-458.

Tangendjaja., B., Johnson, Z.B. and Noland, P.R., 1988b. Effect of cooking, ensiling, water treatments, extrusion and pelleting of rice bran as feed ingredients for pigs. Nutr. Rep. Int. 37: 939-949.

Taverner, M.R., 1975. Sweet lupin seed meal as a protein source for pigs. Anim. Prod. 20: 413-419.

Tillman, A.D., Kidwell, J.F. and Singletary, C.B., 1951. The value of solvent extracted rice bran in the rations of growing-fattening swine. J. Anim. Sci. 10: 837-840.

Warren, B.E., 1986. A nutritional evaluation of defatted and fullfat Australian rice bran. Ph.D. Thesis. University of New England, Armidale, N.S.W.

Warren, B.E. and Farrell, D.J., 1989a. The nutritive value of full-fat and defatted Australian rice bran. 1. Chemical composition. Anim. Feed Sci. Technol. 27: 219-228.

Warren, B.E. and Farrell, D.J., 1989b. The nutritive value of full-fat and defatted rice bran. IV. Egg production on diets with defatted rice bran. Anim. Feed Sci. Technol. 27: 259-268.

Warren, B.E., Gerdes, R.G. and Farrell, D.J., 1981. A preliminary study of defatted rice bran in the diets of pigs and rats. In: D.J. Farrell, ed. Recent Advances in Animal Nutrition in Australia 1987. University of New England, Armidale, N.S.W.

Warren, B.E., Hume, I.D. and Farrell, D.J., 1990. The nutritive value of full-fat and defatted rice bran. V. The apparent retention of minerals and apparent digestibiltiy of amino acids in chickens and adult cockerels. Anim. Feed Sci. Technol. (In press).

CHAPTER 35

Rubber Seeds, Oil and Meal

E. Nwokolo

INTRODUCTION

Continued protein shortages for human and animal nutrition have forced many tropical countries to look seriously at local protein sources with a view to highlighting those which have a potential for future development. Interest in the use of rubber seeds as a livestock feed was kindled in the 1960s and 1970s by researchers in Southeast Asia (Gick et al., 1967; Le Thouc, 1968; Buvanendran and Siriwardene, 1970; Siriwardene and Nugara, 1972) and elsewhere (Orok and Bowland, 1974; Fetuga et al., 1977). In spite of these investigations, rubber seed meal is still not used to any appreciable extent in livestock feeds. Tonnes of rubber seeds are shed annually and germinate on plantation floors. The resulting trees are usually just slashed down and provide little or nothing in the way of animal feed. However, based on its nutrient analyses, it would appear that defatted rubber seed meal has considerable potential for use as a protein supplement in diets fed to both monogastric and ruminant animals.

PRODUCTION OF RUBBER SEED

The para rubber tree (*Hevea brasiliensis*) is a perennial crop, generally cultivated for its industrially important latex which is used in the production of natural rubber. It is estimated that, on a worldwide basis, over 12 million hectares are planted with rubber trees. Malaysia, Indonesia and Thailand have about 5.0, 3.0, and 1.3 million hectares, respectively, planted with rubber trees (FAO, 1985), and these, together with large hectarages in other Southeast Asian countries, account for over 80% of world rubber production.

In addition to its latex, a rubber tree produces hundreds of fruits yearly. Each dehiscent fruit bears four seeds which are dispersed on drying by an explosive mechanism. It is estimated that each rubber tree produces about 5 kg of seeds per annum (Bressani et al., 1983) and that one hectare of rubber trees in a plantation may yield about 1000 kg of fresh seeds (Le Thuoc, 1968).

Rubber seeds can be harvested by picking them up off the plantation floor or mechanically raking them up in situations where seed concentration is moderate to high. The yield of kernels from fresh rubber seeds ranges from 57 to 63%, with a mean of 60.5%. Therefore, 100,000 hectares of rubber plantation possess the potential to produce about 100,000 tonnes of fresh rubber seeds, on an annual basis, and these can be dehulled to produce about 60,000 tonnes of fresh kernels. Fresh kernels have a relatively high moisture content (45%). Consequently, 60,000 tonnes of fresh rubber seed kernels will dry to about 33,000 tonnes of kernels, which may be milled and utilized as whole or full-fat rubber seed meal. Alternatively, the dried kernels may be milled and defatted by a number of methods (mechanically-pressed, solvent-extracted or pre-press solvent-extracted).

NUTRIENT CONTENT OF RUBBER SEEDS AND MEAL

Rubber seeds have to be dehulled before they can be utilized as animal feed and are usually dehulled immediately after they are collected from the field. Although dehulled fresh rubber seeds have a high moisture content, this decreases rapidly in storage. Seeds stored for a month before dehulling have a moisture content of about 30%. Full-fat rubber seed meal has a dry matter content of 84.5 to 94.2% (Table 35.1).

Table 35.1. Proximate Composition of Various Dehulled Rubber Seed Meals (% DM)

	Full-Fat[1]	Solvent Extracted[1]	Mechanically Pressed[2]	Defatted[2]
Dry Matter	84.5	91.4	90.4	91.2
Crude Protein	22.5	36.5	28.8	38.7
Crude Fiber	3.8	4.4	12.2	13.9
Ether Extract	49.5	8.5	18.6	1.0
Ash	3.5	5.3	8.4	8.4
NFE	27.7	45.3	32.0	38.0

[1]Fetuga et al. 1977.
[2]Nwokolo et al. 1987.

With fat extraction, there is an increase in the content of various nutrients. The protein content of rubber seed meal has been shown to vary from 22.5% in whole rubber seed meal to 38.7% in meals that have been exhaustively solvent-extracted (Nwokolo et al., 1987). Crude fiber content is low in the whole meal, increasing to 12.2 and 13.9% in pressed and defatted meals, respectively. Similarly, ash content increases from 3.5% in the whole meal, to 8.4% in the defatted meal. Digestible energy values of full-fat and defatted meals for swine are estimated at 5560 kcal/kg and 3900 kcal/kg, respectively. Detailed chemical composition of full-fat rubber seeds and defatted rubber seed meals was reported by Fetuga et al. (1977), from an analysis of dehulled Nigerian rubber seeds (Table 35.1).

The content of most amino acids is about 20% lower in defatted rubber seed meal than in defatted soybean meal (Table 35.2). Like many oilseed meals, defatted rubber seed meal has a low content of tyrosine, threonine, isoleucine, phenylalanine and lysine and a very low content of cystine and methionine. The availabilities of amino acids in rubber seed meal reported herein have been determined with young broiler chicks, not growing pigs. However, in the absence of amino acid availability data with pigs, those determined with chicks provide the best estimates for swine. Average availability of amino acids in solvent-extracted rubber seed meal is between 75 and 86%, with a mean availability of 80.3%. Histidine is poorly available, while threonine, cystine, lysine and isoleucine are moderately available in rubber seed meal. In spite of this, defatted rubber seed meal is a satisfactory source of dietary amino acids, when compared to other vegetable protein sources like sunflower seed, cottonseed or rapeseed meal.

Table 35.2. Amino Acid Content (%) and Availability (%) in Rubber Seed Meal[1]

	Amino Acid Content		Amino Acid Availability[2]	
	Defatted RSM	Defatted SBM	Defatted RSM	Defatted SBM
Arginine	4.31	4.08	88.9	95.5
Histidine	1.78	1.58	59.5	86.2
Isoleucine	1.38	2.12	75.3	97.6
Leucine	2.69	3.71	85.5	97.4
Lysine	1.66	3.25	66.2	96.6
Methionine	0.64	0.68	83.3	96.9
Cystine	0.34	0.46	68.4	70.1
Phenylalanine	1.52	2.23	89.3	96.9
Threonine	1.30	1.83	77.6	90.1
Valine	2.57	2.17	90.1	97.1

[1]Nwokolo et al., 1987.
[2]Amino acid availabilities were determined with broiler chicks not pigs.

Dried rubber seed kernels contain high levels of fat, estimated at between 48.3 and 49.5% (Nwokolo et al., 1987; Fetuga et al., 1977). Rubber seed oil is a semi-drying oil. Rubber seed lipids are highly unsaturated, the total unsaturated fatty acids comprising about 79% of total lipids, while the polyunsaturated fatty acids comprise about 52% of total lipids. The ratio of polyunsaturated to saturated fatty acids (P/S ratio) is about 2.25, similar to the P/S ratio of corn oil, soybean oil or sunflower oil. Linolenic acid is high, much higher than in most edible oils, while the content of linoleic acid is lower in rubber seed oil than in most edible oils. In general, there is considerable similarity in physicochemical properties between rubber seed oil and other common vegetable oils (Table 35.3).

Table 35.3. Fatty Acid Profile (% of total fatty acids) and Physicochemical Properties of Rubber Seed Oil

	Rubber Seed Oil	Soybean Oil
Fatty Acid Profile		
14:0	0.08	0.11
16:0	9.27	13.07
16:1	0.14	0.14
18:0	10.58	5.53
18:1	26.64	28.16
18:2	34.92	44.44
18:3	17.27	6.45
20:0	0.57	0.49
20:1	0.18	0.19
22:0	0.15	0.51
24:0	0.12	0.27
Physicochemical Properties		
Refractive Index (25°C)	1.41	1.47
Iodine Value (25°C)	138.0	100.2
Saponification Value	192.0	190.0
Unsaponified Matter (%)	4.8	0.9
Saponified Matter (%)	95.2	99.1
Total Satatürated (%)	20.79	20.01
Total Unsaturated (%)	79.15	79.38
Total Essential Fatty Acids (%)	52.20	50.89

Nwokolo (unpublished data).

Comparative mineral analyses of completely defatted rubber seed meal and soybean meal indicate that rubber seed meal has about twice the phosphorus and magnesium content, one and one-half times the potassium content and the same content of calcium and sodium as soybean meal (Table

35.4). Content of zinc and iron is three to four times higher in solvent-extracted rubber seed meal than in soybean meal, while copper is twice as high. All meals have a high content of potassium. While the mineral content of the soil may affect the concentration of many minerals in rubber seed meal (Nwokolo et al., 1987), it is clear that, in general, rubber seed meals have a higher content of most minerals than other common protein concentrates.

Table 35.4. Mineral Composition of Various Rubber Seed Meals

	Full-Fat[1]	Solvent Extracted[1]	Mechanically Pressed[2]	Defatted[2]
Calcium (%)	0.48	0.88	0.20	0.27
Chloride (%)	0.07	0.18	----	----
Magnesium (%)	0.28	0.34	0.35	0.20
Phosphorus (%)	0.64	0.94	0.67	0.52
Potassium (%)	0.96	1.54	1.38	1.34
Sodium (%)	0.09	0.21	0.55	0.76
Iron (mg/kg)	93	147	225	86
Manganese (mg/kg)	23	25	42	35
Zinc (mg/kg)	78	112	102	36
Copper (mg/kg)	25	32	31	16

[1]Fetuga et al., 1977.
[2]Nwokolo et al., 1987.

UNDESIRABLE CONSTITUENTS OF RUBBER SEEDS AND MEAL

A major problem associated with the use of fresh rubber seeds in animal nutrition is the presence in the kernel of a cyanogenic glycoside, linamarin, similar to that in cassava and lima bean. Linamarin by itself is not toxic, but is decomposed to toxic hydrocyanic acid by endogenous linamarinase. The hydrocyanic acid content of fresh rubber seeds has been reported to be as high as 200 mg/100 g (Gick et al., 1967), although the content of cyanide decreases rapidly with storage (Nwokolo and Akpapunam, 1986). Stosic and Kaykay (1981) noted that HCN toxicity was a problem of immense physiological and nutritional importance because when animals ingest small quantities of HCN on a regular basis, quantities ingested may not be large enough to cause mortality, but may be sufficient to affect the general health and productivity of such animals. These workers suggest that while ruminant animals may be able to detoxify rubber seeds in the rumen, this is not the case with monogastric animals.

It is essential to detoxify rubber seeds before they are fed to swine. Milling of rubber seeds followed by fermentation leads to initial decomposition of linamarin to hydrocyanic acid and subsequent breakdown of the hydrocyanic acid to innocuous substances. In Liberia, rubber seeds are routinely detoxified by boiling and soaking in hot water for 12 hours, or roasting at 350°C for 15 minutes (Stosic and Kaykay, 1981), procedures which are reported to be quite effective in detoxifying rubber seeds for poultry and swine feeding. Heat generated by the mechanical pressing of rubber seeds or utilized in solvent-extraction also reduces the HCN content of the resultant rubber seed meal. Ong and Radem (1981) concluded that there was no danger of cyanide poisoning for pigs resulting from the consumption of feed containing reasonable levels of rubber seed meal (up to 30% of the diet). They further suggested that supplementation with methionine and iodine would enable the pig to detoxify the relatively small amount of hydrocyanic acid in rubber seed meal diets.

Rubber seed meal has recently been reported to cause hyperporphyrinurea and a slight decrease in the size of the reproductive organs in rats (Babatunde and Pond, 1987). This has been attributed to the presence of gossypol. The total and free gossypol content of 28 samples of rubber seed meal averaged 538 and 270 mg/kg for solvent-extracted meal and 632 and 338 mg/kg for mechanically-pressed meal, respectively (Abdullah and Hutagalung, 1981).

FEEDING RUBBER SEED MEAL

Mechanically-pressed rubber seed meals are usually available in most tropical countries because they are easier to produce than solvent-extracted meals, which generally require more elaborate and costly equipment. Pressed cakes have a lower protein content (28-30%) and higher residual oil content (18-20%) than completely defatted rubber seed meal. However, these cakes are more prone to rancidity and spoilage in storage.

Only a few experiments have been reported in which rubber seeds or rubber seed meal have been fed to swine. In spite of the paucity of experimental data, there is considerable evidence that rubber seed meal can contribute quite significantly to swine nutrition. Rajaguru and Ravindran (1979) fed weanling pigs (17.5-19.1 kg) diets containing 0, 10, 20 or 30% rubber seed meal for 8 weeks. Body weight gain was depressed and feed conversion ratio elevated by increasing levels of dietary rubber seed meal, weight gains decreasing from 18.1 to 8.4 kg as level of rubber seed meal in the diet rose from 0 to 30%. Similarly, feed efficiency decreased from 3.85 to 8.37 kg feed/kg gain as the level of rubber seed meal increased. However, these workers suggested that the poor performance of pigs fed rubber seed meal was due to a deficiency in dietary lysine and methionine, rather than to the presence of anti-nutritional factors.

Ong and Radem (1981) fed four groups of ten pigs one of four experimental diets containing 0, 10, 20 or 30% rubber seed meal (Table 35.5). The grower diets were isonitrogenous (17.5% CP), as were the fattening diets (15% CP). Weight gain was similar on all diets (420-440 g/day). Feed intake was similar on all diets (1.53-1.57 kg/day) indicating that even with high rubber seed meal diets, palatability was not a problem. Dressing percentage, carcass length, backfat and leaf fat were similar on all diets. On the basis of daily weight gain, it was suggested that rubber seed meal could be fed at levels of up to 30% in pig diets (Table 35.5).

Table 35.5. Influence of Rubber Seed Meal Diets on Growth Rate and Carcass Characteristics of Pigs

	Level of Rubber Seed Meal (%)			
	0	10	20	30
Daily Intake (kg)	1.53	1.51	1.55	1.57
Daily Gain (kg)	0.43	0.43	0.44	0.42
Feed Efficiency	3.59	3.62	3.57	3.75
Dressing Percentage	67.9	67.7	66.4	68.0
Carcass Length (cm)	76.7	76.0	74.5	76.0
Eye Muscle Area (cm^2)	33.0	33.3	32.5	33.7
Backfat Thickness (cm)	2.0	1.9	2.2	2.1
Leaf Fat (g)	875	841	1068	1000

Ong and Radem, 1981.

In Liberia, Stosic and Kaykay (1981) reported feeding experiments with pigs in which rubber seeds, either roasted or soaked in hot ash solution, were incorporated at 40% into pig diets. The control diet was a locally available commercial pig diet. The experiment lasted 6 weeks. Results indicated that pigs fed rubber seed diets performed better than those fed the commercial control diet. Roasted rubber seeds sustained the highest final live weight (38.25 kg), total live weight gain (19.02 kg), daily live weight gain (475 g) and lowest feed/gain ratio (2.90).

SUMMARY

Rubber seed meal is a good source of protein, capable of contributing significant amounts of essential amino acids to swine diets. It is suggested that rubber seed meal can be included in starter diets to a level supplying 30% of the dietary protein. When fed to grower and finishing pigs, the amount of rubber seed meal can be increased to provide up to 50% of the dietary protein. Fifty percent of the protein of sow diets may also be supplied by defatted rubber seed meal. When feeding rubber seed-based diets, supplementation with lysine and methionine is recommended in order to enhance productivity, especially when rubber seed meal is fed to young pigs. Toasting is important to improve the nutritional quality of rubber seed meal.

REFERENCES

Abdullah, A.S. and Hutagalung, R.I., 1981. Gossypol in rubber seed meal. Pertanika 4: 96-98.

Babatunde, G. and Pond, W.G., 1987. Nutritive value of rubber seed (*Hevea brasiliensis*) meal and oil. 1. Rubber seed meal versus soybean meal as sources of protein in semipurified diets for rats. Nutr. Rep. Int. 36: 617-630.

Bressani, R., Elias, I.G., Ayuso, T., Rosal, O., Braham, J.E. and Zuniga, J., 1983. Nutritional value of protein and oil in rubber seed. Turrialba 33: 61-66.

Buvanendran, V. and Siriwardene, J.A., 1970. Rubber seed meal in poultry diets. Ceylon Vet. J. 18: 33-38.

FAO, 1985. Production Yearbook. Food and Agriculture Organization of the United Nations, Rome.

Fetuga, B.L., Ayeni, T.O., Olaniyan, A., Balogun, M.A., Babatunde, G.M. and Oyenuga, V.A., 1977. Biological evaluation of para rubber seeds (*Hevea brasiliensis*). Nutr. Rep. Int. 15: 497-510.

Gick, L.T., Samsudin, M.D., Husaini, B.S. and Tarwotjo, I., 1967. Nutritional value of rubber seed protein. Amer. J. Clin. Nutr. 20: 300-303.

Le Thuoc, 1968. Research on the Use of Rubber Seed Kernels as a Protein Supplement in Livestock Feed. Publication of the Ministry of Land Reforms, Agriculture, Fisheries and Animal Husbandry Development, Vietnam.

Nwokolo, E. and Akpapunam, M., 1986. Content and availability of nutrients in rubber seed meal. Trop. Sci. 26: 83-88

Nwokolo, E., Bragg, D.B. and Sim, J., 1987. Nutritional assessment of rubber seed meal with broiler chicks. Trop. Sci. 27: 195-204.

Ong, H.K. and Radem, J., 1981. Effect of feeding rubber seed meal-based diets on performance and serum thiocyanate level of growing-finishing pigs. Mardi Research Bulletin 9: 78-82.

Orok, E.J. and Bowland, J.P., 1974. Nigerian para rubber seed meal as an energy source for rats fed soybean meal or peanut meal-supplemented diets. Can J. Anim. Sci. 54: 239-246.

Rajaguru, A.S.B. and Ravindran, V., 1979. Rubber seed meal as a protein supplement in growing swine rations. J. Natl. Sci. Council (Sri Lanka) 7: 101-104.

Siriwardene, J.A. and Nugara, D., 1972. Metabolizable energy of rubber seed meal in poultry diets. Ceylon Vet. J. 20: 61-63.

Stosic, D.D. and Kaykay, J.M., 1981. Rubber seeds as animal feed in Liberia. World Anim. Rev. 39: 29-39.

CHAPTER 36

Rye

R.B. Bazylo

INTRODUCTION

Rye (*Secale cereale*) is most commonly grown as a source of flour for use in bread making and it ranks second to wheat for this purpose (Bushuk, 1976). In addition, some rye is grown for the production of rye whisky, while the remainder finds its way into the livestock feed market. Unfortunately, its market potential is limited due to the perception that rye contains toxic factors which reduce its nutritive value. Although some of the reasons for this discrimination are valid, most are unfounded. Therefore, the potential to use rye in swine diets may be greater than has been previously realized.

GROWING RYE

Rye ranks seventh in world wide production relative to other cereal grains (FAO, 1986). The 10 leading rye producing countries are listed in Table 36.1. Poland, the second largest producer of rye, is one of the few countries where rye acreage still exceeds that of wheat (Carmichael and Norman, 1970).

Rye is a cool weather plant. Since it can be sown in the fall, rye takes advantage of early spring moisture and can be harvested earlier than spring sown crops, helping to distribute farm work more evenly. Rye can also be grown on poorer soils than any other grain crop, although it is more tolerant of dry soils than of wet, poorly drained soils (Briggle, 1959). In fact, rye is the most productive of all of the commonly grown cereal grains under conditions of low temperature, low fertility and drought.

Table 36.1. Rye Production by Country

Country	1000 Metric Tonnes	% of Total
U.S.S.R.	15,000	47.2
Poland	7,179	22.6
West Germany	2,403	7.6
East Germany	1,818	5.7
China	1,000	3.1
Canada	670	2.1
Czechoslovakia	547	1.7
Denmark	546	1.7
United States	495	1.6
Turkey	350	1.1
World	31,791	94.4

FAO, 1986.

The total land area used for rye production in the world has declined from an average of 27.8 million hectares in 1961 to 15.5 million hectares in 1986 (FAO, 1986). However, the marked decline in the area of land under rye cultivation has been offset by yield increases and world wide production of rye has dropped only 6% from 33.8 million metric tonnes to 31.8 million metric tonnes. This substantial increase in yield has been achieved through improvements in agronomic practices, especially in the use of chemical fertilizers and crop rotation and through the use of improved varieties and elimination of the use of infertile land (Bushuk, 1976). Yields of 3500 kg/ha are not uncommon.

NUTRIENT CONTENT OF RYE

The most attractive feature of rye as an ingredient for use in swine diets is its relatively high level of digestible energy (3285 kcal/kg). The low fat content of rye (1.5%) is offset by its low crude fiber content (2.2%). Protein contents ranging between 6.5 and 14.5% have been obtained under North American conditions of cultivation, where nitrogen fertilizer applications are generally higher (Kent-Jones and Amos, 1967). The average crude protein (12.0%) falls between that of wheat and barley and is significantly higher than corn (8.5%). The chemical composition of rye is compared to other common feedstuffs in Table 36.2.

Table 36.2. Proximate Analysis of Rye and Other Commonly Used Feedstuffs

	Rye	Barley	Wheat	Corn
Dry Matter (%)	87.0	89.0	88.0	88.0
Protein (%)	12.0	11.5	12.6	8.5
Crude Fiber (%)	2.2	5.0	2.6	2.3
Ether Extract (%)	1.5	1.7	1.6	3.6
D.E. (kcal/kg)	3,285	3,120	3,402	3,530

National Research Council, 1988.

The amino acid composition of rye compares favorably with other cereals (Table 36.3). Rye differs from wheat, barley and most other cereals in having a higher proportion of water and salt soluble proteins, both of which have an improved content of the essential amino-acid, lysine. However, as with most other cereal proteins, lysine and threonine are still the first and second limiting amino acids in rye (Sauer, 1976). Unfortunately, it appears that some factor or factors in rye reduces the availability of these amino acids and the apparent ileal digestibility of lysine (65%) and threonine (59%) are lower than with other cereals (Heartland Lysine, 1988). Taverner (1986) reported that all amino acids in rye were less digestible than those of wheat or triticale.

Table 36.3. Amino Acid Composition of Cereal Grains (%)

	Rye	Barley	Wheat	Corn
Arginine	0.51	0.52	0.65	0.43
Histidine	0.24	0.24	0.30	0.27
Isoleucine	0.46	0.46	0.53	0.35
Leucine	0.67	0.75	0.87	1.19
Lysine	0.41	0.40	0.40	0.25
Methionine	0.17	0.16	0.22	0.18
Cystine	0.19	0.21	0.30	0.22
Phenylalanine	0.58	0.58	0.71	0.46
Threonine	0.35	0.36	0.37	0.36
Tryptophan	0.11	0.15	0.17	0.09
Valine	0.56	0.57	0.58	0.48

National Research Council, 1988.

The vitamin and mineral content of rye is shown in Table 36.4. The vitamin composition of rye compares favorably with that of the other cereal grains although the level of niacin in rye is lower than that found in other cereals. In addition, the calcium content is very low. However, since the cost of vitamin and mineral supplementation is relatively small when compared to other constituents of the diet, their composition in cereal grains is not of great concern.

Table 36.4. Vitamin and Mineral Composition of Rye

Vitamins		Minerals	
Biotin (μg/kg)	60	Calcium (%)	0.06
Choline (mg/kg)	419	Copper (mg/kg)	7.6
Niacin (mg/kg)	14	Iron (mg/kg)	63
Pantothenic Acid (mg/kg)	7.5	Magnesium (%)	0.11
Riboflavin (mg/kg)	1.7	Manganese (mg/kg)	58
Thiamine (mg/kg)	4.1	Phosphorus (%)	0.32
Vitamin E (mg/kg)	14.5	Potassium (%)	0.45
Folacin (mg/kg)	0.6	Selenium (mg/kg)	0.38
Vitamin B_6 (mg/kg)	2.6	Sodium (%)	0.02
		Zinc (mg/kg)	28

National Research Council, 1988.

UNDESIRABLE CONSTITUENTS IN RYE

Rye is susceptible to contamination and infection from several undesirable fungi, the most important of which is ergot. Ergot may contain 10 or more toxic alkaloids including ergotamine, ergotoxine and ergonovine. These alkaloids are highly toxic to livestock and will produce a variety of symptoms including convulsions, muscular incoordination, lameness, difficult breathing and gangrene if extreme (Burfening, 1973). When fed to breeding stock, a high level of ergot can cause abortions, restriction of mammary growth, agalatia and a poor conception rate (Norskog and Clark, 1945). If fed to growing pigs, ergot will reduce feed intake, slow growth and impair feed conversion efficiency (Friend and MacIntyre, 1970). At high enough levels, it can even cause death.

Although ergot has been shown to reduce the performance of pigs fed rye, it is important to realize that ergot is a fungus contaminating the rye and not inherant to the rye itself. Ergot is usually most prevalent in seasons when moisture is readily available at the soil surface during spring and early summer and when showery weather prevails during flowering. Producers should take particular care to watch for ergot during a wet year. Rye fed to livestock should not contain more than 0.1% ergot. If the grain supply contains more than this, it should be diluted with clean grain.

Recent research has confirmed that rye contains high levels of both insoluble and soluble pentosans (Batterham et al., 1988). Pentosans are a normal constituent of the cell wall and are related to hemicellulose but are more soluble. They are large polymers consisting of a long chain of xylose units with frequent side branches of arabinose and occasional side chains of several other sugars and uronic acids. They result in a highly viscous intestinal fluid that may interfere with the digestion process (Antoniou and Marquardt, 1981). Although pentosans are believed to be a major contributor to decreasing the availability of all nutrients, especially amino acids and to a lesser extent fat, other factors may also be involved. Pig performance has been shown to be improved by supplementing rye-based diets with a crude pentosanase enzyme (Thacker et al., 1987).

Another factor found in rye which has been reported to reduce performance is alkyl resorcinol (Wieringa, 1967). The particular compound found in rye is believed to be fat soluble and to consist of a mixture of 5-n-alkyl resorcinol with odd-numbered side chains of 15 to 23 carbon atoms and a smaller amount of 5-n-alkenyl resorcinol (Pond and Maner, 1984). Table 36.5 shows that rye can contain more than twice as much alkyl resorcinol as wheat. However, the most recent work tends to discount the importance of alkyl resorcinols as a factor contributing to the poor performance of pigs fed rye.

Table 36.5. Relative Amounts of Alkyl Resorcinols in Common Cereals

	Number of Varieties	Alkyl Resorcinol Units (Range)
Rye	15	161 (129-192)
Wheat	18	69 (56-79)
Triticale	19	97 (78-124)

Bushuk, 1976.

Other anti-nutritional factors that appear higher in rye as compared to other cereals are trypsin and chymotrypsin inhibitors (Batterham et al., 1988) as well as nitrates. Recent analyses of Australian grains found the following relative values of trypsin inhibitors: triticale (18 samples) 100%, wheat (4 samples) 83% and rye (2 samples) 202% (Taverner 1986). Trypsin inhibitors in triticale have been shown to reduce pig performance (Erickson et al., 1979) so this may also hold true for pigs fed rye.

FEEDING RYE TO PIGS

Starter Pigs

Feeding of rye as a replacement for wheat or maize, for starter diets, has shown varying results. Pond and Yen (1987) found no difference in performance when they replaced 40% of maize with rye. However, in another trial, Harrison et al. (1981b) found pigs fed rye/soybean diets had lower average daily gains (0.40 vs. 0.53 kg/day), lower feed intakes and poorer feed efficiency when compared with pigs fed a corn/soybean diet. Honeyfield et al. (1983) also reported a slight decrease in rate of gain (8-10%) when rye substituted for maize in diets for starter pigs, but reported no difference in feed efficiency. Bazylo (1979) reported a reduction in daily gain and feed intake when rye replaced wheat in diets fed to starter pigs (Table 36.6).

Table 36.6. Performance of Starter Pigs (13-23 kg) Fed Wheat, or Rye as a Complete Replacement for Wheat

	Wheat/Soybean	Rye/Soybean
Daily Gain (kg)	0.38	0.30
Daily Intake (kg)	0.95	0.82
Feed Efficiency	2.53	2.57

Bazylo, 1979.

The most consistent finding in these experiments is a depression in feed intake when rye was substituted for wheat or corn. Dustiness is often a problem when pigs are fed rye. Therefore, producers are advised to utilize a coarser grind (<4-5 mm screen) when processing rye for use in swine diets. Supplementation with low levels (0.5-2.0%) of fat or vegetable oil may also reduce the dustiness. However, since weanling pigs are often reluctant to eat solid feed anyway, a feedstuff of questionable palatability such as rye should be avoided in starter diets.

Growing Pigs

As with starter diets, there are conflicting reports concerning the feeding of rye to grower pigs. Harrison et al. (1981a) reported that the inclusion of 75 to 100% rye, substituted for corn, resulted in a lower average daily gain (Table 36.7). This is in agreement with Hale et al. (1967), who added 2.5% tallow to diets where rye completely substituted for maize. They found pigs grew 10 to 15% slower and required more feed, concluding rye is worth 78 to 80% as much as maize.

Table 36.7. Performance of Grower Pigs Fed a Corn/Soybean (SB) Diet with Rye Substituting for Corn at Various Levels

	Corn/SB	25% Rye	50% Rye	75% Rye	Rye/SB
Daily Gain (kg)	0.71	0.73	0.72	0.66	0.65
Daily Intake (kg)	2.02	1.86	1.84	1.78	1.77
Feed Efficiency	2.83	2.56	2.57	2.70	2.71

Harrison et al., 1981b.

In contrast to the above, Oliveira et al. (1985) fed 50% and 100% rye in substitution for maize and found pigs gained slightly better from 26 to 89 kg liveweight. Growth rates were 591, 628 and 653 g/day for pigs fed maize, 50% rye and 100% rye substituted diets, respectively. Danielson (1972) fed rye at levels up to 60% in replacement of corn for growing-finishing pigs without any decrease in growth rate or feed efficiency.

When rye replaces barley in grower rations, pig performance is quite similar. Experiments with concentrate rations containing 0, 40, 60, and 80% rye at Federal Research Stations (Canada Department of Agriculture, 1974; 1976) indicated that 60 to 80% ergot-free rye can be included in growing and finishing rations without a reduction in intake or gain. Bazylo (1979) reported that levels of up to 84% rye in the diet, as a replacement for barley, reduced feed intake and rate of gain whereas 56% rye in the diet resulted in no difference in performance. Bowland (1966) indicated that rate of gain was not affected by the addition of 25 or 50% rye in replacement for barley. Savage et al. (1978) also found cereal source did not significantly influence growth rate. Kracht et al. (1981) found only a slight decrease in feed intake (5%) and gain (7-10%) when rye was substituted for barley.

It appears from the above data that rye can be used for up to 80% of the grain portion of the diet for growing pigs when it replaces barley. If it replaces either corn or wheat, performance of grower pigs will likely not be as good due to energy differences of these cereals. However, if prices warrant, ergot-free rye can replace up to 50% of the wheat or corn.

Finishing Pigs

Table 36.8. Performance of Gilts (62-100 kg) and Barrows (70-100 kg) Fed Rye, Barley or 50% Barley/50% Rye Diets

	Barley	Barley/Rye	Rye
Barrows			
Daily Gain (kg)	0.87	0.93	0.96
Daily Intake (kg)	3.34	3.55	3.16
Feed Efficiency	3.83	3.81	3.30
Gilts			
Daily Gain (kg)	0.77	0.73	0.73
Daily Intake (kg)	2.86	2.76	2.64
Feed Efficiency	3.72	3.75	3.62

Bazylo, 1979.

Data from trials using rye in diets for finishing pigs is variable. Feed efficiency seems to improve by using rye versus barley but rate of gain is similar (Table 36.8). Horszczaruk et al. (1985) and Lavorel and Grosjean (1983) agree that rye can be used as the only cereal in finisher diets with little or no difference in pig performance. Harrison et al. (1981) and Kracht et al. (1981) found some decrease in performance when rye was used but in the case of the Harrison data, rye was used as a replacement for corn instead of barley. Therefore, it appears that ergot-free rye can be used as the only cereal for finishing pigs. Most data indicate carcass measurements are not influenced if rye is used in the finishing stages (Friend and MacIntyre, 1970; Bazylo, 1979).

Breeding Stock

Very little work has been conducted on the feeding of rye to gestating or lactating sows. If rye is included in the diet of gestating sows, it should be limited to 25% of the diet and care taken to ensure that it is absolutely ergot-free. There is some data supporting the fact the lactating sows fed rye tend to consume less feed. Therefore, rye should be excluded from their diets to ensure maximum feed intake of nursing sows.

SUMMARY

Economics will likely dictate whether or not rye will be used as a feedstuff in swine diets. If prices warrant, ergot-free rye grain can be used as the sole cereal source in finisher diets. Producers who use barley as the only cereal source in grower rations can substitute ergot-free rye in its place. If rye is going to substitute for wheat or corn, limit it to a 50% substitution but expect a reduction in performance. Rye should not be used in the starter stage or for lactating sows, as feed intake in both cases will likely drop. Limit the feeding of ergot-free rye to 25% of the diet for gestating sows.

REFERENCES

Antoniou, T. and Marquardt, R.R., 1981. Influence of rye pentosans on the growth of chicks. Poult. Sci. 60: 1898-1904.

Batterham, E.S., Saini, H.S. and Anderson, L.M., 1988. Nutritional value and carbohydrate content of rye and wheat for growing pigs. Nutr. Rep. Int. 38: 809-818.

Bazylo, R.B., 1979. The effect of rye in rations for growing pigs. M.Sc. Thesis, University of Manitoba, Winnipeg, 73 pp.

Bowland, J.P., 1966. Rye for market pigs. University of Alberta, 45th Annual Feeders Day Report 51: 1-3.

Briggle, L.W., 1959. Growing rye. USDA Farmers Bull. 2145: 16 pp.

Burfening, P.J., 1973. Ergotism. Amer. Vet. Med. Assoc. J. 163: 1288-1290.

Bushuk, W., 1976. Rye: Production, Chemistry and Technology. Amer. Assoc. Cereal Chem. Inc., St. Paul, Minnesota, 181 pp.

Canada Department of Agriculture, 1974. Research Branch Report: Charlottetown, P.E.I.

Canada Department of Agriculture, 1976. Growing fall rye for grain in the Atlantic provinces. Publ. No. 1578. Ottawa, Ontario.

Carmichael, J.S. and Norman, M.A., 1970. Rye production in Canada. Can. Farm Econ. 5: 17-21.

Danielson, D.M., 1972. Effect of different levels of rye for growing-finishing swine. J. Anim. Sci. 35: 1103.

Erickson, J.P., Miller, E.R., Elliott, F.C., Ku, P.K. and Vilroy, D.E., 1979. Nutritional evaluation of triticale in swine starter and grower diets. J. Anim. Sci. 48: 547-553.

FAO, 1986. Production Yearbook. Food and Agriculture Organization of the United Nations, Rome. Vol. 40: 306 pp.

Friend, D.W. and MacIntyre, T.M., 1970. Effect of rye ergot on growth and N-retention in growing pigs. Can. J. Comp. Med. 34: 198-202.

Hale, O.M., Johnson, J.C. and Southwell, B.L., 1967. University of Georgia Agricultural Experimental Station, Athens, Research Bulletin No. 14, p. 17.

Harrison, M.D., Copelin, J.L., Combs, G.E. and Olson, T.A., 1981a. The value of rye (Wesser) in growing-finishing swine diets. J. Anim. Sci. 52 (Suppl. 1): 237.

Harrison, M.D., Copelin, J.L. and Combs, G.E., 1981b. The effect of antibiotic supplementation on rye utilization by starting, growing and finishing swine. J. Anim. Sci. 53 (Suppl. 1): 245.

Heartland Lysine Inc., 1988. Apparent ileal digestibility of crude protein and essential amino acids in feedstuffs for swine. Chicago. (Chart)

Honeyfield, D.D., Froseth, J.A. and McGinnis, J., 1983. Comparative feeding value of rye for poultry and swine. Nutr. Rep. Int. 28: 1253-1260.

Horszczaruk, F., Batorski, M. and Kulisiewicz, J., 1985. Ground rye as an essential energy feed for fattening pigs. Builetyn Informacyiny Przemyslu Paszowego. 24: 21-26.

Kent-Jones, D.W. and Amos, A.J., 1967. Modern Cereal Chemistry. 6th ed. Food Trade Press Ltd., London, 730 pp.

Kracht, W., Otto, E., Matzke, W. and Ohle, H.O., 1981. Use of a discontinuous supply of high-protein concentrate in pig fattening. 3. Use of different types of cereal. Pig News and Information 4: 566 (Abstract).

Lavorel, O. and Grosjean, F., 1983. Comparative utilization of wheat and rye by bacon pigs during growing and finishing. Pig News and Information 4: 501 (Abstract).

National Research Council, 1988. Nutrient Requirements of Domestic Animals, No. 2. Nutrient Requirements of Swine. 9th ed. National Academy of Sciences, Washington, D.C.

Norskog, A.W. and Clark, R.T., 1945. Ergotism in pregnant sows, female rats, and guinea pigs. Amer. J. Vet. Res. 6: 107.

Oliveira, O.E.R., Azovedo, P.M.A. and Fernades, T.H., 1985. A note on the use of rye as a substitute for maize in a diet for growing pigs. World Rev. Anim. Prod. 21: 3-9.

Pond, W.G. and Maner, J.H., 1984. Swine Production and Nutrition. AVI Publishing Company, Wesport, Connecticut, 731 pp.

Pond, W.G. and Yen, J.T., 1987. Effect of supplemental carbadox, an antibiotic combination, or clinoptilolite on weight gain and organ weights of growing swine fed maize or rye as the grain sources. Nutr. Rep. Int. 35: 802-809.

Sauer, W.C., 1976. Factors affecting amino acid availabilities for cereal grains and their components for growing monogastric animals. Ph.D. Thesis. University of Manitoba, 245 pp.

Savage, G., Smith, W.C. and Pickles, J., 1978. The use of rye in the diet of the growing pig. Anim. Prod. 26: 398.

Taverner, M.R., 1986. The digestibility by pigs of amino acids in triticale, wheat and rye. Proc. Intern. Triticale Symp., Sydney. Publ. No. 24: 507-510.

Thacker, P.A., Campbell, G.L. and Grootwassink, J., 1987. The effect of enzymatic treatment of rye on its nutritive value for swine. University of Saskatchewan, Saskatoon, Department of Animal Science, Research Report, pp. 140-145.

Wieringa, G.W., 1967. On the occurrence of growth inhibiting substances in rye. Institute for Storage and Processing of Agricultural Produce, Wageningen, Publ. No. 156, 68 pp.

CHAPTER 37

Safflower Meal

C.S. Darroch

INTRODUCTION

Although soybean meal is the major protein supplement used in swine rations throughout the world, meal from locally grown plants can often replace all or part of the soybean meal in the diet. These alternate plant proteins can be used to improve the overall quality of dietary protein or to reduce the costs associated with feeding high priced protein supplements. Safflower (false or bastard saffron, dyer's saffron, kardiseed, kusum, suff) is a warm-temperature plant which is cultivated throughout tropical climates for its oil. Safflower seed meal, a by-product of the oil seed crushing industry, may have potential as an alternate source of protein for use in swine rations.

GROWING SAFFLOWER

Safflower (*Carthamus tinctorius* L.) belongs to the family *Asteraceae* of the group *Compositae* and represents one of the oldest oilseed crops in the world. Safflower originated in the tropical regions of Asia and Africa and was initially cultivated for its florets, which were used to produce an orange dye. Young plants were occasionally used as a vegetable or fodder. Investigation into the use of safflower as an oilseed crop began in the mid-1900s (Knowles, 1989) and it is now grown primarily for its edible oil.

The safflower plant is a branched, herbaceous, thistle-like annual with yellow flowers and the seed resembles that of the sunflower. It grows best in dry, arid climates on land with a high water table (Kneeland, 1966). In these regions, it is routinely used in crop rotations with other spring-planted grains. The plant is slow to start but grows rapidly in the hot summers in arid regions and is ready to harvest after 120 to 150 days (Knowles, 1989). Although safflower has a high water requirement, it can withstand periods of drought because of its large, deep taproot and this adaptation makes safflower a desirable crop for areas in which rainfall is seasonal. It has also been grown successfully under irrigation. However, in regions of high humidity, the safflower plant is very susceptible to diseases.

The safflower seed is an achene, about the size of a barley seed, and resembles a small sunflower seed. A thick, white hull protects the embryo which is covered by a thin seed coat. The seed is ready to harvest about 35 days after flowering at which time it has reached maximum dry weight, oil and germination percentage and minimum hull percentage (Salunkhe and Desai, 1986). Safflower seed harvested at this time has an average composition ranging from 33 to 45% hulls, 55 to 65% kernels, and 35 to 45% oil. There are varietal differences in seed composition, especially between commercial and experimental varieties (Table 37.1).

The world production of safflower seed in 1985 was 840,000 metric tonnes (FAO, 1985) but this represented only 0.9% of the world soybean production (Mielke, 1989). India and Mexico produce approximately three-quarters of the world's safflower and the United States, Africa, Australia and Spain represent the other major producers (FAO, 1985). Typical safflower seed yields in 1985 in North and Central America, South America, Asia and Australia averaged 847, 800, 571 and 830 kg/ha, respectively (Knowles, 1989).

Table 37.1. Composition of Safflower Seed (%)

	Hull	Kernel	Crude Protein	Crude Fiber	Ash	Oil
Commercial[1,2]	33-44	55-65	15-19	21-22	2.5	37-40
High Linoleic[1]	18-30	66-81	-----	-----	---	39-50
High Oleic[1,2]	54-56	43-44	17.2	35.5	2.3	22.6
High Steric[1,2]	48	51-52	20.0	31.8	2.7	27.8
Thin Hull[2]	24	76	21.1	11.2	3.3	47.2

[1]Applewhite, 1966.
[2]Guggolz et al., 1968.

NUTRIENT CONTENT OF SAFFLOWER SEED AND MEAL

Safflower seed meal and hull are the two major by-products of oilseed crushing, the hull being removed prior to pre-press solvent extraction to increase the efficiency of the crushing equipment. The high proportion of hull in relation to kernel lowers the market value of safflower seed when compared to other oilseeds and restricts the use of whole seed as an ingredient in animal diets.

The hulls consist of 57 to 62% crude fiber (Applewhite, 1966), 70% cellulose and 21% lignin (Vohra, 1989). Only 3% of the carbohydrates in the hull are available to monogastric animals. The breeding of thin-hulled varieties may improve the nutritional value of whole seeds. Hulls from these varieties have a lower fiber content and a higher content of protein and oil (Table 37.1). The hulls are utilized as a fiber source or are incorporated in different amounts with the meal to provide meals with protein levels ranging from 20 to 70%.

Table 37.2. Composition of Safflower Constituents and Meals (% DM)

	Crude Protein	Oil	Crude Fiber	Ash	D.E. (kcal/kg)
Whole Seed[1]	16.44	44.23	21.94	2.68	3814
Dehulled Seed[2]	26.90	61.50	1.60	3.20	----
Kernel[1]	22.38	65.15	2.74	3.13	----
Hull[1]	6.75	13.03	49.38	1.96	----
Expeller Presscake[1]	55.50	1.74	9.95	8.82	----
Partially Dehulled Meal[1]	45.06	1.76	17.33	7.47	3504
Desolventized Meal[3]	24.89	1.30	33.15	---	----
Defatted Meal[2]	24.87	1.20	35.00	4.20	2650
Defatted, Dehulled Meal[2]	66.93	1.12	4.12	7.90	3790
Defatted, Debittered Meal[2]	26.70	0.80	36.20	4.50	----
Defatted, Dehulled Debittered Meal[2]	69.92	0.90	4.80	8.10	----

[1]Betschart, 1975.
[2]Zazueta and Price, 1989.
[3]National Research Council, 1988.

The energy content of safflower meal also varies with the percentage of hulls in the meal and the type of processing used to remove the oil from the seed. Grinding tends to reduce the digestible energy (D.E.) content of whole seed but does not reduce the metabolizable energy (M.E.) value for swine, whereas mechanical (expeller or hydraulic) extraction of the oil reduces both D.E. and M.E. Dehulling prior to oil extraction has a large impact on the energy content of the meal, increasing both digestible and metabolizable energy. Despite a higher content of fiber, partially dehulled (commercial) safflower meal, when compared to SBM, has a similar M.E. content for swine at 2.9 Mcal/kg and a similar protein content (Bell, 1989).

Most commercial meals have protein levels between 20 and 40% (Knowles, 1989) with meal composition dependent on the type of processing and the proportion of hulls in the meal (Table 37.2). Safflower meal protein is a poor source of lysine, methionine and isoleucine for swine (Table 37.3). Both safflower and soybean meal proteins are poor sources of sulfur amino acids for swine and safflower protein has only 27% of the isoleucine found in protein from soybean meal.

Table 37.3. Amino Acid Content (g/16 g N) of Safflower (SFM) and Soybean (SBM) Meals

Amino Acid	SFM	SBM	SFM/SBM
Arginine	8.5	7.3	116
Histidine	2.2	2.5	88
Isoleucine	1.2	4.5	27
Leucine	5.3	7.7	69
Lysine	3.1	6.6	47
Methionine	1.5	1.2	125
Phenylalanine	4.5	4.8	94
Threonine	2.2	3.9	56
Tryptophan	1.2	1.5	80
Valine	4.5	4.6	98

National Research Council, 1988.

In assessing the nutritive value of oilseed proteins, Evans and Bandemer (1967) rated safflower protein at 50 and soybean protein at 96 relative to casein. The low level and availability of lysine was a major factor contributing to the overall poor nutritive value of the protein in safflower meal. However, supplementation of safflower meal with lysine and methionine increased the nutritive value of the protein. In a study with mice, Sarwar et al. (1973) estimated the apparent digestibility and biological value (BV) of the protein from safflower meal at 74.7 and 27%, respectively. The low BV was attributed to low levels of lysine and methionine.

In an attempt to increase the lysine content of safflower meal, Betschart (1975) produced a water soluble protein concentrate from safflower meal that contained approximately 60% more lysine than the meal protein. Although this would improve the nutritional value of the protein for swine, the protein concentrate would probably not be economical.

There are also varietal differences in the amino acid composition of safflower protein. Several experimental varieties, derived from central Asian safflower plants, produced seed 15% higher in lysine than commercial North American varieties (Palter et al., 1969). At the expense of oil content, protein and lysine levels in safflower seed can also be increased by applying high levels of nitrogen fertilizer to the soil (Salunkhe and Desai, 1986). Thus, the possibility exists to improve the nutritive value of safflower through plant breeding and agronomic practices, even though programs at the present time are directed at improving oil yield and quality.

Safflower meal is a rich plant source of iron, having approximately 3.5 times the level found in soybean meal (Table 37.4), but it does not contain selenium. Both meals have comparable levels of calcium and phosphorus.

Table 37.4. Mineral Content of Safflower (SFM) and Soybean (SBM) Meals[1]

Mineral	SFM	SBM	SFM/SBM
Calcium (%)	0.37	0.33	112
Chlorine (%)	nd[2]	0.04	0
Magnesium (%)	0.37	0.32	116
Phosphorus (%)	0.83	0.72	115
Potassium (%)	0.81	2.30	35
Sodium (%)	0.05	0.04	125
Sulphur (%)	0.14	0.47	30
Copper (mg/kg)	10.76	25.50	42
Iron (mg/kg)	540.00	155.50	347
Manganese (mg/kg)	19.78	34.00	58
Selenium (mg/kg)	----	0.11	0
Zinc (mg/kg)	44.60	57.80	77

[1]National Research Council, 1988.
[2]Not detected.

Safflower meal has a relatively poor vitamin profile when compared to other oilseed proteins. It does not contain any detectable levels of thiamine, pyridoxine or vitamin B_{12} and it is low in fat-soluble vitamins and choline (National Research Council, 1988). The meal is a good source of biotin (1.43 mg/kg), riboflavin and niacin when compared to soybean meal (Table 37.5).

Table 37.5. Water Soluble Vitamins in Safflower (SFM) and Soybean (SBM) Meals[1]

	SFM (mg/kg)	SBM (mg/kg)	SFM/SBM
Biotin	1.4	0.3	447
Choline	816	2609	31
Folacin	0.4	0.6	67
Niacin	11	28	39
Pantothenic Acid	37.3	16.3	229
Riboflavin	2.0	2.9	69
Thiamine	nd[2]	6.0	0
Pyridoxine	nd	6.0	0
Vitamin B_{12}	nd	nd	---

[1]National Research Council, 1988.
[2]Not detected.

UNDESIRABLE CONSTITUENTS IN SAFFLOWER

In addition to the high content of fiber and lignin in commercial safflower meals, the meals contain two phenolic glucosides; matairesinol-beta-glucoside which imparts a bitter flavor to the meal and 2-hydroxyarctiin-beta-glucoside which has cathartic properties (Lyon et al., 1979). These glucosides are present in commercial partially dehulled meal at levels of 0.39 and 1.62% of dry matter, respectively. Both glucosides are associated with the protein fraction of the meal but can be removed by a debittering process in which the meal is extracted with methanol 1:5 (w/v) at 50°C for 3 hours (Zazueta and Price, 1989). This process removes approximately 90% of the glucosides without changing the protein content of the meal (Table 37.2). The glucosides can also be removed by a water extraction procedure or by the addition of B-glucosidase (Lyon et al., 1979).

UTILIZATION OF SAFFLOWER MEAL

Whole safflower meal contains a high proportion of hulls and the lignin present in the hulls interferes with nutrient digestibility. In comparison to other oilseed meals, whole safflower meal is low in protein and lysine and high in fiber. In early studies reviewed by Kohler et al. (1966), swine performed poorly when fed rations containing whole meal. Compared to SBM, whole safflower meal has a feeding value for swine of 45 to 50% (Ensminger and Olentine, 1978). Therefore, whole safflower meal is not recommended for use as a protein supplement in swine rations. The lower content of fiber in partially dehulled safflower meal (Table 37.2) improves the nutritive value of the meal for swine and therefore the remaining discussion relates to the use of dehulled meal.

Starter Pigs

There are few reports in the literature on the feeding value of safflower meal for weaner pigs. Based on the high energy and lysine requirements of weaner pigs, their inability to digest fiber, the relatively high fiber and low energy and lysine content of safflower meal, the use of safflower meal in weaner diets is not recommended.

Growing-Finishing Pigs

Williams and Daniels (1973) compared partially dehulled safflower meal (SFM) to soybean meal (SBM) and fish meal (FM) in sorghum based diets for grower and finisher pigs. Daily gains were 35% lower and feed efficiency was inferior for Large White pigs fed a 16.6% SFM, sorghum-based diet in comparison with an isonitrogenous 15% SBM, sorghum-based diet (Table 37.6). Supplementation of the safflower meal diet with synthetic lysine up to 0.3% of the diet improved daily gain in grower and finisher pigs by 22 and

27%, respectively, but both daily gain and feed efficiency remained inferior to the SBM supplemented diet.

Table 37.6. Performance of Pigs (19-82 kg) Fed Sorghum Based Diets with or without Soybean Meal (SBM), Safflower Meal (SFM) and Synthetic Lysine (L)

Growth Parameter	15% SBM	16.6% SFM	16.6% SFM +0.2% L	16.6% SFM +0.3% L
Daily Gain (g/day)				
19 - 45 kg	494	322	372	413
45 - 82 kg	689	449	517	617
Feed Efficiency				
19 - 45 kg	2.55	3.49	3.31	3.18
45 - 82 kg	3.19	4.79	4.32	3.60

Williams and Daniels, 1973.

The replacement of safflower meal and synthetic lysine with fish meal, a naturally high lysine feedstuff, in sorghum-based diets improved daily gain and feed efficiency of both grower and finisher pigs (Table 37.7). The highest daily gains were in finisher pigs fed an 11% FM diet. Diets containing safflower meal plus 0.3% synthetic lysine were inferior to diets containing SBM or FM and combined SFM-FM diets. The response of grower and finisher pigs fed wheat-based diets (Williams and Daniels, 1973) and finisher pigs fed barley-based diets (Williams and O'Rourke, 1974) containing SBM, SFM, FM and synthetic lysine paralleled those of pigs fed sorghum based diets.

In Mexican studies reviewed by Maner (1973), partially dehulled safflower meal decreased weight gains and the efficiency of feed utilization of grower pigs when included in SBM-sorghum based diets and in diets containing cottonseed meal and SBM at levels exceeding 12.5% of the supplemental protein. In follow-up studies by other Mexican researchers (Maner, 1973), weight gains of finisher pigs and feed efficiency decreased in proportion to the level of safflower meal in the diet.

Based on the work of Williams and Daniels (1973), Williams and O'Rourke (1974) and the work reviewed by Maner (1973), safflower meal should not be used as the sole supplementary protein source for swine rations. Maner (1973) recommended that partially dehulled meal not exceed 7 to 8% of the diet for growing pigs and not exceed 10 to 12% of the diet for finisher pigs. However, if sufficient lysine is provided by other feed ingredients to meet the pig's requirement, safflower meal could probably be included in both grower and finisher diets at levels up to 12%.

Table 37.7. Performance of Pigs (23-82 kg) Fed Sorghum Based Diets with or without Soybean Meal (SBM), Safflower Meal (SFM), Fish Meal (FM) and Synthetic Lysine (L)

Growth Parameter	15% SBM	15.7% SFM 0.3% L	11% FM	9% FM 3% SFM	5% FM 9% SFM
Daily Gain (g/day)					
23 - 45 kg	579	323	550	451	416
45 - 82 kg	590	508	657	626	535
Feed Efficiency					
23 - 45 kg	2.53	4.02	2.52	2.99	3.32
45 - 82 kg	3.67	3.98	3.19	3.17	3.73

Williams and Daniels, 1973.

Breeding Swine

Bred gilts, sows and adult boars have lower requirements for energy and lysine when compared to grower and finisher pigs (National Research Council, 1988). Breeding swine are also kept on a restricted feeding regime to maintain proper body condition. One means of controlling their voluntary feed intake is to increase the fiber content of the diet. The inclusion of fiber can reduce the energy density of the diet without affecting overall quality. Safflower meal has a higher fiber content than soybean meal and thus may be used as a high fiber protein supplement in breeding rations to help limit nutrient intake. Since breeding swine have a lower requirement for lysine, performance would not be adversely affected by the relatively low lysine levels in safflower meal. However, safflower meal should be used at very low levels, if at all, in rations fed to lactating sows.

SUMMARY

Safflower meal is a minor oilseed crop and its use will depend on availability and the price of other protein supplements such as soybean meal. In areas where safflower is grown, safflower meal would be a relatively cheap protein supplement for swine. However, the meal must be dehulled prior to feeding.

Partially dehulled safflower meal can be used as a protein supplement in most swine rations. Safflower meal should not be used in weaner diets because of relatively low energy and lysine levels and a relatively high fiber content. In grower and finisher diets, safflower meal can replace up to 50% of the supplemental protein without reducing growth performance. If particular attention is paid to lysine, methionine and fiber levels, safflower meal can be included in grower and finisher diets at levels up to 12%. Safflower meal can be used at levels up to 15% in bred gilt, sow and adult boar diets, fed restrictively to control voluntary feed intake and help maintain body condition.

REFERENCES

Applewhite, T.H., 1966. The composition of safflower seed. J. Amer. Oil Chem. Soc. 43: 406-408.

Bell, J.M., 1989. Nutritional characteristics and protein uses of oilseed meals. In: G. Robbelen, R.K. Downey and A. Ashri, eds. Oil Crops of the World: Their Breeding and Utilization. McGraw-Hill Publishing Company, Toronto, pp. 192-207.

Betschart, A.A., 1975. Factors influencing the extractability of safflower protein (*Carthamus tinctorius* L.). J. Food Sci. 40: 1010-1013.

Ensminger, M.E. and Olentine, C.G., 1978. Protein supplements. In: Feeds and Nutrition. Ensminger Publishing Company, Clovis, California, 1417 pp.

Evans, R.J. and Bandemer, S.L., 1967. Nutritive values of some oilseed proteins. Cereal Chem. 44: 417-426.

FAO, 1985. Production Yearbook. Food and Agriculture Organization of the United Nations, Rome. Vol. 39.

Guggolz, J., Rubis, D.D., Herring, V.V., Palter, R. and Kohler, G.O., 1968. Composition of several types of safflower seed. J. Amer. Oil Chem. Soc. 45: 689-693.

Kneeland, J.A., 1966. The status of safflower. J. Amer. Oil Chem. Soc. 43: 403-405.

Knowles, P.F., 1989. Safflower. In: G. Robbelen, R.K. Downey and A. Ashri, eds. Oil Crops of the World: Their Breeding and Utilization. McGraw-Hill Publishing Company, Toronto, pp. 363-374.

Kohler, G.O., Kuzmicky, D.D., Palter, R., Guggolz, J. and Herring, V.V., 1966. Safflower meal. J. Amer. Oil Chem. Soc. 43: 413-415.

Lyon, C.K., Gumbmann, M.R., Betschart, A.A., Robbins, D.J. and Saunders, R.M., 1979. Removal of deleterious glucosides from safflower meal. J. Amer. Oil Chem. Soc. 56: 560-564.

Maner, J.H., 1973. Investigation of plants not currently used as major protein sources. In: Alternate Sources of Protein for Animal Production. National Academy of Sciences, Washington, D.C., pp. 96-98.

Mielke, T., 1989. World vegetable protein supply and demand. Proceedings of the World Congress. In: T.H. Applewhite, ed. Vegetable Protein Utilization in Human Foods and Animal Feedstuffs. Amer. Oil Chem. Soc., Singapore, pp. 1-9.

National Research Council, 1988. Nutrient Requirements of Domestic Animals. No. 2. Nutrient Requirements of Swine. 9th ed. National Academy of Sciences, Washington, D.C.

Palter, R., Kohler, G.O. and Knowles, P.F., 1969. Survey for a high lysine variety in the world collection of safflower. J. Agric. Food Chem. 17: 1298-1300.

Salunke, D.K. and Desai, B.B., 1986. Safflower. In: Postharvest Biotechnology of Oilseeds. CRC Press, Boca Raton, Florida, pp. 93-103.

Sarwar, G., Sosulski, F.W. and Bell, J.M., 1973. Nutritional evaluation of oilseed meals and protein isolates by mice. J. Inst. Can. Sci. Technol. Aliment. 6: 17-21.

Vohra, P., 1989. Carbohydrate and fiber content of oilseeds and their nutritional importance. In: G. Robbelen, R. K. Downey and A. Ashri, eds. Oil Crops of the World: Their Breeding and Utilization. McGraw-Hill Publishing Company, Toronto, pp. 208-225.

Williams, K.C. and Daniels, L.J., 1973. Decorticated safflower meal as a protein supplement for sorghum and wheat based pig diets. Aust. J. Exp. Agric. Anim. Husb. 13: 48-55.

Williams, K.C. and O'Rourke, P.K., 1974. Decorticated safflower meal as a protein supplement in diets fed either restrictively or ad libitum to barrow and gilt pigs over 45 kg liveweight. Aust. J. Exp. Agric. Anim. Husb. 14: 12-16.

Zazueta, A.J.S. and Price, R.L., 1989. Solubility and electrophoretic properties of processed safflower seed (*Carthamus tinctorius* L.) proteins. J. Agric. Food Chem. 37: 308-312.

CHAPTER 38

Salseed Meal

S.S. Negi

INTRODUCTION

Sal (*Shorea robusta*) is a large sub-deciduous tree, seldom completely leafless, which is found extensively in parts of northeastern and central India. The tree is extensively cultivated in the Indian subcontinent as a source of timber. Salseeds, though not very palatable, are sometimes eaten by the human population in times of food scarcity. The seeds, when crushed, yield 19 to 20% of an oil which is suitable for use as a substitute for cocoa butter in the manufacture of chocolates (CSIR 1972). The residual salseed cake, generally termed salseed meal, is diverted for utilization as a livestock feed.

PRODUCTION OF SALSEED MEAL

When mature, salseeds drop from the tree and this process can be assisted by threshing the branches with a stick. The seeds are collected by raking under the trees, each of which produces approximately 100 kg of seeds. The harvest is collected, cleared of extraneous matter and dried in the sun. Seed decortication yields over 70% kernels which are further processed by screwpress or solvent extraction to recover the edible oil. The potential production of salseed meal is estimated to exceed 5 million tons (ICAR,

1967), although estimates of actual production range from 0.2 million tons (CSIR 1972) to 0.7 million tons (Punj, 1988).

NUTRIENT CONTENT OF SALSEED MEAL

Salseed meal is comparable to the cereal grains in chemical composition, containing approximately 9% crude protein, 2.5% crude fiber, 3.3% ash and variable amounts of ether extract. However, it is the high nitrogen free extract fraction which has attracted the most attention and prompted suggestions that salseed may have potential for use as an energy source in swine rations, in place of maize or barley.

Unfortunately, the nitrogen free extract fraction in salseed meal has a low proportion of starch and is associated with a high proportion of lignin and tannins. As a consequence, the energy content of salseed meal is actually fairly low. The metabolizable energy (M.E.) value of salseed meal has been reported as 2.65 Mcal/kg (Verma, 1970), based on a low level of incorporation in chick diets, and therefore, this value may be higher than actual (Sibbald et al., 1962; Kumar et al., 1967). Khan et al. (1986) obtained a value of only 1.86 Mcal M.E./kg of salseed meal in open circuit respiration calorimetry using goats. Typical data on the composition of salseed meal are presented in Table 38.1.

Table 38.1. Typical Composition of Salseed Meal

Constituent	% in DM	Constituent	% in DM
Crude Protein	7.8-9.7	Acid Detergent Fiber	21.6
Ether Extract	0.8-9.4	Cellulose	5.7
Crude Fiber	1.6-3.0	Lignin	11.9
NFE	74.6-84.2	Soluble Carbohydrates	57.8
Ash	3.1-3.6	Starch	30.1
Tannins	6.2-13.7		

Negi, 1982.

Salseed meal has a reasonable amino acid balance (Table 38.2). However, since it has a low protein content, it has little value as a protein supplement. The contents of calcium and phosphorus in the dry matter of salseed meal are low, in the range 0.18 to 0.29 and 0.16 to 0.18%, respectively (ICAR, 1967).

Table 38.2. Amino Acid Composition of Salseed Meal Protein

Amino Acid	% of Protein	Amino Acid	% of Protein
Arginine	6.8	Tyrosine	4.3
Histidine	4.8	Cysteine	2.2
Threonine	7.8	Lysine	6.7
Valine/Methionine	7.6	Phenylalanine/ Tryptophan	4.3
Norleucine/Leucine/ Isoleucine	12.0		

Mitra and Mishra, 1967.

UNDESIRABLE CONSTITUENTS IN SALSEED MEAL

The utilization of salseed as a monogastric feed is limited by the presence of tannins which adversely affect the utilization of other feed proteins. The average tannin content of salseed meal is 8% with reported values ranging from 3.5 to 13.3% (Negi, 1982). About two-thirds of the tannins are hydrolyzable (Vijjain and Katiyar, 1973; Singh and Arora, 1980). These constitute the esters of gallic and ellagic acid (Kumar, 1980) and may be removed by simple water washing (Panda et al., 1969). However, the residual condensed tannins are more harmful (McLeod, 1974). Attempts to detoxify salseed meal have not been successful (Negi, 1982; Sinha and Nath, 1983).

INCORPORATION OF SALSEED MEAL IN RATIONS OF PIGS

The invariable absence of a negative control in experimental designs has led to incongruous interpretation of results which mask the harmful effects of salseed meal as a feed ingredient. Negi (1985) has stressed the need of a negative control in experimental designs, particularly in the evaluation of a waste byproduct as a livestock feed, to safeguard against erroneous conclusions.

Starter Pigs

Salseed meal has a low palatability. It should be fed only as an ingredient of a mixed diet, and is more acceptable to adult than younger animals. As a consequence, it is recommended that salseed meal not be included in starter diets.

Grower Pigs

Das and Mohanty (1971) distributed 16 litter mates from 4 sows into 4 groups and assigned them to the diets shown in Table 38.3 when they reached 2.5 months of age (about 12 kg). The data on liveweight gains and digestibility of nutrients in the ration show the decreasing utilization of the diet by young pigs as the proportion of salseed meal in the diet increases.

Table 38.3. Growth of Young Pigs (12-80 kg) on Diets Containing Salseed Meal

	Level of Salseed Meal (%)			
	0	25	37.5	50
Animal Performance				
Liveweight at 8 Months (kg)	82.3	68.3	55.0	40.8
Average Daily Gain (kg/day)	0.43	0.41	0.33	0.25
Digestibility (%)				
Dry Matter	63	63	60	53
Organic Matter	74	66	64	54
Crude Protein	77	67	58	55
Ether Extract	73	53	60	43
Nitrogen Free Extract	78	74	71	65
Total Digestible Nutrients	78	60	58	52

Das and Mohanty, 1971.

Reddy (1972) randomly distributed 20 Yorkshire pigs of 8 to 12 weeks of age into four treatments. The treatments incorporated salseed meal at 0, 5, 10 and 15%, respectively, and replaced corresponding quantities of maize in the ration. Data on weight gains and feed efficiency of rations during a short experimental period are presented in Table 38.4. The feed efficiency of pigs fed salseed meal-containing rations was lower and pigs on these rations recorded lower weight gains than on the control ration, though the differences

Table 38.4. Performance of Grower Pigs (10-45 kg) Fed Salseed Meal

	Level of Salseed Meal (%)			
	0	5	10	15
Daily Gain (kg)	0.40	0.37	0.31	0.33
Daily Intake (kg)	1.99	1.89	1.90	1.88
Feed Efficiency	4.97	5.10	6.13	5.71

Reddy, 1972.

were nonsignificant. The observed decrease in the digestibility of protein with increasing level of salseed meal in the ration was attributed to the presence of tannins in salseed meal.

Finisher Pigs

Pathak and Ranjhan (1973) studied the performance of three groups of Large White x Yorkshire female pigs, fed diets with half or all of the maize moiety replaced by salseed meal for 112 days. The feed intakes and rates of gain of pigs are presented in Table 38.5. There was no significant difference in growth rates between the three diets and it was concluded that during the finishing period salseed meal could replace some of the maize in the diet. However, the authors noted a detrimental effect on the kidneys. These degenerative changes in vital organs of pigs on salseed meal diets have not received specific attention. However, their presence cannot be discounted because of extensive toxic, degenerative changes in kidney, liver, spleen and intestines observed in cattle (Gupta et al., 1977) and poultry (Panda et al., 1975; Mohapatra and Panda, 1978). In addition, the depression in the utilization of nutrients is quite obvious. As a consequence, the economic benefit from utilization of salseed meal, if any, is nullified.

Table 38.5. Performance of Female Finisher Pigs (37-90 kg) Fed Salseed Meal

	Level of Salseed Meal (%)		
	0	50	100
Performance			
Daily Gain (kg)	0.46	0.45	0.45
Daily Intake (kg)	2.90	3.15	3.15
Feed Efficiency	6.3	7.0	7.0
Nitrogen Balance (g/day)	37.2	38.6	21.4
Digestibility (%)			
Dry Matter	76	72	63
Crude Protein	77	68	57
Ether Extract	79	53	68
Total Carbohydrates	79	76	73
Total Digestible Nutrients	74.8	68.5	64.6

Pathak and Ranjhan, 1973.

Breeding Stock

There are no recorded observations on the feeding of salseed meal to gestating or lactating sows. However, the above observations on the feeding of salseed meal to pigs confirm the contention of Robb (1975) who, reviewing the role of unconventional raw materials including salseed meal in livestock rations, concluded that it was unlikely that there would be any cheap alternatives to the use of cereals in swine rations.

SUMMARY

The recorded observations on the feeding of salseed meal to pigs fail to indicate much potential for use in swine production.

REFERENCES

CSIR 1972. The Wealth of India. Raw Materials. Vol IX. A dictionary of Indian raw materials and industrial products. Publication and Information Directorate, Council of Scientific and Industrial Research, New Delhi.

Das, B.K. and Mohanty, G.P., 1971. Annual Progress Report of Investigations on Agricultural Byproducts and Industrial Waste Materials for Evolving Economic Rations for Livestock. Coordinated project of ICAR at Orissa University of Agriculture and Technology, Bhubneshwar, 45 pp.

Gupta, B.S., Gupta, B.N., Saha, R.C., Bhandari, A.B. and Mathur, K.L., 1977. Studies on long term feeding of sal (*Shorea robusta*) seed meal to adult cattle. Indian Vet. J. 54: 463-469.

ICAR, 1967. Report of the Study Group for Exploring the Possibilities of Using Food Byproducts as a Poultry Feed. Indian Council of Agricultural Research, New Delhi.

Khan, M.Y., Lal, M. and Kishan, J., 1986. Nutritional evaluation of deoiled salseed (*Shorea robusta*) meal for goats. Indian J. Anim. Nutr. 3: 29-32.

Kumar, D., 1980. Tannin composition of salseed. J. Food Sci. Technol. India 17: 144-145.

Kumar, R. Agrawal, O.P., Rao P.V., Sidhu, G.S. and Negi, S.S., 1967. Determination of metabolizable energy of some poultry feeds available in India. Indian J. Poult. Sci. 2: 13-18.

McLeod, M.N., 1974. Plant tannins - Their role in forage quality (A review). Nutr. Abstr. Rev. 44: 803-815.

Mitra, C.R. and Mishra, P.S., 1967. Amino acid composition of processed oilseed meal proteins. J. Agric. Food Chem. 15: 697-699.

Mohapatra, H.C. and Panda, N.C., 1978. Effect of feeding sal (*Shorea robusta*) seed and sal oil meal to chicks. Indian Vet. J. 55: 559-566.

Negi, S.S., 1982. Tannins in salseed (*Shorea robusta*) and salseed meal limit their utilization as livestock feeds. Anim. Feed Sci. Technol. 7: 161-183.

Negi, S.S., 1985. A review on the utilization of agricultural wastes as feeds for livestock and experimental safeguards against erroneous conclusions. Agricultural Wastes 13: 93-103.

Panda, B., Jayaraman, M., Rammurty, N.S. and Nair R.B., 1969. Processing and utilization of salseed (*Shorea robusta*) as a source of energy in poultry feed. Indian Vet. J. 46: 1073-1077.

Panda, N.C., Mitra, A., Kedary, S. and Parichha, S.N., 1975. Salseeds and sal oil meal as cereal substitute in starter mashes. Indian Vet. J. 52: 186-194.

Pathak, N.C. and Ranjhan, S.K., 1973. Nutritional studies with salseed meal as a component of finisher rations in Large White Yorkshire pigs. Indian J. Anim. Sci. 43: 424-427.

Punj, M.L., 1988. Agricultural byproducts and industrial wastes for livestock and poultry feeding. Major recommendations of the All-India Coordinated Research Project, Indian Council of Agricultural Research, New Delhi.

Reddy, K.J., 1972. Studies on supplementation of deoiled salseed meal in growing swine. Andhra Pradesh Agriculture University, Hyderabad *vide* Thesis Abstracts (1978), 4(2) 92.

Robb, J., 1975. Alternatives to conventional cereals. In: H. Swan and D. Lewis eds., Feed Energy Sources for Livestock, Butterworths, London, pp. 13-27.

Sibbald, I.R., Slinger, S.J. and Ashton, G.C., 1962. Factors influencing the metabolizable energy content of poultry feeds. Poult. Sci. 41: 107-114.

Sinha, R.P. and Nath, K., 1983. Digestibility of untreated, lime treated and lime-sodium hydroxide treated deoiled salseed meal and its influence on growth in calves. Indian J. Anim. Sci. 47: 8-12.

Singh, K. and Arora, S.P., 1980. Studies on tannins of sal (*Shorea robusta*). Indian J. Anim. Sci. 50: 1043-1051.

Verma, S.V.S., 1970. Studies on metabolizable energy value of salseed and salseed cake by chemical assay and biological evaluation in chicks. M.V.Sc. Thesis, Agra University.

Vijjain, V.K. and Katiyar, R.C. 1973. Types of tannins in deoiled salseed meal and oak kernels and their metabolic behaviour in sheep. Indian J. Anim. Sci. 43: 398-401.

CHAPTER 39

Screenings

R.M. Beames

INTRODUCTION

In most countries, grain which is destined either for human consumption or for use as animal feed is cleaned by removal of straw, soil and foreign seeds. This is achieved by the use of seed cleaners. They consist of a series of blowers and sloping, shaking screens, which separate particles on the basis of density and size. The contaminants, collectively known as screenings, are generally defined on the basis of the allowable content of broken grain, fiber, weed seeds (some specifically mentioned) and dust. Definitions for the United States and Canada are presented in an Appendix on page 404 (Canada Grains Act, 1984; Assoc. Am. Feed Control Officials, 1989). Although screenings are most commonly used as an ingredient in diets fed to ruminants, they may also find use in swine production.

BOTANICAL AND CHEMICAL COMPOSITION

Number 1 Feed Screenings

The major category of screenings which result from the cleaning of wheat under Canadian conditions is defined as Number 1 Feed Screenings. They has been described by various authors to contain levels of wild buckwheat ranging from 4 to 38% with the balance being predominantly cracked wheat (Biely and Pomeranz, 1975; Stapleton et al., 1980; Beames et al., 1986). As a consequence, the nutritive value of Number 1 Feed Screenings is heavily dependent on the chemical composition and nutritive value of wild buckwheat (Table 39.1). However, on a dry matter basis the screenings contain 14.8% crude protein, 3.1% ether extract, 8.7% acid detergent fiber and 2.2% ash (Beames et al., 1986; Table 39.2). The digestible energy (D.E.) content of Number 1 Feed Screenings is calculated at 3100 kcal/kg, based on a D.E. for wild buckwheat of 2206 (Harrold et al., 1980a), a D.E. of 3402 for wheat (National Academy of Sciences, 1969) and a ratio of wheat to wild buckwheat of 75:25 (Tait et al., 1986). The lysine content (0.48 g/100 g DM) is similar to that of barley (National Research Council, 1988).

Uncleaned Screenings

Another category of screenings is "uncleaned screenings" (Appendix I) which are highly variable and are so defined because they do not meet the requirements of Number 1 Feed Screenings. These screenings generally contain less wild buckwheat but higher levels of other seeds such as wild oats and rapeseed. Their high variability makes nutritional characterization even more difficult than with many other types of screenings.

Refuse Screenings

Refuse screenings are grain screenings which do not meet the grade requirements of Number 1 or uncleaned feed screenings because of their high content of weed seeds, chaff or dust. In Canada, all dust is incorporated into refuse screenings. Dust from other screenings (e.g. rapeseed screenings) may also be added to this material at elevators. In the United States, dust is handled separately. Botanical analysis of refuse screenings does little to elucidate the true nature of the material because of the high and variable content of unidentifiable chaff, grass (35-67%) and dust (2-36%; Beames et al., 1986). Because of this, the chemical composition of refuse screenings is very variable.

A chemical analysis of refuse screenings is presented in Table 39.2. On average, they contain 13.1% crude protein, 5.4% ether extract, 28.8% acid detergent fiber and 10.6% ash. The fiber content of Canadian refuse screenings (28.8%) is considerably higher than the fiber content of dust (13.7% crude fiber; Behnke and Clark, 1979), even after correcting for the fact that acid detergent fiber values are consistently higher than corresponding crude fiber values. This difference reflects the higher content of chaff and grass present in refuse screenings. No D.E. values are available for this material. However, a crude estimate of 2700 kcal/kg DM is based on an attributed gross energy of 4400 kcal/kg and a dry matter digestibility of 61.3% (Tait et al., 1986).

Table 39.1. Chemical Composition of Some Weed Seeds Found at High Levels in Screenings (% DM)

	Wild Buckwheat		Stink-weed	Lamb's Quarters		Wild Oats
Crude Fat	3.72[1]	2.10[2]	34.00[3]	8.40[2]	4.63[1]	6.30[3]
Crude Protein	12.40	10.00	10.00	22.70	21.40	13.20
Crude Fiber	---	9.00	---	12.80	---	16.90
Arginine	0.91	0.78	1.64	1.50	0.96	0.64[2]
Histidine	0.38	0.21	0.56	0.46	0.67	0.36
Isoleucine	0.40	0.39	0.89	0.62	0.46	0.35
Leucine	0.70	0.58	1.52	1.04	0.93	0.78
Lysine	0.48	0.40	1.23	0.83	0.74	0.47
Methionine	0.28	0.22	0.26	0.33	0.29	0.35
Cystine	0.16	0.16	0.55	0.36	0.21	0.22
Phenylalanine	0.48	0.37	0.97	0.63	0.61	0.56
Tyrosine	0.27	0.20	0.59	0.45	0.53	0.34
Threonine	0.42	0.32	1.04	0.58	0.62	0.41
Tryptophan	0.25	0.08	0.27	0.27	0.33	0.16
Valine	0.45	0.48	1.27	0.76	0.55	0.48

[1]This and values below from Harrold and Nalewaja, 1977.
[2]This and values below from Tkachuk and Mellish, 1977.
[3]This and values below from National Academy of Sciences, 1969.

Dust

In the Canadian cleaning system, dust is incorporated into refuse screenings, which is the residue after other definable screening fractions have been separated, and contains the straw or chaff. In the United States, the dust is utilized separately (Behnke and Clark, 1979; Hubbard et al., 1982). According to Baird (1980), the dust fraction consists of dust, dirt, broken grain, weed seeds and grass. It is an inexpensive commodity (approximately $8 per tonne U.S. in April 1988) and at times even has a negative value.

Dust has been analyzed to contain 92.3% DM, 8.4% crude protein, 12.9% crude fiber, 1.7% ether extact, 6.6% ash, 62.7% NFE (Baird, 1980) and is estimated to provide a digestible energy level of 3053 kcal/kg DM. Similar values for mixed grain dust (Midwest United States origin, presumably mostly corn) but with a higher ash content (7.4-11.1%) were reported by Behnke and Clark (1979). These latter workers reported a lysine content in the dust of 0.34 g/100 g DM, which was higher than the 0.20 g/100 g DM listed by National Academy of Sciences (1969). Dust separated from refuse screening derived from Canadian wheat and barley was shown in one report (Beames et al., 1986) to contain 28.8% ash, indicating a high soil content. These authors used this value to caution on the extrapolation on the use of results obtained with corn or mixed dust to dust derived exclusively from wheat and barley.

Mixed Feed Oats

Mixed feed oats are predominantly wild oats, which, in the Canadian definition (Canada Grains Act, 1984) contain not less than 50% wild oats (*Avena fatua*). The balance consists of other seeds (whole and broken) as well as straw and chaff, with a maximum allowance of 0.25% ergot (*Claviceps purpurea*). Although this fungus is not common in wild oats, it is prevelant in the seeds of many native grasses and cultivated grains (Seaman, 1980). A proximate analysis of wild oats is presented in Table 39.1, while a chemical analysis of mixed feed oats is presented in Table 39.2. On average, mixed feed oats contain 13.2% crude protein, 4.3% ether extract, 22.5% acid detergent fiber and 4.3% ash. Their digestible energy content, calculated using a gross energy of 4680 kcal/kg (Harrold and Nalewaja, 1977) and a dry matter digestibility of 62.5% (Tait et al., 1986) is 2925 kcal/kg. Lysine content is 0.47% (Harrold and Nalewaja, 1977).

Rapeseed Screenings

Rapeseed screenings are the general classification for cleanings from rapeseed and canola. Unseparated screenings include all the material (both coarse and fine) separated in the cleaning process, but the bulk of the screenings is traded as fines or, in North American, canola fines, which do not include the coarse material (McKinnon, 1988). Botanical composition of the screenings (unseparated) were reported by Beames et al. (1986) to contain 23.8 (SD 3.5)% barley, 10.2 (SD 2.9)% rapeseed, 6.8 (SD 2.4)% wheat, 2.1 (SD 0.9)% lamb's-quarters, 8.4 (SD 2.5)% stinkweed, 2.4 (SD 0.8)% wild buckwheat, 24 (SD 7.1)% wild oats and 17.4 (SD 3.4)% chaff, grass and dust. The fines, after removal of most of the large grain seeds, chaff and grass, consist of fine particles of less than 1.5 to 2.0 mm, and include the smaller and broken grains of rapeseed, the fine chaff, dust and soil as well as other small seeds, predominantly lamb's-quarters and stinkweed. Botanical analyses reported by Bell and Shires (1980) showed the rapeseed content to range from 25 to 55%; lamb's-quarter, 6 to 20%; stinkweed, 0 to 15% and small unidentifiable particles, 16 to 46%, with the last mentioned often containing considerable rapeseed fragments. In a grain cleaning facility, where a lowering of the level of contamination with small seeds takes precedence over minimizing the level of dockage, the fines would contain a higher content of rapeseed.

The chemical content of rapeseed screenings is shown in Table 39.2. They contain approximately 19.6% crude protein, 22.5% ether extract, 28% acid detergent fiber and 7.4% ash. The digestible energy content is estimated at 4035 kcal/kg DM, using a gross energy of 6023 kcal/kg (Keith and Bell, 1983) and a digestibility of 67%. The protein quality of rapeseed screenings, in terms of content of amino acids (g amino acid/16 g N), indicates the lysine content to be somewhat lower than that present in canola seed (4.58-5.21 vs. 5.28, respectively) but methionine plus cystine levels are higher (3.48-3.57 vs. 3.59; Bell and Shires, 1980; Clandinin, 1981).

Because of the high content of rapeseed in the screenings, the product contains a proportionate level of the anti-nutritional factors contained in the separated rapeseed, with this being combined with the anti-nutritional factors present in the small seeds. The most important components are the toxic glucosinolates which yield isothiocyanates and oxazolidinethione on hydrolysis by myrosinase (Hill, 1979), which is released on rupture of the seed. Stinkweed contains 7 to 8% glucosinolate, of which approximately 90% is sinigrin (Beames et al., 1986). The oil of stinkweed is high in erucic acid (35%; Appelqvist, 1971) which is similar to the level found in high erucic acid rapeseed oils.

Table 39.2. Chemical Composition (%) of Number 1 Feed Screenings, Refuse Screenings, Rapeseed Screenings and Mixed Feed Oats

	As Received	Dry Matter Basis			
	Dry Matter	Crude Protein	Ether Extract	ADF	Ash
Number 1 Feed Screening (n=30)					
Mean	89.8	14.8	3.1	8.7	2.2
Range	87.7-91.6	13.5-16.7	1.9-4.1	6.8-13.4	1.1-2.7
SD	0.7	0.7	0.6	1.3	0.3
Refuse Screening (pelleted) (n=16)					
Mean	92.2	13.1	5.4	28.8	10.6
Range	89.0-94.7	10.6-19.3	3.2-8.6	22.8-37.2	7.6-12.8
SD	1.3	2.0	1.5	4.6	1.5
Rapeseed Screenings (n=15)					
Mean	91.8	19.6	22.5	28.0	7.4
Range	90.6-93.0	17.3-21.0	14.9-25.0	23.4-33.4	6.0-9.2
SD	0.6	0.9	2.8	2.7	0.8
Mixed Feed Oats (n=23)					
Mean	93.0	13.4	4.3	22.5	4.3
Range	91.6-95.0	12.0-15.3	3.6-5.3	17.2-29.6	3.3-4.7
SD	1.0	0.7	0.5	2.5	0.3

Beames et al., 1986.

NUTRITIVE VALUE OF GRAIN SCREENINGS

Number 1 Feed Screenings

There has only been limited work on the evaluation of Number 1 Feed Screenings for pigs. Tait et al. (1986), in an experiment in which Number 1 feed screenings were fed to growing-finishing pigs in combination with a barley-soybean meal basal diet (60:40 screenings to basal diet ratio), reported the DM digestibility of the basal diet (80.4%) to be virtually the same as that of the combination (80.7%) which gave a predicted DM digestibility value of 80.9% for the screenings. The digestibility of energy would have been of a similar magnitude, as DM digestibility values of conventional diets have been shown to be not markedly different from energy digestibility values (Taverner et al., 1975). As the greatest component of Number 1 Screenings is cracked wheat (52-76%) (Biely and Pomeranz, 1975; Stapleton et al., 1980; Beames et al., 1986), with the balance almost entirely wild buckwheat, the D.E. concentration should be inversely related to the content of the buckwheat. With the lower D.E. value and higher protein quality (i.e. lysine/protein) of wild buckwheat than of wheat, it would appear to be reasonable to ascribe a nutritive value to Number 1 Screenings approximately equal to that of barley when the screenings are being used in pig diets. It is suggested that Number 1 Feed Screenings should be able to substitute for barley at levels up to complete replacement in all types of swine diets.

There has been much more work done with broiler chickens (Biely and Pomeranz, 1975; Stapleton et al., 1980; Wolde-Tsadick and Bragg, 1980; Proudfoot and Hulan, 1986). In broiler diets, results generally indicate that these screenings can substitute for wheat at levels ranging from 45 (Proudfoot and Hulan, 1988) to 100% (Stapleton et al., 1980) of the diet.

Refuse Screening

Because of the high levels of fiber and ash, and only moderate levels of lipid, refuse screenings have not been considered seriously as a dietary component for pigs. The blending of dust with the coarse screenings to form refuse screenings produces a feedstuff which is more suited for feeding to cattle and sheep than to pigs. However, the protein content (10.6-19.3%) reported by Beames et al. (1986) reflects the fact that a considerable amount of this material originates from the kernel. Tests with rats by these authors have shown a true protein digestibility (TD) of 77.4%, a biological value (BV) of 83.8% and a net protein utilization (TD x BV) of 64.8%. In growing pigs, these same workers obtained an organic matter digestibility of 63.8% and an apparent nitrogen digestibility of 65.4%.

Green foxtail (*Setaria viridis*) is a contaminant which has been reported as a component (0-12%) of refuse screenings (Beames et al., 1986). In growth studies with pigs, Castell (1984) found no effect on performance of growing-finishing pigs with 10% replacement of the grain portion of a wheat-barley-soybean meal diet with green foxtail. Harrold et al. (1976) ascribed a digestible energy value of 3.0 kcal/g to yellow foxtail (*Setaria lutescens*), which is a common contaminant in grain in North Dakota. The low level of digestible energy would make refuse screenings an unsuitable dietary component for growing pigs. However, screenings would be suitable for inclusion in a gestation diet where energy dilution may even be beneficial. Because of the reasonably good protein quality, no upper limit for inclusion need be set, providing digestible energy requirements are met.

Grain Dust

Baird (1980) tested grain dust at levels of 25%, 50%, 75% and 100% as a replacement for corn in a corn-soybean meal diet for self-fed Yorkshire pigs from weaning to 95 kg bodyweight (Table 39.3). The author's conclusion from these results were that "under most feeding conditions . . . it would appear that levels under 25% could successfully be fed without a significant loss in production." Such a statement could be somewhat misleading, with a non-significant 8% reduction in growth rate at the 20% level of replacement. However, with the low cost of dust, even if its inclusion caused a slight reduction in performance it could still prove to be economical to incorporate in pig diets. The maximum economic level of replacement of conventional feedstuffs would require the use of a complete economic model. The profitable level of inclusion would be inversely related to the ratio of fixed costs to total costs.

The results of Behnke and Clark (1979) illustrate the problems resulting from the application of a generalization to the nutritive value of a highly variable product. In an experiment of a somewhat similar design to that of Baird (1980), they tested 25 and 50% replacement of corn with corn dust in corn-soybean meal grower (16% CP) and finisher (14% CP) diets. Their results (Table 39.3) are at variance with those of Baird (1980) showing a slight improvement in daily gain and only a negligible, non-significant reduction in feed efficiency with 25% replacement of corn and a further small non-significant reduction with 50% replacement. These findings would reinforce the essentiality of analyzing each batch of such material before it is used as a replacement for conventional grains in swine diets. Even with such analyses, no idea is gained on levels of available nutrients. Because of the high lysine content of the dust used by Behnke and Clark (1979) in relation to that of corn (0.35 and 0.20 g/100 g DM, respectively), it is considered that an analysis for this amino acid, although expensive, would be desirable to gain some idea of nutritive value. Fiber and ash analysis would be of value in the assessment of energy content.

Table 39.3. Utilization of Corn Grain Dust as a Replacement for Corn in an 18% CP Grower-Finisher Soybean Meal Diet (Expt. 1) and a 16% CP Grower, 14% CP Finisher Diet (Expt. 2)

	Level of Replacement for Corn (%)[3]				
	0	25	50	75	100
Experiment 1[1]					
Daily Intake (kg)	2.34[a]	2.65[b]	2.77[c]	2.80[c]	2.71[b]
Daily Gain (kg)	0.78[a]	0.72[ab]	0.70[b]	0.66[bc]	0.59[c]
Feed Efficiency	3.01[a]	3.72[b]	4.09[c]	4.49[d]	4.58[d]
Dressing Percent	72.1[a]	71.4[a]	68.90[b]	70.3[ab]	69.0[b]
Experiment 2[2]					
Daily Intake (kg)	2.06[a]	2.25[ab]	2.35[b]		
Daily Gain (kg)	0.74	0.79	0.77		
Feed Efficiency	2.81	2.86	3.05		

[1]Adapted from Baird, 1980.
[2]Adapted from Behnke and Clark, 1979.
[3]Means without a common superscript are different ($P<0.05$).

Mixed Feed Oats

Wild oats have been evaluated for both pigs (Harrold et al., 1980b; Tait et al., 1986) and rats (Barranco et al., 1978; Harrold et al., 1980a; Tait et al., 1986). The wild oats which were investigated by Tait et al. (1986) containing 18.2% ADF, were estimated to have a DM digestibility of 62.5% (cf. 80.4% DM digestibility for the barley-soybean meal basal diet). Energy digestibility should be approximately the same as DM digestibility (Henry, 1976). Using a high density wild oat, Harrold et al. (1980b), who tested up to 40% inclusion in a corn-soy diet, showed that 20% inclusion had no effect on daily gain but caused a 5.6% reduction in feed efficiency in 25 kg pigs, and a 3.7% reduction in 43 kg pigs.

Rapeseed Screenings

From a superficial examination of the chemical composition of rapeseed fines, the product would appear to have a high nutritive value. With approximately 65% consisting of rapeseed, lamb's-quarter and stinkweed, with respective lysine levels as a percent of protein being 6.0, 4.5 and 5.4 (Table 39.1), provided that the protein is of a reasonable digestibility, rapeseed screenings have the potential of being a good source of supplementary protein for wheat and barley. However, measurements of digestibility have been somewhat difficult to make. First, the small size of the seed requires grinding with a hammermill with a 1.0 mm screen in order to break all the grain. The high oil content dictates a slow rate of throughput. With undiluted screenings, 44% passes through a 2 mm screen of a commercial hammermill without being ground. With this same screen size, premixing at a level of 15% with a standard ground diet, although overcoming potential clogging problems, has no effect on the percentage of fine seed remaining unground (Beames and Tait, 1987). Once stinkweed is ground, the appropriateness of its name becomes obvious. The bad aroma and (presumably) bad taste lower the acceptability of the screenings, with the result that only a low level (approximately 15%) can be added to a basal diet without a marked reduction in intake. The precision and accuracy of the measurement of digestibility by difference is thus reduced.

Pelleting is a requirement for Canadian canola screenings for export (McKinnon, 1988) as it has been shown that pelleting prevents germination of all seeds (Wood, 1957). Similar results can be achieved by the application of dry heat at 120°C for 1.5 hours (Beames and Tait, 1987). It is thus fortuitous that pelleting improves digestibility of rapeseed screenings to a degree similar to that achieved by fine grinding (1.0 mm screen). The results presented in Table 39.4 (Beames, unpublished) demonstrate this advantage.

Table 39.4. Digestibility of Dry Matter, Crude Protein and Crude Fat (After Hydrolysis) of Canola Fine Screenings in Different Physical Forms in Diets for Growing Pigs (6 Pigs/Diet)[1]

	Diet Composition (% air dry)			
Ingredients				
Basal (Barley-SBM)	100	85	85	85
Whole Screenings[2]	--	15	15	--
Ground Screenings (1 mm)[2]	--	--	--	15
Pelleted	No	No	Yes	No
Digestibility (%)				
Dry Matter	80.1	71.9	79.9	78.0
Crude Protein	77.1	68.3	80.2	77.6
Crude Fat[3]	39.8	22.1	59.6	48.3
Screenings Dry Matter[4]	--	27.2	68.8	66.2

[1]From Beames, unpublished data.
[2]Major components (by weight): stinkweed 10.5%, lamb's-quarter 23.6%, canola (whole and broken) 31.9%, straw, dust, hulls 23.2%.
[3]Calculated by method described by Thorbek and Henkel, 1977.
[4]Calculated by difference.

Somewhat parallel responses to pelleting and to grinding have been obtained with canola seed by Bell et al. (1985). However, the above improvement from processing rapeseed screenings appears to be somewhat at variance with the results of Bell and Shires (1980). The latter workers investigated five ground rapeseed screenings after fat extraction, with each replacing 12% of the canola meal in a grain plus 15% canola meal diet (level of inclusion was thus 1.8%). They showed a reduction in apparent protein digestibility of the diets containing the five samples of screenings meals from 78.3% in the control diet to 76.8, 76.6, 74.2, 73.1 and 72.9% and in apparent energy digestibility from 79.8% in the control diet to 79.3, 78.0, 77.3, 78.2 and 78.9%, with three of the former and four of the latter reductions being significant. Such results, with such low levels of inclusion of the test material, make interpretation somewhat difficult. In an associated trial in which pigs were fed from 34 to 57 kg bodyweight with 4, 8 and 12% replacement of the 15% canola meal, by each of five rapeseed dockage meals in a grain plus 15% canola meal diet, there was no significant adverse effect of inclusion of screenings on daily feed intake or efficiency of feed utilization.

Keith and Bell (1983) have investigated the possible beneficial effect of processing canola screenings by boiling, with or without ammoniation. Boiling plus ammoniation reduced the glucosinolate content in the screenings. Although ammoniation was shown to improve the intake of growing pigs, the adverse effects of ammoniation would seem to outweigh any benefits, as its application reduces lysine utilization and the efficiency of feeds utilization (Keith and Bell, 1983; Bell et al., 1987).

SUMMARY

The names and composition of the various types of screenings vary from country to country. Screenings discussed in this chapter include Number 1 Feed Screenings (mostly cracked wheat and buckwheat), refuse screenings (chaff, dust, small seeds, soil), dust, wild oats and rapeseed screenings (mostly small seeds). For growing pigs, Number 1 Feed Screenings should be attributed a value approximately that of barley. Refuse screenings, with a high fiber content, are too low in digestible energy to be seriously considered for use in pig diets, but could be of some value in gestation diets. Dust, which is widely available in the U.S., often at a very low cost, has been shown to have no adverse effect on growth when replacing corn at levels of up to 50%, but causes a reduction in efficiency of feed utilization. Even so, the low price of dust could make it attractive to feed at high levels. Experiments with wild oats have shown satisfactory performance at dietary levels of 20%, with only a 4 to 6% reduction in feed efficiency. Rapeseed screenings are unpalatable for pigs and normally will not be tolerated at dietary levels in excess of 10%. Optimum utilization is obtained with pelleting, which gives energy and protein digestibility values similar to those obtained with barley.

REFERENCES

Appelqvist, L.A., 1971. Lipids in cruciferae. VIII. The fatty acid composition of seeds of some wild or partially domesticated species. J. Amer. Oil Chem. Soc. 48: 740-744.

Association of American Feed Control Officials Incorporated, 1989. Official Publication, 1989. pp. 185-186. See Appendix, p. 405.

Baird, D.M., 1980. Feeding and energy value of elevator grain dust. Feedstuffs 52 (49): 17-21.

Barranco, A., Morgan, D.J., Capper, B.S. and Ogilvie, L.M., 1978. A note on the nutritive value of wild oats. Anim. Prod. 27: 129-132.

Beames, R.M., Tait, R.M. and Litsky, J., 1986. Grain screenings as a dietary component for pigs and sheep. I. Botanical and chemical composition. Can. J. Anim. Sci. 66: 473-481.

Beames, R.M. and Tait, R.M., 1987. Maximizing the nutritional value of Canola screenings in pig and sheep diets. In: Research on Canola Seed, Oil, Meal and Meal Fractions. 8th Progress Report, Canola Council of Canada, Winnipeg, pp. 100-104.

Behnke, K.C. and Clark, H.M., 1979. Nutritional utilization of grain dust by monograstrics. In: B.S. Miller and Y. Pomeranz eds. Proceedings of the International Symposium on Grain Dust, Kansas State University, Manhattan, Kansas, pp. 219-233.

Bell, J.M. and Shires, A., 1980. Effects of rapeseed dockage content on the feeding value of rapeseed meal for swine. Can. J. Anim. Sci. 60: 953-960.

Bell, J.M., Keith, M.O. and Kowalenko, W.S., 1985. Digestibility and feeding value of frost-damaged canola seed (low glucosinolate rapeseed) for growing pigs. Can. J. Anim. Sci. 65: 735-743.

Bell, J.M., Keith, M.O., Darroch, C.S. and McGregor, D.I., 1987. Effects of ammoniation of canola seed contaminated with wild mustard seed on growth, feed utilization and carcass characteristics of pigs. Can. J. Anim. Sci. 67: 113-125.

Biely, J. and Pomeranz, Y. 1975. The amino acid composition of wild buckwheat and No. 1 wheat feed screenings. Poult. Sci. 54: 761-766.

Canada Grains Act, 1984. Off grades and grades of screenings order. In: Canada Grains Act. Regulations PC 1975-1530 (Amended to 1 Aug. 1984). See Appendix, p. 404.

Castell, A.G., 1984. Response of growing-finishing pigs to dietary inclusion of green foxtail (*Setaria viridis*) seeds. Can. J. Anim. Sci. 64: 1063-1066.

Clandinin, D.R., 1981. Canola Meal for Livestock and Poultry. Publ. No. 59. Canola Council of Canada, Winnipeg, 25 pp.

Harrold, R.L., Craig, D.L., Nalewaja, J.D. and North, B.B., 1980a. Nutritive value of green or yellow foxtail, wild oats, wild buckwheat or redroot pigweed seed as determined with the rat. J. Anim. Sci. 51: 127-131.

Harrold, R.L., Zimprich, R.C. and Johnson, J.N., 1980b. Utilization of wild oats (*Avena fatua*) by growing swine. J. Anim. Sci. 51 (Suppl. 1): 74.

Harrold, R.L., Johnson, J.N. and Dinusson, W.E., 1976. "Pigeon grass" screenings for non-ruminants. J. Anim. Sci. 42: 1354-1355.

Harrold, R.L. and Nalewaja, J.D., 1977. Proximate, mineral and amino acid composition of 15 weed seeds. J. Anim. Sci. 44: 389-394.

Henry, Y., 1976. Prediction of energy values of feeds for swine from fibre content. 1st International Symposium on Feed Composition, Animal Nutrition Requirements and Computerization of Diets. Utah State University, Logan, Utah, pp. 270-281.

Hill, R., 1979. A review of the "toxic" effects of rapeseed meals with observations in meal from improved varieties. Brit. Vet. J. 135: 3-16.

Hubbard, J.D., Lai, F.S., Martin, C.R., Miller, B.S. and Pomeranz, Y., 1982. Amino acid composition of grain dust. Cereal Chem. 59: 20-22.

Keith, M.O. and Bell, J.M., 1983. Effects of ammonia and steam treatments on the composition and nutritional value of canola (low glucosinolate rapeseed) screenings in diets for growing pigs. Can. J. Anim. Sci. 63: 429-441.

McKinnon, P.J., 1988. Canola fines - what are they? Proc. 23rd Annual Pacific North West Animal Nutrition Conference, Spokane, Washington, pp. 159-170.

National Academy of Sciences, 1969. United States-Canadian Tables of Feed Composition, 2nd revision. Publ. No. 1684. National Research Council, Washington, D.C.

National Research Council, 1988. Nutrient Requirements of Domestic Animals, No. 2. Nutrient Requirements of Swine. 9th ed. National Academy of Sciences, Washington, D.C.

Proudfoot, F.G. and Hulan, H.W., 1986. The nutritive value of wheat screenings as a feed ingredient for adult leghorn hens. Can. J. Anim. Sci. 66: 791-797.

Proudfoot, F.G. and Hulan, H.W., 1988. Nutritive value of wheat screenings as a feed ingredient for broiler chickens. Poult. Sci. 67: 615-618.

Seaman, W.L., 1980. Ergot of grains and grasses. Publ. 1438. Agriculture Canada, Ottawa.

Stapleton, P., Bragg, D.B. and Biely, J., 1980. The botanical and chemical composition and nutritive value of wheat feed screenings. Poult. Sci. 59: 333-340.

Tait, R.M., Beames, R.M. and Litsky, J., 1986. Grain screenings as a dietary component for pigs and sheep. II. Animal utilization. Can. J. Anim. Sci. 66: 483-494.

Taverner, M.R., Rayner, C.J. and Biden, R.S., 1975. Amino acid content and digestible energy value of sprouted, rust-affected and sound wheat in pig diets. Aust. J. Agric. Res. 26: 1109-1113.

Thorbek, G. and Henkel, S., 1977. Apparent digestibility of crude fat determined in trials with calves and pigs in relation to analytical methods applied. Zeit. Tierphysiol, Tierernahr. u. Futtermittelkde 39: 48-55.

Tkachuk, R. and Mellish, J.V., 1977. Amino acid and proximate analyses of weed seeds. Can. J. Plant Sci. 57: 243-249.

Wolde-Tsadick, M.S. and Bragg, D.B., 1980. Utilization of wheat screenings in the broiler diet as a major source of energy. Poult. Sci. 59: 1674-1675.

Wood, A.J., 1957. Memorandum to J.E. Gage, Pacific Elevators, Vancouver from A.J. Wood, University of British Columbia. Quoted by Beames and Tait (1986).

APPENDIX

A. Description of Screenings as Defined in the Regulations of the Canada Grain Act (Anon., 1984).

Number 1 feeding screenings+ (a) shall be grain screenings (b) shall be cool and sweet (c) shall contain not less than 35% in the aggregate of broken or shrunken grain; not more than 2% hare's-ear mustard; not more than 1% hulls; not more than 3% in the aggregate of small weed seeds that can pass through a 4-1/2/64 inch round hole sieve, chaff, hulls, dust; not more than 6% in the aggregate of small weed seeds that can pass through a 4-1/2/64 inch round hole sieve, chaff, hulls dust, wild and domestic mustard seed, ball mustard and rapeseed; nor more than 8% wild oats and not more than 1% of the seeds designated as injurious in the "Feed Regulations"; and may contain wild buckwheat and not more than 10% of other large seeds named in this section.

Uncleaned screenings++shall be grain screenings that do not meet the specifications for Number 1 or No. 2 feed screenings because of the content of weed seeds, hulls, chaff or dust but that contain (a) at least 35% of material that if separated would meet the grade requirements for Number 1 feed screenings; and not more than 1% of excessive wild oat hulls.

Refuse screenings+++shall be grain screenings that do not meet the grade requirements for Number 1 or Number 2 feed screenings or uncleaned screenings because of the content of weed seeds, chaff or dust.

Mixed feed oats (1) A mixture of western wild oats and western cereal grain because of being so mixed shall, if it has the qualities set out in subsection (2), be graded "mixed feed oats" (2). Mixed feed oats shall (a) contain not less than 50% wild oats (b) contain not more than 1% of material that passes through a 4-1/2/64 inch round hole sieve and (c) contain in whole, broken, hulled or hulless form not more than (i) 5% heated kernels (ii) 5% wild buckwheat (iii) 5% wheat heads (iv) 4% knuckles, straw and chaff (v) 5% in the aggregate of wild buckwheat, wheat heads, knuckles, straw and chaff (vi) 5% flax seed (vii) 0.1% stones or (viii) 0.25% ergot.

Rapeseed screenings: There is no such category defined in the regulations. However, these are known as refuse screenings (rapeseed) and are produced in the cleaning of rapeseed.

+ International Feed No. 4-02-154.

++ International Feed No. 4-02-153.

+++ International Feed No. 4-02-151.

B. Description of Screenings as Defined by the Association of American Feed Control Officials (Anon., 1989).

Screenings is obtained in the cleaning of grains which are included in the United States Grain Standard Act and other agricultural seeds. It may include light and broken grains and agricultural seeds, weed seeds, hulls, chaff, joints, straw, elevator or mill dust, sand, and dirt. It must be designated as Grain Screenings, Mixed Screenings and Chaff and/or Dust.

No grade of screenings may contain any seeds or other material in amount that is either injurious to animals or will impart an objectionable odor or flavor to their milk or flesh. The screenings must contain not more than four whole prohibited noxious weed seeds per pound and must contain not more than 100 whole restricted noxious weed seeds per pound. The prohibited and restricted noxious weed seeds must be those named as such by the seed control law of the state in which the screenings is sold or used.

All grades of screenings must bear minimum guarantees of crude protein and crude fat and maximum guarantees of crude fiber and ash. (Adopted 1953, Amended 1959, 1960.)

81.1 Grain Screenings must consist of 70% or more of grains, including light and broken grains, wild buckwheat, and wild oats. It must contain not more than 6.5% ash. (Adopted 1953, Amended 1959, 1960.)

IFN 4-02-156 Cereals grain screenings.

81.2 Mixed Screenings is screenings excluded from the preceding definition. It must contain not more than 27% crude fiber and not more than 15% ash. (Adopted 1953, Amended 1954, 1960.)

IFN 4-02-157 Cereals mixed grain screenings.

81.3 Chaff and/or Dust is material that is separated from grains and seeds in the usual commercial cleaning processes. It may include hulls, joints, straw, mill or elevator dust, sweepings, sand, dirt, grains, seeds. It must be labeled, "chaff and/or dust". If it contains more than 15% ash the words "sand" and "dirt" must appear on the label. (Adopted 1953.)

IFN 4-02-149 Cereals-legumes chaff and/or dust.

T81.4 _______ Screenings must consist of 70% or more grains including light and broken grains, wild buckwheat and wild oats. It must contain not more than 6.5% ash. The predominating grain must be declared as the first word in the name. (Proposed 1974.)

IFN 4-00-542 Barley screenings
IFN 4-20-687 Maize screenings
IFN 4-03-329 Oats screenings
IFN 4-08-085 Rice screenings
IFN 4-27-721 Sorghum screenings
IFN 4-05-216 Wheat screenings

CHAPTER 40

Seaweed

R.M. Beames

INTRODUCTION

Seaweed is the popular name for the large attached (benthic) marine algae found in the groups *Chlorophyaceae, Rhodophyaceae* and *Phaeophyaceae*, or the green, red and brown algae, respectively (Chapman and Chapman, 1980). Kelp is a term of unknown origin dating to the 14th century and is defined as "any large brown seaweed, especially any species of Laminaria."

Seaweed has been used for a wide variety of commercial purposes (Neushel, 1987). It has been used as a source of potash and acetone for gunpowder production and has also been used for the production of soda and mulch. Seaweed was a major source of iodine in the early 1800s, but was replaced by Chile saltpetre in the latter half of the 19th Century. Today, seaweed continues to be widely utilized for a variety of purposes, with the total annual production valued in excess of $780 million (Jensen, 1979).

Vast quantities of seaweed are located throughout the world. For example, it is estimated that there are about 4 million tons of seaweed produced around the coast of Scotland (Black, 1955b) and 1.5 million tons of annual growth produced on the west coast of British Columbia (Whyte and Englar, 1974). Since it is so plentiful, a considerable amount of research has been conducted to evaluate the potential to use seaweed as a feed ingredient and approximately $10 million worth of seaweed meal is used annually around the world (Jensen, 1979). The nutritive value of seaweed is based mainly on its content of macro and micro elements but there may be some potential to increase its utilization in swine rations, predominately as a bulking agent.

GROWING SEAWEED

Seaweeds are very specific in their requirements for growth, each having a narrow margin of tolerance for exposure and immersion in water. This feature dictates the occurrence in the various tidal zones and the zone below low tide (sub-littoral). The plant is attached to a base (eg. rock) by a holdfast, to which the stalk or stipe is joined. Above the stipe is the thallus which further divides into fronds, which include the reproductive organs. Air bladders are incorporated into the structures to aid with buoyancy.

Harvesting of seaweed may be done in a variety of ways but the use of a cutter is recommended where regeneration is considered desirable (Whyte and Englar, 1974). After the material is drained, drying can be carried out by conventional methods including natural drying and drum drying. Some chopping may occur before drying after which the material is ground to a meal.

NUTRIENT CONTENT OF SEAWEED

The nutritive value of seaweed, as with any other dietary component, must be assessed on the basis of its ability to provide energy, amino acids, minerals and vitamins and on the presence of any other growth stimulating or inhibiting factors. Considerable information has been published on the proximate composition of seaweed meals. Unfortunately, much of the work on the nutritive value of seaweed has not provided details on the harvesting and drying conditions under which it was obtained. It has been shown (Woodward, 1951; Black, 1950) that the chemical composition of seaweed varies over the growing season. Consequently, a contributory factor in the differences in estimates of nutritive value obtained by various research workers is the time of harvest.

Values presented in Table 40.1 show that most meals are low in crude protein and high in ash. Fats do not appear to have been measured in great detail (Woodward, 1951) but the total amount has been reported to vary from less than 1% in sub-littoral seaweeds to 9% in some exposed weeds (Black, 1955a).

The most significant carbohydrate components of seaweed consist of mannitol, alginic acid and its salts, laminarin, fuciodin and algal cellulose. The level of mannitol varies from 5 to 25% of dry matter, depending on time of harvest. In large doses, it can have a pronounced laxative effect (Whittemore and Percival, 1975). Alginic acid is the main structural carbohydrate in the brown seaweeds, taking the place of cellulose in land plants. It is a straight-chain polymannuronide containing β-D mannopyruronic acid residues linked through the 1:4-positions (molecular weight of approximately 185,000). It is thought to form a complex with proteins, which may explain the negative protein digestibility values obtained in some trials (Black, 1955b). Laminarin is the most important reserve food material (corresponding to starch in land plants) and would be expected to be the main source of energy in seaweed for monogastric animals. Laminarin is composed of β-D-glucopyranose units linked through the 1:3-positions and having a chain length of approximately 20 units. Fucoidin is a fucose sulphate ester polymer and is a cell wall constituent. Algal cellulose is the insoluble residue which remains after the removal of alginic acid from seaweed.

Table 40.1. Proximate Composition of Seaweed Meals (%)

Species	Dry Matter	Crude Protein	Ether Extract	Ash	Crude Fiber	NFE
Fucus vesiculosus[1]	87.6	5.0	2.0	13.1	5.5	62.0
Fucus vesiculosus[2]	88.7	5.2	4.2	16.3	9.4	53.6
Fucus serratus[1]	87.7	4.4	0.8	16.0	5.7	68.9
Ascophyllum nodosum[3]						
Unwashed	95.1	8.4	3.1	20.9	3.0	59.7
Washed	94.1	4.9	6.1	17.7	2.5	62.9
Laminaria digitata[1]						
Rich in laminarin	81.4	5.8	0.6	11.3	26.7	36.6
Poor in laminarin	81.4	9.0	0.6	11.3	26.6	47.0
Laminaria saccharina[3]						
October sample	96.0	7.3	1.7	26.7	3.9	56.4
January sample	94.1	10.5	0.8	40.4	5.9	36.5
Algit[2]	83.7	11.4	1.1	16.8	8.6	45.8
Nereocystis luetkeana[4]						
Frond	100	15.3	1.9	42.7	1.3	38.8
Stipe	100	7.3	2.4	43.1	6.2	41.0

[1]Chapman and Chapman, 1980.
[2]Ringen, 1939.
[3]Barta et al., 1981.
[4]Black, 1955a.

Nitrogen, besides being a component of proteins and their constituent amino acids, is present in seaweed as ammonia and as methylamine and triethylamine, both in the salt form (Woodward, 1951). The reports of Johnston (1972) and Whyte and Englar (1975) indicate the major portion of the nitrogen in seaweed is tightly bound and unavailable for endogenous enzymatic hydrolysis. In two species of Japanese seaweed used for human food, Johnston (1972) found only 12 to 16% of the nitrogen to be present as amides plus amino acids.

The protein content of seaweed varies considerably among species in average value and within species throughout the year. According to Black and Woodward (1957) "only in the Spring of the year is the protein reasonably high (12-15% of the dry matter), and this falls rapidly as growth proceeds, so that for the greater part of the year it is 5 to 10%, consequently seaweed cannot be regarded as a source of protein." This would appear to be a rather sweeping statement, because the authors, in the same article, list five seaweed meals with protein levels of 7.6, 8.4, 12.6, 23.4 and 30.5%, with all samples being collected in Scotland from November to January (at a time when protein content, according to the authors, normally should be relatively low). Samples of *Macrocystis integrifolia* collected on the West Coast of British Columbia, Canada also showed changes in the level and distribution of nitrogen throughout the season (Whyte and Englar, 1975).

Some of the reports from the 1950s on the amino acid composition of seaweed are of limited value for nutritional purposes (e.g. Coulson, 1953) since the amino acids were determined only qualitatively. Amino acid content, determined by ion-exchange chromatography, has been reported by Beames et al. (1977) and Barta et al. (1981) for *Nereocystis luetkeana*, while values for other seaweeds, e.g. *Macrocystis pyrifera*, are available in commercial publications (Stauffer Chem. Co., 1975). The above reports indicate lysine (g/100 g DM) to range from 0.2 to 1.7 and threonine from 0.12 to 1.1 with a similar range in the remaining essential amino acids. However, it would be of little value, and even perhaps misleading, to present a detailed listing of amino acids in a review such as this, because of the overriding influence of the low availability of the total protein.

Table 40.2. Macro Element Levels in Seaweed (g/kg DM)

Species	Harvest Time	P	K	Ca	Mg	Na	Cl
Cladosporum rupestris	Jan	2.7	32.8	15.2	7.3	2.5	6.3
Rhodymenia palmata	Jan	5.6	79.1	7.2	3.9	20.7	97.0
Laminaria cloustoni	Nov						
(frond)		2.8	52.5	10.4	5.8	28.8	59.2
(stipe)		2.5	81.5	18.0	7.3	13.5	124.0
Ascophyllum nodosum	Dec	0.9	22.6	21.6	8.2	2.9	18.9
Nereocystis luetkeana	Aug	2.1	117.1	10.1	10.7	91.3	222.0
Nereocystis luetkeana	May						
(frond)		--	13.0	5.1	--	50.8	--
(stipe)		--	14.5	6.5	--	33.7	--
(bulb)		--	24.6	6.8	--	45.0	--
Algit (unknown)	?	1.3	15.3	47.1	2.3	11.6	9.8
Fucus vesicutosus	?	0.6	19.7	14.5	6.6	42.7	16.8

Ringen, 1939; Black and Woodward, 1957; Jensen, 1972; Beames et al., 1977 and Barta et al., 1981.

Historically, seaweed has earned its reputation as a food and as a fertilizer largely on the basis of its content of macro and micro elements (Booth, 1961). Many of the macro elements are present at a concentration within or markedly above the normal range for land plants (Table 40.2). Levels of calcium, phosphorus and magnesium are similar to those found in land plants, whereas sodium, potassium and chlorine are present at much higher concentrations. Of these elements, the only one which could be of reasonable economic importance is phosphorus, but the level of 0.2% to 0.3% found in most seaweeds is still low when compared with the dietary requirement of 0.4 to 0.7% for the pig (National Research Council, 1988).

Of the essential trace elements, it would seem that only iodine and iron are present in concentrations either approaching or in excess of the requirements of monogastric animals (Table 40.3). Chapman and Chapman (1980) state that the trace elements in seaweed are present in an organic form. They contended that this made them more readily assimilable than trace elements in the inorganic form. However, even if this were so, such a feature would be of little economic significance in modern animal production where trace elements, as mixtures of inorganic salts, normally are added to diets at very little cost.

Table 40.3. Micro Element Levels in Seaweed (mg/kg)

Species	Harvest Time	Co	Fe	Cu	Zn	Mn	I
Cladophora rupestris	Jan	16.2	4,400	31	92	1260	1,100
Rhodymenia palmata	Jan	2.6	1,355	48	200	110	300
Laminaria cloustoni							
(frond)	Nov	0.4	437	5	170	<20	5,000
(stipe)	Nov	0.5	446	5	59	47	3,300
Ascophyllum nodosum	Dec	1.4	1,132	61	110	45	500
Nereocystis luetkeana	Aug	--	1,677	<2	6	5	377
Nereocystis luetkeana							
(frond)	May	<0.1	78	<5,000	30	--	--
(stipe)	May	<0.1	31	<4,000	12	--	--
(bulb)	May	<0.1	103	<5,000	9	--	--
Algit (unknown)	?	--	380	3	--	16	1,639

Ringen, 1939; Black and Woodward, 1957; Jenson, 1972; Beames et al., 1977 and Barta et al., 1981.

Vitamins and precursors present include thiamin, riboflavin, vitamin B_{12}, beta-carotene and the brown pigment fucoxanthin. Several of the green seaweeds contain vitamin B_{12} at levels as high as that found in liver. Seaweed has been shown to have an antirachitic effect for chickens (Black, 1955b). Vitamin C has been reported at levels up to 8,000 mg/kg, vitamin E at levels of 10 to 350 mg/kg, while pantothenic acid and folic acid have also been found (Black, 1955b). Nebb and Jensen (1966) concluded that 3% seaweed meal can satisfy the vitamin A and B_2 requirements of bacon pigs, but the nature of the basal diet used by these authors made such conclusions equivocal.

EFFECT OF PROCESSING

For seaweed to be used as a dietary component for animals other than those with access to the seashore or access to freshly harvested material, it must be processed by drying, ensiling or by the use of a preservative. Most seaweed used as a dietary component is dried and ground. Although a large proportion of the nitrogen in the fresh material may be strongly bound chemically and thus already of low availability (Whyte and Englar, 1975), the remaining protein and amino acids in the hydrophilic component would be vulnerable to binding and would have a reduced availability as a result of any high temperatures occurring during the drying process. Such binding has been demonstrated in terrestrial plants like forages (Van Soest, 1965) with temperatures as low as 50°C, where protein is bound by non-enzymatic browning to produce "artifact lignin". Therefore, the temperature and rate of removal of moisture during drying must be carefully controlled to ensure that it does not have an adverse effect on the availability of the protein not associated with the polymeric residue. It is suggested that the drying temperature be kept below 120°C if protein availability is to be a primary consideration.

The effect of ensiling appears to be quite variable depending on species and time of harvest, with some studies reporting protein being broken down to inorganic nitrogen during ensiling and other studies suggesting that protein is synthesized from inorganic nitrogen with the utilization of mannitol. Regardless, ensiling appears to have an adverse effect on palatability. In trials reported by Black (1955c), sheep would not accept ensiled *Ascophyllum nodosum* and *Laminaria cloustoni*, although the unensiled seaweeds were readily consumed. Bender et al. (1953) also reported poor palatability of *Laminaria cloustoni* where rats completely refused to consume diets containing this material. Therefore, ensiling would not appear to be applicable for seaweed used as a swine feed.

UNDESIRABLE CONSTITUENTS

The high sodium, potassium and chlorine content of seaweed, with their resultant diuretic effect (Beames et al., 1977), has been the main reason that the dietary inclusion level has been limited to 5 to 10% in most research trials conducted with seaweed. In the trials of Beames et al. (1977), the urine volume of grower pigs (36-57 kg) increased from 81 ml/kg $BW^{0.73}$ to 470 ml/kg $BW^{0.73}$ as the dietary seaweed level increased from 0 to 20%. In these experiments, the chloride concentration in the urine remained in the range of 0.47 to 0.84 g/100 g urine, as the dietary seaweed level rose from 5 to 20%, indicating that pigs are able to tolerate high dietary ion concentrations by increasing water intake. However, they must have free access to low-mineral water. With tolerance levels to dietary potassium of approximately 3% (National Research Council, 1988), most seaweed meals could be included at levels up to 20% without potassium levels becoming a problem.

Of the micro-elements, only the concentration of iodine would appear to be of possible concern and then only in growing pigs, where 800 mg/kg diet can result in reduced performance (National Academy of Sciences, 1980). Sows are more tolerant, showing no adverse effects when receiving diets containing up to 2500 mg/kg diet (National Academy of Sciences, 1980).

NUTRITIVE VALUE OF SEAWEED

Starter Pigs

Of the research results collected by the author, none was designed to evaluate the suitability of seaweed for classes of pig other than grower-finishers. However, the low level of digestible protein and the high level of structural carbohydrate make seaweed meal unsuitable for use in diets fed to weanling pigs.

Growing-Finishing Pigs

Cameron (1954) obtained an 8% increase in feed intake but no improvement in growth rate as a result of adding 6% seaweed meal (predominantly *Ascophyllum nodosum*) to a 78% barley, 16% tankage diet fed to growing-finishing pigs (Table 40.4). The only additive was fish oil as a source of vitamins A and D. The major conclusion of the authors was that 2 to 6% of this type of seaweed meal could be safely added to a hog ration.

Table 40.4. Performance of Growing Pigs Fed Graded Levels of Seaweed

	Level of Seaweed (%)			
	0	2	4	6
Daily Gain (kg)	0.67	0.69	0.67	0.67
Daily Intake (kg)	2.56	2.65	2.61	2.77
Feed Efficiency	3.82	3.84	3.89	4.13

Cameron, 1954.

Similar trials with growing-finishing pigs conducted at the East of Scotland College of Agriculture with *Ascophyllum meal* showed that barley meal could be replaced by seaweed at a level of 5%, increasing to 12% without an adverse effect on growth rate but with a reduction in the efficiency of feed utilization (Black, 1955b). Experiments with *Eklonia cava* and *Ascophyllum nodosum* (Murakami et al., 1984) showed a reduced weight gain and poorer efficiency in finishing pigs with a 20% level of dietary inclusion of the seaweed. The conclusion to be drawn from the various research results is that the upper level of inclusion of seaweed meal in a grower diet should be 10%.

Digestibility studies with *Eklonia cava* showed that when the level of dietary inclusion increased from 10 to 40%, digestibility of dry matter decreased from 84.7 to 54.8% and of crude protein from 65.4 to 43.8%. Ringen (1939) fed two commercially prepared meals - "Algit" made from desalted kelp (undefined species) and "Neptun" prepared from rockweed (*Fucus vesiculosus*), both containing 16 to 17% ash, with 11.4 and 5.2% crude protein, respectively (all figures on DM basis). Organic matter digestibility values for Algit and Neptun were 41 and 31%, respectively. Nitrogen digestibility of the Algit was "low" but with Neptun, "for each 100 g DM provided, 7 to 8 g of the digestible protein in the other feed was lost." A similar negative protein value for seaweed has been reported by Beames et al. (1977), where inclusion of 5 to 20% of dried *Nereocystis luetkeana* meal in the diet of growing pigs (36-57 kg bodyweight) resulted in a negative effect on nitrogen digestibility ranging from 1.9 to 4.7 g faecal nitrogen loss/100 g dry kelp intake. Whittemore and Percival (1975) examined the residue of *Ascophyllum nodosum* after the removal of alginic acid. The test diet of 50% dried seaweed residue and 50% basal diet was fed to six, 40 kg pigs for a 3-day pre-collection and a 3 to 4 day collection period. Digestibility of energy and protein, measured by difference, was 12% and 25%, respectively.

Experiments demonstrating a substantial protein or energy contribution by seaweed are scarce. One of the few which showed a moderate to high utilization was conducted by Sheehy et al. (1942). The meal was produced from *Laminaria digitata* harvested in Ireland in September, and contained (DM basis) 16.38% ash, 2.55% chlorine, 2.0% calcium, 7.82% crude protein, 0.40% ether extract and 6.93% crude fiber. Both unhydrolyzed and hydrolyzed (boiled with 5% sulphuric acid for one hour, then neutralized with sodium hydroxide) seaweed were evaluated. Digestibility was measured by difference with two 107 kg pigs, with the seaweed incorporated at a level of approximately 25%. Agreement between pigs was high, with a mean dry matter digestibility of 67.4% and mean crude protein digestibility of 27.3% for the unhydrolysed meal, with figures for the hydrolysed meal being slightly lower. Differences between the results obtained by various workers could be attributed not only to the use of different species of seaweed, but also to differences in the time of harvesting (Woodward, 1951; Whyte and Englar, 1975).

Seaweed has been shown to be of value in the control of parasites. Jensen (1972) reported that 3 to 5% inclusion of seaweed meal in the diet markedly reduced the extent of liver condemnations in pigs at slaughter.

Breeding Stock

For sows, the high level of structural carbohydrate found in seaweed could be advantageous in improving the satiety value of diets during gestation.

The laxative effect could be of value in sow diets during the last week before farrowing. However, the economics of using seaweed as an energy diluent would be debatable. Also, the two to three fold increase in urine volume (and water consumption) with 5 to 10% levels of dietary inclusion (Beames et al., 1977) need to be considered. Taking these various factors into account, the upper inclusion level of seaweed meal in sow diets is suggested as 5%.

SUMMARY

There are large variations in the composition of seaweed and seaweed meal, even within species over the growing season. Seaweeds contain high levels of vitamins and inorganic constituents. Although this feature may be of considerable value in health food supplements for human consumption, it is of little economic value in animal diets, where the convenience, consistency and low cost of purified vitamins and inorganic mineral supplements make these latter materials the nutrient sources of preference. All but a few of the research projects which have been designed to evaluate the energy and protein content of seaweed have not been promising. Most protein utilization measurements have produced very low figures, with some even being negative, the reason for this being that much of the nitrogen is closely bound to cell wall structures. Seaweed has a laxative and a diuretic effect, which has prompted some research workers to suggest a 10% upper limit to dietary inclusion. A marked reduction in the incidence of internal parasites as a result of seaweed feeding has been reported. This, and other health related features, could prove to be the major value of seaweed as a dietary component.

REFERENCES

Barta, E.S., Branen, A.L. and Leung, H.K., 1981. Nutritional analysis of Puget Sound Bull Kelp (*Nereocystis luetkeana*). J. Food Sci. 46: 494-497.

Beames, R.M., Tait, R.M., Whyte, J.N.C. and Englar, J.R., 1977. Nutrient utilization experiments with growing pigs fed diets containing from 0 to 20% kelp (*Nereocystis luetkeana*) meal. Can. J. Anim. Sci. 57: 121-129.

Bender, A.E., Miller, D.S. and Tunnah, E.J., 1953. Biological value of algal protein. Chem. Ind. 38: 1340-1341.

Black, W.A.P., 1950. The seasonal variation in weight and chemical composition of the common British Laminariaceae. J. Marine Biol. Assoc. 29: 45-72.

Black, W.A.P., 1955a. Seaweeds and their constituents in foods for man and animal. Chem. Ind. pp 1640-1645.

Black, W.A.P., 1955b. Seaweed in animal foodstuffs. I. Availability and composition. Agriculture 62: 12-15.

Black, W.A.P., 1955c. The preservation of seaweed by ensiling and bactericides. J. Sci. Food Agric. 6: 14-23.

Black, W.A.P. and Woodward, F.N., 1957. The value of seaweeds in animal feedingstuffs as a source of minerals, trace elements and vitamins. J. Exp. Agric. 25: 51-59.

Booth, E., 1961. Trace elements and seaweeds. In: A.D. DeVirville and J. Feldmann, eds. 4th International Seaweed Symposium, Pergamon Press Ltd., Oxford, pp. 385-392.

Cameron, C.D.T., 1954. Seaweed meal in the ration for bacon pigs. Can. J. Agric. Sci. 34: 181-186.

Chapman, V.J. and Chapman, D.J., 1980. Seaweeds and Their Uses. Chapman and Hall, London, pp. 1-334.

Coulson, C.B., 1953. Proteins of marine algae. Chem. Ind. 38: 997-998.

Jensen, A., 1972. The nutritive value of seaweed meal for domestic animals. In: 7th International Seaweed Symposium, University of Tokyo Press, Tokyo, Japan, pp. 7-14.

Jensen, A., 1979. Industrial utilization of seaweeds in the past, present and future. In: A. Jenson and J.R. Stein, eds. 9th International Seaweed Symposium, pp. 17-34.

Johnston, H.W., 1972. A detailed chemical analysis of some edible Japanese seaweeds. In: 7th International Seaweed Symposium, University of Tokyo Press, Tokyo, Japan, pp. 429-435.

Murakami, Y., Kazutosi, N., Kyozo, A., Shinichi, S. and Shuhei, I. 1984. Utilization of burst algal meal as feed for domestic animals and fowls. In: 11th International Seaweed Symposium, Dr. W. Junk, Publisher, Dordrecht, pp. 101-105.

National Academy of Sciences, 1980. Mineral Tolerance of Domestic Animals. National Academy of Sciences, Washington, D.C.

National Research Council, 1988. Nutrient Requirements of Domestic Animals, No. 2. Nutrient Requirements of Swine. 9th ed. National Academy of Sciences, Washington, D.C.

Nebb, H. and Jensen, A., 1966. Seaweed meal as a source of minerals and vitamins in rations for dairy cows and bacon pigs. In: 5th International Symposium on Seaweed. Pergamon Press, Oxford, pp. 387-393.

Neushel, P., 1987. Energy from marine biomass: The historical record. In: K.T. Bird and P.H. Benson, eds. Seaweed Cultivation for Renewable Resources. Elsevier, Amsterdam, pp. 1-381.

Ringen, J. 1939. Forverdien av. tangmel. Norges Landbruk Inst. Husdyr. og Forings. Flyveblad nr. 3. pp 8.

Sheehy, E.J., Brophy, J. Dillon, T. and O'Muineachian P., 1942. Seaweed (*Laminaria*) as stock food. Royal Dub. Soc. Econ. Proc. 3: 150-159.

Stauffer Chemical Company, 1975. Dehydrated California Kelp. Fact sheet. Los Angeles, California.

Van Soest, P.J., 1965. Use of detergents in analysis of fibrous feeds. III. Study of effects of heating and drying on yield of fiber and lignin in forages. J. Assoc. Offic. Agric. Chem. 48: 785-790.

Whittemore, C.T. and Percival, J.K., 1975. A seaweed residue unsuitable as a major source of energy or nitrogen for growing pigs. J. Sci. Food Agric. 26: 215-217.

Whyte, J.N.C. and Englar, J.R., 1974. Commercial kelp drying operation at Masset. Fisheries Research Board of Canada, Ottawa, Technical Report No. 453: 1-30.

Whyte, J.N.C. and Englar, J.R., 1975. Basic organic chemical parameters of the marine alga *Nereocystis luetkeana* over the growing season. Fisheries and Marine Service, Ottawa, Technical Report No. 589: 1-42.

Woodward, N., 1951. Seaweeds as a source of chemicals and stock feed. J. Sci. Food Agric. 2: 477-487.

CHAPTER 41

Sesame Meal

V. Ravindran

INTRODUCTION

Sesame (*Sesamum indicum* L.), also known as benniseed, gingelly, simsim, til and ajonjoli, is probably the most ancient oilseed crop known by man. It is often referred to as the "queen of the oilseed crops," owing to the excellent culinary properties of its oil (Salunkhe and Desai, 1986).

Due to an increasing demand for sesame oil in non-European countries, world production of sesame has increased during the past twenty years (Weiss, 1983). The major producers of sesame are India, China, Sudan, Burma and Mexico. These countries account for over 60% of the annual world production of 2.4 million tonnes (FAO, 1986). The estimated world production of sesame meal, the residue remaining after oil extraction of sesame seeds, was 1.0 million tonnes in 1980 (Salunkhe and Desai, 1986).

GROWING SESAME SEED

Sesame is considered a crop of the tropics and subtropics, but its extension into temperate zones is possible through breeding of suitable varieties. The plant is typically an erect, branched annual with a wide range of ecological adaptation. It is multi-flowered and the fruit is a cylindrically shaped capsule containing a number of small oleaginous seeds. The 1000-

seed weight varies from 2 to 4 g. The seeds may be black, white, yellow, reddish brown or grey in color, depending on variety. Light colored seeds yield a better quality oil than the dark ones and are also preferred by consumers (Weiss, 1983).

Sesame grows well on a variety of soil types, but thrives on those which are moderately fertile and free-draining. It normally requires a temperature of 25 to 27°C to produce maximum yields (2000-2500 kg seeds/ha). If the temperature falls below 20°C for any length of time, germination and growth will be delayed (Salehuzzaman and Pasha, 1979). Sesame's drought-resistant qualities are among its chief virtues, since it can be planted in relatively arid zones and still produce a good crop. Localities suitable for sorghum and dryland cotton are considered suitable for non-irrigated sesame.

Sesame is normally ready for harvesting 80 to 150 days after sowing. The capsules tend to split open and shed when ripe. Thus, harvesting time is critical for sesame, otherwise large yield reductions may occur. With the introduction of non-shattering varieties, it is now possible to harvest the sesame crop mechanically using either a reaper-binder or combine harvester. The cultural practices, fertilizer requirements, pests and diseases of the crop have been described by Weiss (1971, 1983).

NUTRIENT CONTENT OF SESAME MEAL

The nutrient composition of sesame meal compares favourably with that of soybean meal but varies widely depending on the variety used, the degree of decortication and the processing method employed (Lease and Williams, 1967; Johnson et al., 1979). For example, Lease and Williams (1967) analyzed sesame meal from five different varieties and found that the crude protein content ranged from 41 to 58%. An average protein content of 42% and fiber content of 6.5% is typical for dehulled expeller-extracted sesame meal (Table 41.1). Solvent-processed meals contain slightly higher protein (45%) and lower fat (1%) levels than those produced by expeller extraction. The digestible energy content of sesame meal is lower than in soybean meal and appears to be related to its high ash content (12.0%).

Table 41.1. Proximate Analysis of Sesame Meal and Soybean Meal (as fed)

	Soybean Meal	Sesame Meal
Dry Matter (%)	90.0	93.0
D.E. (kcal/kg)	3483	3130
Protein (%)	42.6	42.0
Crude Fiber (%)	6.2	6.5
Ether Extract (%)	4.0	7.0

National Research Council, 1979.

The thin hull of the sesame seed accounts for 15 to 20% of the whole seed (Johnson and Raymond, 1964). The hull can be separated from the kernel in decorticating machines or by soaking and rubbing the seed. Removal of the hull results in a reduction in fiber content of approximately 50% and increases the protein content, digestibility and palatability of the meal. Occasionally, the seed is milled without decortication to improve the efficiency of oil extraction. However, the meal resulting from such processing is of relatively poor nutritive quality.

Sesame meal is an excellent source of methionine, cystine and tryptophan, but is low in lysine (Table 41.2). Smith and Scott (1965) found a low level of free lysine in the plasma of sesame meal-fed chicks and concluded that lysine was the first limiting amino acid in sesame meal. This lysine deficiency precludes the use of sesame meal as the sole protein supplement in non-ruminant rations (Squibb and Braham, 1955). Grau and Almquist (1944) found chicks fed a basal ration containing 20% protein supplied from sesame meal showed poor growth. The growth response was improved when the sesame meal-based ration was supplemented with 0.5% lysine. Similar responses to lysine supplementation have been reported by Patrick (1953) and Daghir et al. (1967). Smith and Scott (1965) reported a marginal deficiency of threonine in sesame protein, based on plasma amino acid analysis. Responses to threonine supplementation have been reported (Kik, 1960; Cuca and Sunde, 1967b), indicating that threonine may be the second limiting amino acid in sesame meal-based rations.

Table 41.2. Essential Amino Acid Profile of Soybean Meal and Sesame Meal (% of Meal)

	Soybean Meal	Sesame Meal
Arginine	3.3	4.2
Histidine	1.4	1.1
Isoleucine	2.8	2.1
Leucine	3.9	3.3
Lysine	1.7	1.3
Methionine	0.7	1.2
Cystine	0.7	0.6
Phenylalanine	2.9	2.2
Threonine	2.1	1.6
Tryptophan	0.7	0.8
Valine	3.2	2.4

National Research Council, 1979.

The amino acid profile of sesame meal complements most other oilseed proteins. This complementary value was well demonstrated by the early studies of Almquist and Grau (1944) who fed soybean meal, sesame meal and combinations of these two to chicks. Best gains were obtained at the soybean meal/sesame meal ratio of about 2:1. However, subsequent studies showed that a ratio of 1:3 was sufficient to support satisfactory growth of chicks (Cuca and Sunde, 1967b).

Almost 75 to 80% of the sesame protein is reported to be digestible (Ellis and Bird, 1951; Ravindran, 1982). However, prolonged heating may severely depress the availability of amino acids (Aherne and Kennelly, 1985). Processing of sesame meal at high temperatures can also result in the destruction of cystine, and the sulphur-containing amino acids may actually become deficient under these conditions (Ravindran et al., 1982).

Sesame meal contains a higher fat content than does soybean meal (Table 41.1). The principle fatty acids of sesame oil are oleic and linoleic acids (Table 41.3). The saturated fatty acids account for only about 12% of the fatty acids in sesame oil and consist mainly of palmitic and stearic acid. The high degree of unsaturation in sesame oil can result in soft carcass fat in pigs fed high levels of sesame meal.

Table 41.3. Fatty Acid Composition of Sesame Oil (%)

Oleic Acid	45.3 - 49.4
Linoleic Acid	37.7 - 41.2
Palmitic Acid	7.8 - 9.1
Stearic Acid	3.6 - 4.7
Archidonic Acid	0.4 - 1.1
Hexadecenoic Acid	0.0 - 0.5
Myristic Acid	0.1 - 0.2

Budowski and Markley, 1951.

Although sesame meal is not a rich source of vitamins, its vitamin content can be considered comparable with soybean meal (Table 41.4) and other oilseed meals (Aherne and Kennelly, 1985).

Table 41.4. Vitamin Composition of Soybean Meal and Sesame Meal

	Soybean Meal	Sesame Meal
Biotin (ug/kg)	330.0	340.0
Choline (mg/kg)	2703.0	1690.0
Niacin (mg/kg)	37.0	30.0
Pantothenic Acid (mg/kg)	14.0	6.0
Riboflavin (mg/kg)	3.7	3.6
Thiamin (mg/kg)	1.7	2.8

National Research Council, 1979.

The mineral composition of sesame meal is presented in Table 41.5. In general terms, sesame meal is particularly rich in phosphorus, magnesium and microminerals. The calcium content of sesame meal is seven times higher than that in soybean meal. However, Cuca and Sunde (1967a) studied the availability of calcium in sesame meal and reported that the calcium supplied by corn-sesame meal rations was not as available as the calcium supplied by corn-soybean meal rations. The presence of phytates and oxalates may have contributed to the relatively low calcium availability in sesame meal-based rations. However, the high level of calcium present in sesame meal more than compensates for its low availability.

Lease et al. (1960) showed that chicks fed a sesame meal-based ration developed leg deformities similar to those reported in zinc deficiency, even though they received adequate amounts of zinc in the ration. Addition of 60 mg/kg zinc improved growth rates and greatly reduced leg deformities. In a subsequent study (Lease, 1966), autoclaving sesame meal for two hours decreased leg abnormalities and improved growth. Although no evidence of zinc deficiency has been reported in the literature, it is recommended that additional zinc be supplied when sesame meal forms a significant part of swine rations.

Table 41.5. Mineral Composition of Soybean Meal and Sesame Meal[1]

	Soybean Meal	Sesame Meal
Macrominerals (%)		
Potassium	1.83	1.20
Calcium	0.27	1.99
Magnesium	0.26	0.86
Phosphorus	0.61	1.37
Sodium	0.27	0.04
Microminerals (mg/kg)		
Zinc	60	100
Manganese	31	48
Iron	140	320[2]
Copper	18	35[2]

[1]National Research Council, 1979.
[2]Ravindran et al., 1982.

UNDESIRABLE CONSTITUENTS IN SESAME MEAL

Sesame meal can vary in color from light yellow to greyish-black, depending on the predominant testa color of the seeds used. Dark-colored meals are often less palatable than lighter ones, and the latter are preferred where there is a choice. The bitter taste of the darker meals is related to the high oxalic and phytic acid levels present in the hull (Aherne and Kennelly, 1985). Sesame meal contains approximately 5% phytate and 35 mg/100 g of oxalates and these compounds may interfere with mineral availability (Toma et al., 1979). Decortication of seeds almost completely removes the oxalates, but has little effect on phytate. Phytates are known to lower the availability of calcium, zinc, phosphorus, magnesium and possibly iron (Reddy et al., 1982).

FEEDING SESAME MEAL

Starter Pigs

There is no published information on the use of sesame meal in diets fed to starter pigs. However, based on its relatively high crude fiber content (6.5%) and the possibility of palatability problems due to the presence of phytates and oxalates, it would seem wise to limit the use of sesame meal to 5% in starter rations until more research data is available.

Growing-Finishing Pigs

Squibb and Salazar (1951) were probably the first to evaluate the possible use of sesame meal in swine rations. Their studies indicated that satisfactory performance of growing-finishing swine can be obtained with 15% of the ration supplied by sesame meal. Later reports (Maner and Gallo, 1963; Gohl, 1981) have shown that sesame meal should be blended with high-lysine sources such as soybean meal, fish meal or meat meal when used as a protein supplement in swine rations based on cereal grains. Sesame meal can replace one-half or all of the soybean meal in corn-soybean meal rations containing 4-5% fish meal or meat meal (Maner and Gallo, 1970; Gohl, 1981) and can be substituted for soybean meal at dietary levels as high as 10% in corn-soybean meal rations which do not contain animal protein (Gallo and Maner, 1970).

Because of the low levels of lysine in sesame meal and cottonseed meals, or blends of these two feed ingredients, they are inadequate for supplementing cereal-based growing-finishing swine rations. However, 50:50 blends of these protein sources fed along with cereal grains and 5-6% fish meal produce adequate gains and feed efficiency (Hervas et al., 1965; cited by Pond and Maner, 1984). Use of too high levels of sesame meal in finisher rations should be avoided, since it tends to produce soft pork (Gohl, 1981).

Breeding Stock

There is no published information on the use of sesame meal in diets fed to breeding stock. However, there does not appear to be any reason why sesame meal could not be used successfully in diets fed to sows in gestation and lactation.

SUMMARY

The use of sesame meal as a swine feed under practical conditions is governed almost entirely by price and availability. The scarcity of sesame meal, in relation to the large quantities of other oilseed meals available and the total production of livestock feed, has relegated its use to high-price or specially compounded feeds. However, sesame meal can be profitably utilized in swine diets provided that it can be purchased at a cost lower than that of soybean meal.

In practical diets, sesame meal can be used to provide a portion or all of the supplemental protein in cereal-based swine diets for all ages of swine except the starter pig. The level of substitution will depend upon the type and quantity of other protein supplements used in the formulation. For example, with 4 to 6% fish meal in the formulation, sesame meal can replace all of the soybean meal in cereal based diets. Additional mineral supplementation is recommended when sesame meal is used at levels of more than 10%.

REFERENCES

Aherne, F.X. and Kennelly, J.J., 1985. Oilseed meals for livestock feeding. In: D.J.A. Cole and W. Haresign, eds. Recent Developments in Pig Nutrition, Butterworths, London, pp. 278-315.

Almquist, H.J. and Grau, C.R., 1944. Mutual supplementary effect of proteins of soybean meal and sesame meal. Poult. Sci. 23: 341-343.

Budowski, P. and Markley, K.S., 1951. The chemical and physiological properties of sesame oil. Cereal Chem. 48: 121-151.

Cuca, M. and Sunde, M.L., 1967a. The availability of calcium from Mexican and Californian sesame meals. Poult. Sci. 46: 994-1002.

Cuca, M. and Sunde, M.L., 1967b. Amino acid supplementation of a sesame meal diet. Poult. Sci. 46: 1512-1516.

Daghir, N.J., Ullah, M.F. and Rottensten, K., 1967. Lysine supplementation of sesame meal broiler rations. Trop. Agric. (Trinidad) 44: 235-242.

Ellis, N.R. and Bird, H.R., 1951. By-products as feed for livestock. In: Crops in Peace and War. Yearbook of Agriculture. U.S. Department of Agriculture, Washington, D.C.

FAO, 1986. Production Yearbook. Food and Agriculture Organization, Rome, Vol. 40.

Gallo, J.T.C. and Maner, J.H., 1970. Sesame meal for pigs. I. Nutritive value of sesame meal for growing and finishing pigs. Revista Instituto Colombiano Agropecuario 5: 107-112.

Grau, C.R. and Almquist, H.J., 1944. Sesame meal protein in chick diets. Proc. Soc. Exp. Biol. Med. 57: 187-189.

Gohl, B., 1981. Tropical Feeds. Food and Agriculture Organization, Rome.

Johnson, R.H. and Raymond, W.D., 1964. The chemical composition of some tropical food plants. III. Sesame. Trop. Sci. 6: 173-179.

Johnson, L.A., Suleiman, T.M. and Lucas, E.W., 1979. Sesame protein: A review and prospects. J. Amer. Oil Chem. Soc. 56: 463-468.

Kik, M.C., 1960. Effects of amino acid supplements, vitamin B_{12} and buffalo fish on the nutritive value of proteins in sesame seed and meal. J. Agric. Food Chem. 8: 327-330.

Lease, J.G., 1966. The effect of autoclaving sesame meal on the phytic acid content and on the availability of its zinc to the chick. Poult. Sci. 45: 237-241.

Lease, J.G. and Williams, W.P., Jr., 1967. Availability of zinc and comparison of in vitro and in vivo zinc uptake of certain oilseed meals. Poult. Sci. 46: 233-241.

Lease, J.G., Barnett, B.D., Lease, E.J. and Turk, D.E., 1960. The biological unavailability to the chick of zinc in a sesame meal ration. J. Nutr. 72: 66-70.

Maner, J.H. and Gallo, J.T., 1963. Valor nutritivo de la torta de ajonjoli comoreemplazo de la terta de soya en dietas para cerdos en crecimiento y acabado. Primer Congreso Nacional de la Industria Porcina, Bogata, Colombia, pp 1-3.

Maner, J.H. and Gallo, J.T., 1970. La torta de ajonjoli en alimentacion de cerdos. II. Efecto de la supplementation de las tortas de soya y ajonjoli con metionina, en raciones para cerdos. Revista Instituto Colombiano Agropecuario. 5: 113-119.

National Research Council, 1979. Nutrient Requirements of Domestic Animals, No. 2. Nutrient Requirements of Swine. 8th ed. National Academy of Sciences, Washington, D.C.

Patrick, H., 1953. Deficiency in a sesame meal type ration for chicks. Poult. Sci. 32: 744-745.

Pond, W.G. and Maner, J.H., 1984. Swine Production and Nutrition. AVI Publishing Company, Westport, Connecticut, 731 pp.

Ravindran, V., 1982. Methodology to Evaluate the Nutritive Value of Feedstuffs for Poultry and Swine. M.Sc. Thesis, Virginia Tech. University, Blacksburg.

Ravindran, V., Kornegay, E.T., Webb, J.R., K.E. and Rajaguru, A.S.B., 1982. Nutrient characterization of some feedstuffs of Sri Lanka. J. Natl. Agric. Soc. Ceylon. 19: 19-32.

Reddy, N.R., Sathe, S.K. and Salunkhe, D.K., 1982. Phytates in legumes and cereals. Adv. Food Res. 28: 1-92.

Salehuzzaman, M. and Pasha, M.K., 1979. Effects of high and low temperatures on the germination and growth of flax and sesame. Indian J. Agric. Sci. 49: 260-261.

Salunkhe, D.K. and Desai, B.B., 1986. Postharvest Biotechnology of Oilseeds. CRC Press, Boca Raton, Florida.

Smith, R.E. and Scott, H.M., 1965. Use of free amino acid concentrations in blood plasma in evaluating the amino acid adequacy of intact proteins for chick growth. II. Free amino acid patterns of blood plasma of chicks fed sesame and raw, heated and overheated soybean meals. J. Nutr. 86: 45-50.

Squibb, R.L. and Braham, J.E., 1955. Blood meal as a lysine supplement to all-vegetable protein rations for chicks. Poult. Sci. 34: 1050-1053.

Squibb, R.L. and Salazar, E., 1951. Value of corozo palm nut and sesame oil meals, bananas, A.P.F. and cow manure in rations for growing and fattening pigs. J. Anim. Sci. 10: 545-550.

Toma, R.B., Tabekhia, M.M. and Williams, J.D., 1979. Phytate and oxalate contents in sesame seed. Nutr. Rep. Int. 20: 25-31.

Weiss, E.A., 1971. Castor, Sesame and Safflower. Leonard Hill, London.

Weiss, E.A., 1983. Oilseed Crops, Tropical Agricultural Series. Longman, London.

CHAPTER 42

Single Cell Protein

D.G. Waterworth

INTRODUCTION

During the 1960s, it was postulated that the rapidly increasing world population and the increasing consumption of meat by both industrially developed and developing countries would create an ever increasing demand for quality protein materials to supplement animal diets and human food. For this reason, there was a considerable amount of activity during the 1960s and 1970s directed towards the development of supplies of single cell protein.

There are a number of reasons why most of the projects based on the fermentation of feedstock originating from fossil fuels have not continued (Sharp, 1989). Firstly, the global requirements for protein have been reduced by changes in the method by which these requirements are assessed. The world's supply of soya protein has also increased dramatically and this increased supply has reduced the price of soya protein in real terms. In addition, the energy crisis of the 1960s and 1970s increased feedstock costs and thereby reduced the profit margin for manufacturers of single cell protein. Finally, research to more closely define our knowledge of amino acid requirements and the increased use of synthetic amino acids has helped to improve the efficiency of utilization of animal feeds. This, together with the development of more efficient breeds of pigs and poultry has helped to minimize the demand for protein.

Despite these changes, the original premise still holds that global demands for protein will increase with time. Alternatives must be available to allow the continued development of a profitable livestock sector. This chapter is principally a description of "Pruteen", a bacterial single cell protein (SCP) developed by Imperial Chemical Industries. However, other single cell proteins that have potential as protein sources for pigs are also included.

SOURCES OF SINGLE CELL PROTEIN

The main criteria for selecting a microorganism and a process for the development of an single cell protein suitable for inclusion in animal diets are:

a) The single cell protein must have a high protein content and a good balance of essential amino acids.

b) It must be palatable, digestible by the animal species and free from toxic components.

c) The micro-organism must grow economically on a chosen substrate.

There are a number of organisms that have been used to produce single cell protein including various forms of algae, fungi, yeasts and bacteria. Examples of the major groups of organisms used for the production of single cell protein and their chemical analyses are given in Table 42.1. The algae that have been investigated grow slowly, require light and generally do not lend themselves to high tonnage production while the fungi generally have low protein contents. However, some fungi have the advantage of utilizing waste cellulosic substrates. "Pekilo" was a product developed in Finland using fungi grown on sulphite liquor effluent.

Table 42.1. Chemical Composition (%) of Single Cell Proteins Derived from Various Organisms, Compared with Conventional Protein Sources

	Algae	Fungi	Yeast	Bacteria	Soybean Meal	White Fish Meal
Crude Protein	52.0	32.0	60.0	74.0	45.0	64.0
Fat	15.0	5.0	9.0	8.0	1.0	9.0
Ash	7.0	2.0	6.0	8.0	6.0	18.0
Fiber	11.0	28.0	---	---	6.0	----
Lysine	2.4	1.5	4.2	4.1	2.8	4.7
Methionine + Cyst.	1.7	0.8	1.7	2.3	1.3	2.8

Van der Wal, 1979.

Yeasts are the oldest source of single cell protein used in animal feeds. *Saccharomyces cerevisae*, a by-product of the brewing industry, has been used at low levels in animal feeds for centuries. In addition, some *Saccharomyces* species can be grown on whey (Moebus and Teuber, 1979) and a process was developed by Bel Industries in France to make an ingredient for calf milk replacers. Candida yeasts, utilizing paraffins and gas oil residues from the oil industries, were the basis of the product "Toprina" made by British Petroleum. "Toprina" was a high protein yeast (62% crude protein and 4.1% lysine) and was shown to be an excellent protein supplement for pigs of all ages.

Bacteria have very high cell growth rates in optimum fermentation conditions, a characteristic which when put together with the wide variety of species and utilizable substrates, makes them attractive for single cell protein production. A species of Pseudomonas utilising methane as the substrate was investigated by Shell (Wilkinson et al., 1974) but the choice of potential bacteria able to use methane for energy was very restricted. Imperial Chemical Industries developed a single cell protein based on the fermentation of methanol by *Methylophilus methylotrophus*, the basis of "Pruteen". Similar organisms were chosen by Hoechst (*Methylomonas clara*) and by Norsk Hydro (*Methylomonas methanica*). The names of products developed or investigated from these groups of organisms are given in Table 42.2 (Waterworth, 1981).

Table 42.2. Substrates and Micro-Organisms of Commercial Potential

Organisms	Substrate	Process	Trade Name
Algae			
Chlorella spp.	water, sewage		
Scenedesmus spp.			
Fungi			
"an Ascomycete"	woodpulp	Paper Industry	"Pekilo"
Bacteria			
Pseudomonas and other spp.	methane	Shell	
Methylophilus methylotrophus	methanol	ICI	"Pruteen"
Methylomonas clara	methanol	Hoechst	"Probion"
Methylomonas methanica	methanol	Norsk-Hydro	"Norprotein"
Pseudomonas spp.	methanol	Mitsubishi	
Yeasts			
Candida spp.	n-paraffin	Dainippon	"Ronipron"
Candida spp.	n-paraffin	Liquichimica	"Liquipron"
Candida lipolytica	n-paraffin	British Petroleum	"Toprina"
Candida tropicalis	gas oil	British Petroleum	
Pichia	methanol	Phillips Petroleum	"Provesteen"
Saccharomyces cerevisiae	maltose etc	Brewing Industry	Feed yeast
Saccharomyces cerevisiae	whey	Bel Industries	"Bel yeast"
Torulopsis sp.	sulphite liquor	Paper Industry	Torula yeast

Waterworth, 1981.

SAFETY TESTING

The guidelines for the toxicological and safety testing of industrially derived single cell proteins intended for inclusion in animal feeds were drawn up by the International Union of Pure and Applied Chemists and the Protein Advisory Group of the United Nations (Protein Advisory Group, 1974; Hoogerheide et al., 1979). Single cell proteins have been subjected to extensive toxicological testing to ensure their safety to the consumer, the process operator and any animals which might consume the product. The toxicological testing program for "Pruteen" was described by Stringer (1982) and for "Toprina" by Shaklady and Gatumel (1972).

THE PRODUCTION PROCESS

"Pruteen" differs from conventional protein sources in that it is produced using the growth of a microorganism in a fermenter. A novel pressure cycle fermenter was developed by Imperial Chemical Industries in the 1970s (Craig and Trotter, 1983). The bacterium *Methylophilus methylotrophus* uses methanol as a carbon and energy source and ammonia for nitrogen. Fermentation takes place in a pressure fermenter under monoculture conditions which exclude the ingress of unwanted pathogens. After fermentation, the organism is killed during harvesting and concentrated to a cream by flotation and centrifugation before drying to 10% moisture in flash driers designed to minimise heat damage. The standard product is granular in texture.

The process lends itself to the addition of soluble nutrients for the animal, such as minerals, via the liquid phase of the process or added after the drying stage, e.g. supplementary fat. The controlled production means that the product is consistent in composition, free from salmonella and other pathogens and production is unaffected by seasonal factors. The low moisture content, low unsaturated fatty acids and the virtual absence of reducing sugars makes it chemically very stable. Therefore "Pruteen" stores well over long periods. The process for the production of "Toprina" from Gas Oil or n-paraffin involves fermentation in stirred fermenters (Levi et al., 1979, Midgley, 1979).

NUTRIENT CONTENT OF "PRUTEEN"

The proximate analysis of the standard granular product is given in Table 42.3. The mean digestible energy determined for pigs by Imperial Chemical Industries is 4150 kcal/kg of product as received (van Weerden and Hausman, 1978; Wiseman et al., 1977; Whittemore et al. 1976a). The average protein content is about 72% with about 10% ash. The lipid is principally phospholipid measured by extraction with chloroform and methanol by the method of Bligh and Dyer (1959).

Table 42.3. Nutrient Composition of "Pruteen" Granules (as fed)

Crude Protein (%)	72
Lipid (%)	8
Ash (%)	10
Moisture (%)	8
D.E. (kcal/kg)	4150

Imperial Chemical Industries (unpublished).

"Pruteen" has a high content of some of the essential amino acids including lysine, threonine, tryptophan and the sulphur amino acids (Table 42.4). The biological availability of these amino acids is high. D'Mello et al. (1976) measured the digestibility of these amino acids using semi-synthetic diets fed to 43 kg pigs. The digestibility of both lysine and threonine was 96% and methionine 97%. In an experiment using barley based diets, Whittemore et al. (1976a) found the digestibility of the protein to be 91%. Similar values were obtained by Braude et al. (1977) and van Weerden and Hausman (1978). The digestibilities of the protein and energy in other single cell proteins are given by Roth (1980).

Table 42.4. Essential Amino Acid Composition of "Pruteen" Granules (% as fed)

Arginine	3.7	Cystine	0.5
Histidine	1.3	Phenylalanine	2.6
Isoleucine	3.3	Tyrosine	2.2
Leucine	5.2	Threonine	3.3
Lysine	4.1	Tryptophan	1.0
Methionine	1.4	Valine	4.0

Imperial Chemical Industries (unpublished).

The mineral and vitamin contents are given in Table 42.5. The phosphorus originates from the phospholipid and nucleic acid components of the cell and from phosphoric acid used in the production process. The availability of phosphorus is regarded as 100%. The high availability of the phosphorus in "Pruteen" allows for cost savings in diet formulation by reducing the need for inclusion of calcium disphosphate. The calcium contained in "Pruteen" derives from $Ca(OH)_2$ used for neutralization in the harvesting process. The sodium and potassium content are low. The chloride level is very low because neither the sodium nor potassium originates from salts. This is a favorable value for nutritionists wishing to optimise the ionic balance using the formula Na + K minus Cl in poultry or pig diets. Sodium selenite is usually included to provide an evenly distributed level of 0.4 mg/kg Se in the final product to equate with the average, but variable, level in fish meal.

Shacklady and Gatumel (1972) highlighted the need to consider the vitamin content of single cell protein. The vitamins in "Pruteen" are not major contributors with the exception of biotin, which at 2.9 mg/kg is one of the richest sources of this essential vitamin. In contrast, the extraction of lipid material from "Toprina" with a polar solvent significantly reduces the choline content. Since synthetic choline chloride is normally added at a set level in the formulation of pig diets, high dietary inclusion of "Toprina" with corresponding removal of another material naturally high in choline could lead to a deficiency of this vitamin in the diet.

Table 42.5. Mineral and Vitamin Content of "Pruteen" Granules

	%		mg/kg		mg/kg
Phosphorus	2.5	Iron	350	Biotin	2.9
Calcium	1.3	Copper	<30	Thiamin	5
Sodium	0.2	Zinc	<50	Riboflavin	37
Potassium	0.18	Selenium	0.4	Pyridoxine	2
Magnesium	0.22	Folic Acid	14	Vitamin B_{12}	0.03
Chlorine	0.03	Nicotinic Acid	52	Pantothenic Acid	10

"PRUTEEN" AS A PROTEIN SOURCE FOR PIGS

Most of the developmental work on "Pruteen" was carried out by Imperial Chemical Industries at the Jealotts Hill Research Station, Bracknell, England. The external research on pigs was conducted mainly in Europe at universities, research centres and commercial companies. The nutrition reports and formulation recommendations are given in the Imperial Chemical Industries brochures. In general, "Pruteen" should be formulated in pig diets on the basis of the digestible energy, principal essential amino acids, available phosphorus and calcium with appropriate adjustments for sodium and chloride if required.

Suckling Piglets

The high digestibility of the energy and protein and the absence of any undesirable flavors makes "Pruteen" an appropriate material to replace the protein in skimmed milk powder in piglet prestarter diets. Newport and Keal (1980) replaced 15% skim milk powder by "Pruteen" and whey in diets for piglets weaned at only 2 days of age and showed equivalent growth and reduced health problems for the animals on the "Pruteen" diet. Hutagalung (1982) replaced 10% skim milk powder with various levels up to 10% "Pruteen" in prestarter diets fed to 280 piglets and concluded that "Pruteen" is a satisfactory protein material for piglets.

"Pruteen" can be substituted for skim milk powder by normal linear programming formulation or using a general substitution formula based on two parts of skim milk being equal to one part "Pruteen" and one part whey powder. Other energy sources acceptable to piglets such as starch, lactose and fats can be used, in part, instead of whey powder. The general recommendation is to replace all or part of the skim milk powder by up to 7.5% "Pruteen".

Starter Pigs

Fast growing starter pigs are laying down lean meat very rapidly and require high quality protein in their diets. In a summary of 13 experiments (Imperial Chemical Industries, 1978) involving 3300 pigs, "Pruteen" replaced fish meals of various origins in diets with balanced protein, lysine, methionine, energy, calcium and phosphorus. Overall, the results and responses were as given in Table 42.6 showing a statistically significant ($P<0.01$) improvement for the "Pruteen" fed pigs over the controls. The reason for these improvements is unknown but higher amino acid availability or a more favorable balance of amino acids, e.g. threonine, was suggested.

Table 42.6. Growth Responses of Starter Pigs Fed Diets in Which "Pruteen" Replaced Fish Meal

	Daily Gain (g/d)	Daily Intake (g/d)	Feed Efficiency
Control	409	964	1.93
"Pruteen"	522	978	1.87

In six trials conducted in the United Kingdom and Germany, Imperial Chemical Industries reported equivalent growth performance and feed intake when 4 to 10% "Pruteen" was used to replace up to 25% of the skim milk powder in diets fed to 574 piglets from 18 days of age to approximately 49 days (Imperial Chemical Industries, 1978). Control piglets grew satisfactorily at 282 g/day with a feed efficiency of 1.28. O'Grady (1978) replaced a mixture of skim milk powder, fish and soybean by "Pruteen" in the diets of six week old pigs. Over the subsequent five weeks the efficiency of conversion and growth were improved significantly on the diets with "Pruteen" inclusion.

Growing-Finishing Pigs

Braude et al. (1977) compared "Pruteen" at 3 and 7% in the diet as a replacement of white fish meal fed to pigs up to 60 or 90 kg. No difference in growth performance was found. The carcasses were extensively tested for physical measurements, dissected into lean and fat and taste panel and cooking quality tests were performed (Braude and Rhodes, 1977). No differences were found between pigs fed "Pruteen" or fish meal. Imperial Chemical Industries reported that in a number of pork and bacon trials, "Pruteen", at levels which replaced all non-cereal protein from soybean meal or fish meal (i.e. 5-15%) gave the same growth performance and carcass backfat measurements. Whittemore et al. (1976a,b) obtained equivalent growth when "Pruteen" replaced white fish meal in diets fed to pigs between 23 and 60 kg.

Breeding Pigs

Diets containing 0 or 8% "Pruteen" were fed to sows for various periods up to four years of age (Waterworth and Heath, 1981). In total, over 7000 piglets were born to both the control and "Pruteen" fed sows and no abnormalities in reproductive performance of the sows, or in the number or weights of piglets born and reared, were recorded as a result of feeding "Pruteen". A similar breeding study under practical farm conditions was carried out on the same diets (Burt and Heath, 1980). There was no effect of "Pruteen" on the number of piglets born and weaned and there was no effect on breeding performance, fecundity of the sows or spermatogenesis of the boars. Similarly designed pig production experiments also gave no adverse effect with "Toprina" (Van der Wal, 1979).

SUMMARY

"Pruteen" is a high protein and energy raw material suitable for inclusion in the diets of pigs of all ages. The controlled production and monitoring of single cell proteins can ensure consistent quality, good storage properties and continuity of supply. "Pruteen" supplies 4150 kcal D.E. per kg, 72% crude protein, 4.1% lysine, 1.9% sulfur amino acids, 3.3% threonine and 1.0% tryptophan. The availability of these and most other essential amino acids is over 95%. The phosphorus content at 2.5% is 100% available, as is the 2.9 mg/kg biotin.

"Pruteen" is recommended at dietary inclusion levels of 6% in prestarter and 7.5% in starter diets to achieve maximum production performance. Total substitution of the supplementary protein can generally be achieved by 10% "Pruteen" in these diets and give equivalent growth. In grower-finisher and breeding diets, any inclusion rate up to 15% "Pruteen" will give equivalent performance.

REFERENCES

Bligh, E.G. and Dyer, W.J., 1959. A rapid method of total lipid extraction and purification. Can. J. Biochem. Physiol. 37: 911-917.

Braude, R., Hosking, Z.D., Mitchell, K.G., Plonka, S. and Sambrook, I.E., 1977. "Pruteen", a new source of protein for growing Pigs. I. Metabolic experiment: Utilisation of nitrogen. Livest. Prod. Sci. 4: 79-89.

Braude, R. and Rhodes, D.N., 1977. "Pruteen", a new source of protein for growing pigs. II. Feeding trial: Growth rate, feed utilisation and carcass and meat quality. Livest. Prod. Sci. 4: 91-100.

Burt, A.W. and Heath, M.E., 1980. "Pruteen" as a replacement for soybean meal and fish meal in the diet of breeding pigs over three generations. Anim. Prod. 30: 470 (Abstract).

Craig, J.B. and Trotter, S.G., 1983. ICI Technology for SCP production from Methanol. Proceedings International Symposium on Single Cell Protein from Hydrocarbons, Algiers, Algeria, pp. 17-34.

D'Mello, J.P.F., Peers, D.G. and Whittemore, C.T., 1976. Utilization of dried microbial cells grown on methanol in a semi-purified diet for growing pigs. Brit. J. Nutr. 36: 403-410.

Hoogerheide, J.C., Yamada, K., Littlehailes, J.D. and Ohno, K. 1979. Guidelines for testing of single cell protein destined as protein source for animal feed. Pure Appl. Chem. 51: 2537-2560.

Hutagalung, R.I., 1982. Dried microbial cells as a source of protein for swine. Second Animal Science Congress of Australasian Association of Animal Production, Manila, Philippines.

Imperial Chemical Industries, 1978. A New Protein Source. Billingham, Cleveland, England.

Levi, J.D., Shennon, J.L. and Ebbon, G.P., 1979. Biomass from Liquid Alkanes. Economic Microbiology Vol. 4. Academic Press, pp. 362-419.

Midgley, G., 1979. A Process-Protein from Petroleum. Joint IVA/CBMPE Symposium, Gothenberg, Sweden.

Moebus, O. and Teuber, M., 1979. Bundesanstalt fur Milchforschung Ann. Rep. B. 35-46.

Newport, N.J. and Keal, H.D., 1980. Artificial rearing of pigs. Effect of replacing dried skimmed milk by a single cell protein ("Pruteen") on performance and digestion of protein. Brit. J. Nutr. 44: 161-170.

O'Grady, J., 1978. Protein for young pigs. Irish Farmers J. (Feb. 25) p. 7.

Protein Advisory Group, 1974. Guideline No. 15 on Nutritional Safety Aspects of Novel Protein Sources for Animal Feeding, United Nations, Geneva.

Roth, F.X., 1980. Micro-organisms as a Source of Protein for Animal Nutrition. Anim. Res. Dev. 12: 7-19.

Shacklady, C.A. and Gatumel, E., 1972. Proteins From Hydrocarbons. Academic Press, New York.

Sharp, D.H., 1989. Bioprotein Manufacture: A Critical Assessment, Publ. Horwood, p. 15.

Stringer, D.A., 1982. Industrial development and evaluation of new protein sources: Micro-organism. Proc. Nutr. Soc. 41: 289-300.

Van der Wal, D., 1979. Perspective on Bioconversion of Organism Residues for Rural Communities. United Nations University, Geneva, Food and Nutr., Bull. Suppl. 2: 3-24.

van Weerden, E.J. and Hausman, J., 1978. The digestibility of protein and amino acids and energy value of "Pruteen" for pigs. Anim. Feed Sci. Technol. 2: 377-383.

Waterworth, D.G. and Heath, M.E., 1981. "Pruteen" in the diet of breeding pigs: Reproductive performance. Anim. Feed Sci. Technol. 6: 297-307.

Waterworth, D.G., 1981. Single Cell Protein. Outlook on Agriculture 10: 403-408.

Whittemore, C.T., Moffat, I.W. and Taylor, A.G., 1976a. Evaluation by digestibility, growth and slaughter of microbial cells as a source of protein for young pigs. J. Sci. Food Agric. 27: 1163-1170.

Whittemore, C.T. and Moffat, I.W., 1976b. The digestibility of dried microbial cells grown on methanol in diets for growing pigs. J. Agric. Sci. (Camb.) 86: 407-410.

Wilkinson, T.G., Topwala, H.H. and Hamer, G., 1974. The Utilization of Methane by Bacteria. Biotechnol. Bioeng. 19: 45-59.

Wiseman, J., Cole, D.J.A. and Lewis, D., 1977. The energy yielding value of two microbial proteins in diets for growing pigs. Anim. Prod. 24: 136-137.

CHAPTER 43

Soybeans: Full-Fat

A.C. De Schutter and J.R. Morris

INTRODUCTION

Solvent extracted soybean meal, a by-product of the oil-seed industry, remains the single largest source of supplemental protein used by livestock producers in North America. Its popularity can be attributed to several factors including its widespread availability, its high protein, lysine and energy content as well as its palatability. Although soybean meal tends to be cost competitive, alternative protein sources must continue to be examined.

In the last 10 to 15 years, the use of heat-treated whole or "full-fat" soybeans in swine rations has increased dramatically. Much of the initial interest was generated as the cost of full-fat soybeans became competitive with soybean meal. This occurred due to increases in feed grain prices and decreases in vegetable oil prices on the world market (Pfost, 1984). However, the economics of using this alternative have fluctuated considerably with changes in the price of oil and meal.

Other factors have also contributed to the popularity of the full-fat soybean. One main factor is that full-fat soybeans allow the opportunity for swine producers to utilize a home-grown protein supplement in their swine rations. In addition, the use of high fat rations in swine feeding programs have become increasingly popular due to advantages such as increased energy density, improved palatability, reduced dust and an overall enhanced animal appearance. Whole soybeans have a high oil content and their inclusion in swine rations can be a means of increasing the fat content of the feed. The use of whole soybeans is preferred in on-farm mixing situations compared to fat alternatives such as tallow, or pure soybean oil, due to its relative ease of handling.

NUTRIENT CONTENT OF FULL-FAT SOYBEANS

The nutrient content of the full-fat soybean is compared to 48% SBM, the standard protein supplement, in Tables 43.1 and 43.2. Whole soybeans are substantially lower in protein content than soybean meal, containing only 36 to 38% crude protein. Since the oil content (18%) of the full-fat soybean is approximately three times higher than that found in 48% SBM, there is an obvious potential for a corresponding increase in digestible energy content.

Table 43.1. Proximate Analysis of Cooked Full-Fat Soybeans and Solvent Extracted Soybean Meal (SBM)

	Solvent 44% SBM	Dehulled Solvent 48% SBM	Cooked Full-Fat Soybeans
Dry Matter %	90	90	90
D.E. (kcal/kg)	3490	3680	4035
Crude Protein (%)	44.0	48.5	36.7
Crude Fiber (%)	7.3	3.4	5.2
Ether Extract (%)	1.1	0.9	18.8

National Research Council, 1988.

There is considerable variation in the nutrient content of full-fat soybeans, with protein values ranging from 32% (Danielson, 1985) to 41% (Manitoba Agriculture, 1987) and oil from 17% (Castell and Cliplef, 1988) to 21.3% (Manitoba Agriculture, 1987). Most of this variation can be attributed to either soybean variety or weather conditions during the growing season. As the protein content is quite variable, chemical analysis of the whole soybean is recommended. Other factors, such as the method and conditions of processing, can also have considerable effect on protein and energy availability.

Table 43.2. Essential Amino Acid Content of Cooked Full-Fat Soybeans and Solvent Extracted Soybean Meal (SBM; % of protein)

	Dehulled Solvent Extracted SBM	Solvent Extracted SBM	Cooked Full-Fat Soybeans
Isoleucine	4.4	4.5	4.4
Leucine	7.5	7.7	7.2
Lysine	6.4	6.6	6.1
Methionine	1.5	1.2	1.2
Phenylalanine	4.9	4.8	4.9
Threonine	3.9	3.9	3.8
Tryptophan	1.4	1.5	1.5
Valine	5.1	4.6	4.4

National Research Council, 1988.

UNDESIRABLE CONSTITUENTS IN FULL-FAT SOYBEANS

For over fifty years, it has been recognized that pigs cannot utilize the nutrients in raw soybeans efficiently. The major cause of this reduced performance is the presence of a group of compounds known as trypsin inhibitors. As their name implies, these compounds reduce the activities of trypsin and chymotrypsin which are pancreatic enzymes involved in protein digestion (Yen et al., 1977). While the precise mechanism by which these compounds exert their effect has not been well established, it is certain that the inhibitors reduce protein digestibility and appear to interfere with systemic protein utilization (Myer et al., 1982; Myer and Froseth, 1983).

Surprisingly, there are relatively few studies which have examined the direct relationship between trypsin inhibitor level and performance. The acceptable levels of trypsin inhibitor are within the range of 5.3 mg/g to 18.6 mg/g (Hansen et al., 1984; Pontif et al., 1987). The level of trypsin inhibitor found in raw soybeans has been reported to be well above this range, at 21.1 mg/g (Dale et al., 1987) and 31.1 mg/g (Vandergrift et al., 1983), and can be expected to vary with the source of soybean.

While trypsin inhibitor content is believed to be the ideal analysis for predicting hog performance, the procedure itself is rather difficult and consequently expensive to perform. Fortunately, measurement of other heat-liable compounds which are also found in raw soybeans, such as urease, can be correlated to trypsin inhibitor content (Albrecht et al., 1966; Carter and Cox, 1984). Urease activity is relatively simple to determine and consequently has become the standard test used to evaluate the effectiveness of the heating process of full-fat soybeans (Vandergrift, 1985). The units used to describe urease activity are given in change in pH, which reflects the decrease in pH. Urease activity levels of 0.20 to 0.05 change in pH indicate adequate denaturation of trypsin inhibitor (Smith, 1977; Kohlmeier, 1988).

Plant geneticists have examined the potential for producing soybean varieties which have low trypsin inhibitor contents, and which could thus be incorporated raw into swine rations. Hymowitz (1986) reported that germ plasm, from a soybean variant of low trypsin inhibitor content, could be successfully incorporated into commercially-grown cultivars. Unfortunately, when fed in the raw form to swine these modified varieties, while superior to conventional varieties, still produced inferior performance compared to soybean meal (Cook et al., 1988).

Other anti-nutritional factors present in raw soybeans are substances such as tannins, alkaloids, saponins, glucosides and hemagglutinins (lectins) (Liener, 1969). While hemagglutinins are known to cause agglutination of red blood cells, the exact influence that these substances and others exert on the feeding value of the unprocessed soybean has not been well established (Rackis, 1974).

PROCESSING FULL-FAT SOYBEANS

Although the anti-nutritional factors in raw soybean greatly depress performance, even the most detrimental factor can be readily destroyed through the use of heat (Rackis, 1974). Any method of heat treatment from boiling to autoclaving can be used to achieve the necessary denaturation. Those procedures which are performed commercially include dry roasting, jet-sploding, micronization and a fourth process, extrusion, which may be performed with or without the addition of moisture.

To date, the most common process used in North America has been dry roasting. This process involves passing the beans through a rotating chamber during which time they are directly exposed to a flame. Both the temperature of the chamber and the time the beans spend in the chamber can be used to adjust the degree of heating achieved. As a general rule, when dry roasting whole soybeans for swine, the beans should be heated for 3 to 5 minutes and should have an exit temperature of 125°C. Roasting can be accomplished on the farm through farmer-owned or mobile units, as well as by larger units owned by feed mills.

Jet-sploding is a technique which also involves dry heat. Instead of being exposed directly to a flame, the soybeans are introduced into an air stream which has been pre-heated to extremely high temperatures (260°C to 280°C). The intense heat produces molecular vibration causing the soybeans to heat from the inside out. The soybeans reach temperatures of 90 to 95°C. As they exit the heating chamber, the beans are immediately passed through a roller mill which is believed to ensure maximum energy availability.

The third dry-heating process, micronization, has been designed to utilize infra-red or radiant heat. This process results in a very consistent heat treatment as the beans are actually heated from the inside out, and is also very energy efficient. In this process, ceramic plates are heated by gas jets which then radiate heat to beans passing through a chamber on a vibrating conveyer below. These beans are also passed through a roller before cooling to maintain subsequent starch availability. Micronizers are generally limited to off-farm milling operations due to the expense of the equipment involved.

Extrusion heats through the application of shear forces and high pressure. Extruders are designed with flights along a shaft encased in a cylinder. The flights force the material forward through the cylinder, generating heat through friction. Additional heat or moisture may be added to the material at different stages of extrusion. In a recent review, Vandergrift (1985) suggested that an exit temperature of approximately 140°C is desirable when extruding whole soybeans to ensure adequate destruction of the trypsin inhibitors. As the material is extruded through a die at the end of the shaft, the sudden decrease in pressure causes the beans to expand and rupture, converting the material to a semi-fluid state. As the product is cooled, the oil is absorbed into the meal. Although all heat treatment processes improve energy availability of the soybean through a gelatinization of starch, extrusion achieves a further increase in energy availability due to a complete rupturing of the oil cells. However, due to the expense of the equipment and the electric input requirements, its use is limited to large, generally off-farm, milling operations.

The effectiveness of the heat treatment can be affected by a number of factors, the main one being the actual degree of heating or "total heat load" which the beans have received. The total heat load can be altered in virtually all processes by adjusting the duration, temperature, and pressure (if any) applied during the cooking process, as well as the rate of cooling. More research is required to determine the amount of heat required to provide optimum processing of the soybean for swine.

Furthermore, characteristics of the raw soybean product such as initial moisture content, oil content, particle size (such as whole versus flaked) and maturity, will influence the total heat load required to achieve the necessary denaturation (Seerley et al., 1974). Presoaking or the addition of steam at the time of heat processing also greatly influences the length of cooking time required (Rackis, 1974). Above all, uniformity of the cooking process is critical to ensure adequate denaturation of anti-nutritional factors and maximum retention of nutrient availability.

One crude, but frequently used, means of testing the adequacy of heat treatment is evaluate the final product on appearance and taste. In processes where the bean remains intact, the majority of the beans should be split and be fairly uniform in color. Proper heat processing improves the palatability of soybeans, producing a bean with a "nutty" flavor and eliminating the bitterness. To provide more concise evaluation of heating adequacy, several analytical techniques are available.

It is critical in assessing the adequacy of the heat treatment of full-fat soybeans for use in swine rations that the product be analyzed to evaluate the possibility of both under- and over-heating. Under-heating will result in inadequate denaturation of the anti-nutritional factors discussed previously and a subsequent decrease in performance. While the risk of over-heating is less likely, due to the high energy cost of processing, it is none the less also a concern as excessive heat treatment will result in reduced amino acid availability. The amino acids most susceptible to over-heating are lysine (Mauron, 1981) and arginine, cystine, methionine and isoleucine (Chang et al., 1984). Further problems associated with over-heating include a decrease in feed intake (Noland et al., 1976) or performance (Hansen et al., 1984).

Urease activity, as measured by pH change, can give some indication of processing adequacy as samples that test over 0.20 change in pH are considered under-processed for swine and samples testing below 0.05 considered over-processed. To identify soybeans which have been over-processed, protein solubility analysis can also be conducted. Quick test procedures, which translate these analyses to color indicators, are commercially available and provide processors with some indication of nutritional quality.

In addition, cooked whole soybeans are less susceptible to rancidity than raw soybeans due to the denaturation of naturally occurring lipase enzymes upon exposure to heat (Pfost, 1984). It is critical that the beans be adequately cooled before putting in storage. Once ground, the storage life of the heat processed whole soybean is decreased appreciably as its susceptibility to oxidation is increased. To determine the extent of oxidation that has occurred, samples can be submitted for peroxide analysis. In general, a low peroxide value indicates freshness and minimal oxidation.

FEEDING FULL-FAT SOYBEANS

Starter Pigs

The response of starter pigs to diets containing heat processed full-fat soybeans has been variable, with the greatest variation seen in young pigs fed full-fat soybeans processed by the dry roasting method. As can be seen from the results presented in Table 43.3, starter rations containing roasted soybeans and balanced to contain 18% CP, do not result in any consistent improvements in either feed conversion or growth rate when compared to rations containing soybean meal. Conversely, starter pigs fed rations containing extruded soybeans have consistently demonstrated 5 to 10% improvements in feed conversion, with some improvements in growth rate.

The difference in performance observed with these two methods of processing can be attributed to the digestibility of various nutrients after processing. During the extrusion process, the soybean is completely ruptured, rendering both oil and protein readily available. Conversely, after roasting and grinding, the oil cells still remain essentially intact. Young pigs lack the ability to completely digest the fibrous cell wall and consequently the availability of the oil within the cell wall is reduced. This has been confirmed by work from several researchers which have reported that energy and protein digestibility of roasted beans is inferior to either soybean meal plus oil, or extruded soybeans (Faber and Zimmerman, 1973; Noland et al., 1976).

Table 43.3. Performance of Starter Pigs Fed Corn-Based Diets Containing Heat-Processed Full-Fat Soybeans[1]

	Source of Dietary Protein		
	Soybean Meal	Roasted Soybeans[2]	Extruded Soybeans[2]
Daily Intake (kg)	0.93	0.92	0.93
Daily Gain (kg)	0.46	0.43	0.50
Feed Efficiency	2.05	2.17	1.87

[1]De Schutter, unpublished.
[2]Full-fat soybeans included at 34.1% of the total ration.

Others have suggested that the decreased performance of starter pigs fed roasted soybeans is due to a decrease in feed intake which has been attributed to the pig's attempt to maintain a constant energy intake. This is based on the reduction in feed intake which is noted when starter pigs are fed rations containing 5 to 10% added oil (Liebbrant et al., 1975). If other nutrient levels in the feed, such as protein, vitamins and minerals, have not been adjusted, the reduction in feed intake results in deficiencies of these nutrients. However, this explanation does not seem to apply with the use of full-fat soybeans, as a reduction in feed intake is not always seen when full-fat soybeans are incorporated into starter rations.

To summarize, extrusion achieves the highest utilization of full-fat soybeans by starter pigs. The extruded soybeans can be used to replace all the soybean meal in the ration, with improvements in feed conversion in the order of 5 to 10% expected. If other processes are used, the expected performance is equal to soybean meal at best. The relative cost of the full-fat soybeans should determine their usage in starter feeds.

Grower-Finisher Pigs

As with starter rations, full-fat soybeans cannot be substituted into grower-finisher rations on an equal weight-for-weight basis with soybean meal if maximum performance is to be expected. A weight-for-weight substitution with soybean meal will produce a 14% CP ration. Higher rates of inclusion must be used when feeding full-fat soybeans to account for the reduced percentage of crude protein in the whole soybean. Grower-finisher rations containing full-fat soybeans should not contain less than 16% CP (Table 43.4).

Table 43.4. Effect of Dietary Level of Crude Protein on the Growth Performance of Grower/Finisher Pigs Fed Full-Fat Soybeans[1]

	16% CP SBM[2]	12% CP RSBM[3]	14% CP RSBM	16% CP RSBM
Daily Intake (kg)	2.44	2.21	2.36	2.30
Daily Gain (kg)	0.85	0.72	0.82	0.83
Feed Efficiency	2.88	2.21	2.36	2.30

[1]Morris, 1988.
[2]SBM=corn and 47% SBM based ration
[3]RSBM=corn and roasted soybean ration

If rations are balanced to contain adequate crude protein, fairly consistent improvements in feed conversion are noted when full-fat soybeans are substituted for soybean meal. The improvements in feed conversion increase linearly up to the maximum of 25.5% of the ration, at which point improvements in the order of 5 to 10% can be expected (Table 43.5). This is due to the increased energy density of such rations.

Table 43.5. Effect of Heat Processing Method on the Growth Performance of Grower/Finisher Pigs Fed Full-Fat Soybeans

	SBM[3]	RSBM[4]	ESBM[5]	MSBM[6]
Feeder Fed:[1]				
Daily Intake (kg)	2.48	2.49	2.39	2.32
Daily Gain (kg)	0.80	0.82	0.82	0.79
Feed Efficiency	3.10	3.04	2.93	2.93
Floor Fed:[2]				
Daily Intake (kg)	2.04		2.03	
Daily Gain (kg)	0.72		0.77	
Feed Efficiency	2.84		2.64	

[1]De Schutter and Usborne, unpublished.
[2]De Schutter et al., 1988.
[3]SBM=corn and 47% SBM based ration
[4]RSBM=corn and roasted soybean ration
[5]ESBM=corn and extruded soybean ration
[6]MSBM=corn and micronized soybean ration

While the response in feed conversion is consistently observed, improvements in growth rate due to the inclusion of properly heat processed full-fat soybeans in grower-finisher rations is much less consistent. One major factor contributing to this variation is the effect of the full-fat soybean on feed intake. As can be expected, if feed intake is reduced, as has been observed in some instances where high-fat rations are used, then little, if any, improvement in growth rates is likely to occur. The decrease in feed intake has been associated with the pig's tendency to maintain a constant energy intake. However, this appears to apply only in situations where energy intake is already maximized. In situations where energy intake is restricted, either through the use of low energy rations or in floor feeding situations, then an increase in energy intake and, consequently, growth rate is likely to occur when soybean meal is replaced with full-fat soybeans.

Due to the improvements in feed efficiency, with or without corresponding reductions in feed intake, concentrations of vitamins and minerals in the ration should be increased in this proportion to ensure that adequate nutrients are available for maximum growth and metabolism. Over and above this need to adjust for differences in energy density, absolute daily nutrient requirements are also affected by the inclusion of full-fat soybeans in swine rations. In particular, requirements for vitamin E are increased due to the high concentration of polyunsaturated fatty acids (PUFA) in whole soybeans (Leat, 1983). While limited research has been done correlating levels of full-fat soybeans and vitamin E requirements, current recommendations suggest 2.5 mg vitamin E per gram PUFA in the ration to ensure optimal growth performance and pork quality (Weiser and Salkeld, 1977). Based on these recommendations, a grower-finisher ration containing 20% full-fat soybeans would contain about 3.0% PUFA and should, therefore, contain 75 mg/kg of vitamin E.

Much concern has been expressed regarding the effect of the full-fat soybeans on the quality of hog carcasses, with one of the primary concerns being their effect on the carcass index. Some reduction in carcass index, which is related to the percentage lean yield, has occurred in instances where hogs were feeder-fed corn based rations containing 25.5% full-fat beans. Floor feeding hogs similar rations did not result in reductions in lean yield or carcass index. In the feeder-fed studies, pigs fed full-fat soybeans also had higher live fat probes, indicating that the additional energy from the full-fat soybean was deposited as fat. Whether pigs could utilize more of this energy for lean tissue growth if amino acid levels were increased above those provided in a 16% ration has not been well established.

Another concern relates to the impact of the full-fat soybean on fat firmness. In pigs, as in many other monogastric animals, the fatty acid composition of fat tissue is greatly influenced by incidence and proportion of fatty acids present in the feed (Castell and Falk, 1980). The fatty acids in soybean oil are predominately oleic (23-26%), linoleic (50-54%) and linolenic (7-9%) (Castell and Cliplef, 1988). As a result, it is not surprising to discover that the concentration of these fatty acids is increased in the carcass fat of hogs fed rations containing full-fat soybeans (De Schutter and Usborne, unpublished; Seerley et al., 1974). These changes in fatty acid composition, particularly the increase in linoleic acid (C18:2), are believed to be directly responsible for the observed increase in softness of pork from hogs fed full-fat soybeans (Berschauer, 1984).

Although obvious decreases in fat firmness are observed, the significance of these changes to the processor are as yet undetermined. There is some indication that softness has no negative effect on the processing quality of the carcass (De Schutter and Usborne, unpublished). However, the acceptability of this pork from a consumer standpoint must also be considered. In addition to the softness of the fat, there are other distinct changes in the carcass. For instance, a paler color of lean has been noted which may also present problems, particularly for export markets.

Consumer acceptability as it relates to taste of the pork product must also be considered. While some studies have reported that the level of full-fat soybean in the ration should be restricted to 10% in order to prevent negative effects on taste of the pork (Usborne et al., 1986; Castell and Cliplef, 1988), others have indicated that levels as high as 25.5% of the ration could be fed without negatively affecting the taste of the pork (Wahlstrom et al., 1971; Seerley et al., 1974; De Schutter and Usborne, unpublished). It is possible that one factor contributing to this discrepancy in results is the nutrient composition of the diets fed.

In summary, properly heat processed full-fat soybeans will result in equal or better growth performance when included in well balanced rations for grower-finisher pigs. The economic analysis shows that when it is profitable to incorporate full-fat soybeans into grower-finisher rations, they are most profitable at the highest inclusion rates (25.5% of the ration). However, there are some definite changes in the carcass with the use of full-fat soybeans, the full impact of which is not fully understood.

Breeding Stock

The advantages of using whole soybeans in sow rations appear to be quite similar to the advantages noted for other types of high energy density sow rations. The use of whole soybeans in gestation rations results in an improved number of piglets born and percent survival rate of piglets, particularly those with low birth weights (Table 43.6). This has been attributed to an increase in the fat and energy content of sow's milk (Boyd et al., 1978). Similarly, a 2.6% improvement in piglet survival has been noted in pigs fed gestation rations containing added fat or oil (Moser and Lewis, 1981). Sows, due to their maturity, appear to be able to utilize the nutrients from raw soybeans more effectively than younger animals.

Table 43.6. Performance of Sows Fed Full-Fat Soybeans During Gestation

	Soybean Meal	Full-Fat Soybean
Number of Sows	88	88
Pigs Born Alive	10.0	10.5
Birth Weight (kg)	1.5	1.6
Pigs Weaned	8.3	8.7
Weaning Weight (kg)	5.4	5.4
Survival (%)	82.5	82.7
Sow Feed Intake (kg/day)	5.5	5.4
Lactation Weight Loss (kg)	7.0	7.0

Crenshaw and Danielson, 1985.

SUMMARY

As properly heat-treated full-fat soybeans produce equal or superior performance relative to soybean meal, economics should be the major consideration determining their usage in swine rations. Factors affecting the profitability of this alternative include prices of soybean meal, raw beans, soy oil, grains, weaner pig and market pig prices.

REFERENCES

Albrecht, W.J., Mustakas, G.C. and McGhee, J.E., 1966. Rate studies on atmospheric steaming and immersion cooking of soybeans. Cereal Chem. 43: 400-407.

Berschauer, F., 1984. In: J.D. Wood, ed. Fat Quality in Lean Pigs. Commission of European Community, Brussels, pp. 74-82.

Boyd, R.D., Moser, B.D., Peo, E.R. and Cunningham, P.J., 1978. Effect of energy source prior to parturition and during lactation on piglet survival and growth and on milk lipids. J. Anim. Sci. 47: 883-892.

Carter, R.R. and Cox, P.J., 1984. Urease and trypsin inhibitor activity in soybeans as affected by extrusion. Proc. Aust. Soc. Anim. Prod. 15: 661.

Castell, A.G. and Falk, L., 1980. Effects of dietary canola seed on pig performance and backfat composition. Can. J. Anim. Sci. 60: 795-797.

Castell, A.G. and Cliplef, R.L., 1988. Performance and carcass responses to dietary inclusion of raw soybeans (cv. maple amber) by boars fed ad libitum from 30 to 95 kilograms liveweight. Can. J. Anim. Sci. 68: 275-282.

Chang, C.J., Tanksley, T.D., Knabe, D.A., Zebrowska, T. and Gregg, E.J., 1984. Effect of different heat treatments during processing on nutrient digestibilities of soybean meal by growing swine. J. Anim. Sci. 59: 303 (Abstract).

Cook, D.A., Jensen, A.H., Fraley, J.R. and Hymowitz, T., 1988. Utilization of growing and finishing pigs of raw soybeans of low Kunitz trypsin inhibitor content. J. Anim. Sci. 66: 1686-1691.

Crenshaw, M.A. and Danielson, D.M., 1985. Raw soybeans for gestating sows. J. Anim. Sci. 60: 163-170.

Dale, N.M., Araba, M. and Whittle, E., 1987. Protein solubility as an indicator of optimum processing of soybean meal. Proceedings Georgia Nutrition Conference, Atlanta, Georgia, pp. 88-95.

Danielson, M., 1985. Raw soybeans in pig feeding. Pig News and Information 6: 35-41.

De Schutter, A.C., Dinnissen, S., Morris, J.R. and Usborne, W.R., 1988. Effect of inclusion rate of full-fat soybeans and ration crude protein on growth performance and carcass quality of swine. Ontario Swine Research Review. Ontario Agricultural College Publication No. 689, pp. 86-93.

Faber, J.L. and Zimmerman, D.R., 1973. Evaluation of infra-red roasted and extruded processed soybeans. J. Anim. Sci. 36: 902-907.

Hansen, B.C., Flores, E.R., Tanksley, T.D. and Knabe, D.A., 1984. Effect of different heat treatments during processing of soybean meal on performance of pigs weaned at four weeks of age. J. Anim. Sci. 59: 304 (Abstract).

Hymowitz, T., 1986. In: M. Friedman, ed. Nutritional and Toxicological Significance of Enzyme Inhibition in Foods. Plenum, New York.

Kohlmeier, R.H., 1988. Methods of processing full-fat soybeans. Canadian Feed Manufacturers Conference, pp. 6-16.

Leat, W.M.F., 1983. Nutritional deficiencies and fatty acid metabolism. Proc. Nutr. Soc. 42: 333-342.

Leibbrandt, V.D., Hays, V.W., Ewan, R.C. and Speer, V.C., 1975. Effect of fat on performance of baby and growing pigs. J. Anim. Sci. 40: 1081-1085.

Liener, I.E., 1969. Toxic Constituents of Plants Foodstuffs. Academic Press, New York, N.Y. 500 pp.

Manitoba Agriculture, 1987. Field crop variety recommendations for Manitoba. Manitoba Agriculture. Winnipeg, Manitoba, 35 pp.

Mauron, J., 1981. The maillard reaction in food; a critical review from the nutritional standpoint. Proc. Food Nutr. Soc. 5: 5-35.

Morris, J.R., 1988. Roasted soybeans - watch the protein level. South West Ontario Pork Producers Conference, Ridgetown, pp. B1-3.

Moser, B.D. and Lewis, A.J., 1981. Fat additions to sow diets: A review. Pig News and Information 2: 265-269.

Myer, R.O., Froseth, J.A. and Coon, C.N., 1982. Protein utilization and toxic effects of raw beans (*Phaseolus vulgaris*) for young pigs. J. Anim. Sci. 55: 1087-1098.

Myer, R.O. and Froseth, J.A., 1983. Extruded mixtures of beans (*Phaseolus vulgaris*) and soybeans as protein sources in barley-based swine diets. J. Anim. Sci. 57: 296-306.

National Research Council, 1988. Nutrient Requirements of Domestic Animals, No. 2. Nutrient Requirements of Swine. 9th ed. National Academy of Sciences, Washington, D.C.

Noland, P.R., Campbell, D.R., Cage, R.K., Sharp, R.N. and Johnson, Z.B., 1976. Evaluation of heat processed soybeans and grains in diets for young pigs. J. Anim. Sci. 43: 763-769.

Pfost, H., 1984. The effects of processing on protein quality and fat stability of full-fat soybeans. Feedstuffs 56 (June 25) pp. E1-E2.

Pontif, J.E., Southern, L.L., Combs, D.F., McMillin, K.W., Bidner, T.D. and Watkins, K.L., 1987. Gain, feed efficiency and carcass quality of finishing swine fed raw soybeans. J. Anim. Sci. 64: 177-181.

Rackis, J.J., 1974. Biological and physiological factors in soybeans. J. Amer. Oil Chem. Soc. 51: 161A-174A.

Seerley, R.W., Emberson, J.W., McCampbell, H.C., Burdick, D. and Grimes, L.W., 1974. Cooked soybeans in swine and rat diets. J. Anim. Sci. 39: 1082-1091.

Smith, K.J., 1977. Soybean meal: Production, composition and utilization. Feedstuffs 49 (Jan. 17): p 22.

Usborne, W.R., Gullet, E.A. and Haworth, C.R., 1986. Evaluation of fresh and cured pork from pigs fed whole soybeans. Ontario Swine Research Review, Guelph, Ontario.

Vandergrift, W.L., Knabe, D.A., Tanksley, T.D. Jr. and Anderson, S.A., 1983. Digestibility of nutrients in raw and heated soyflakes for pigs. J. Anim. Sci. 57: 1215-1224.

Vandergrift, W.L., 1985. Use of soybeans in pig diets. Pig News and Information. 6: 282-285.

Wahlstrom, R.C., Libal, G.W. and Berns, R.J., 1971. Effect of cooked soybeans on performance, fatty acid composition and pork carcass characteristics. J. Anim. Sci. 32: 891-894.

Weiser, H. and Salkeld, R.M., 1977. Acta Vitamin Enzymol. (Milano) 31: 143-155.

Yen, J.T., Jensen, A.H. and Simon, J., 1977. Effect of dietary raw soybean and soybean trypsin inhibitor on trypsin and chymotrypsin activities in the pancreas and in small intestinal juice of growing swine. J. Nutr. 107: 156-165.

CHAPTER 44

Sugar Beet

A.C. Longland and A.G. Low

INTRODUCTION

Intensively-reared pigs are traditionally fed diets where cereals provide most of the energy and a substantial part of the dietary protein. However, there is increasing interest in using feedstuffs which are unsuitable for human consumption in swine diets. These ingredients may be cheaper and in some areas are more readily available than cereal grains. Sugar beet (*Beta vulgaris*) is one such crop.

Sugar beet is chiefly grown as a source of sucrose for human consumption. They were developed from the white fodder beet in Europe during the late 1800s, largely as a result of sugar cane shortages during the Napoleonic Wars. The early varieties of sugar beet contained 5 to 6% sucrose, while modern cultivars yield nearer 20% on a fresh-weight basis (British Sugar Bureau, 1976).

The pulp produced during sugar extraction has a substantial non-starch polysaccharide (NSP) content and is very widely used as a highly fermentable feedstuff for ruminants but can also be used for pigs. In addition, sugar beet is sometimes grown as a forage crop, again mainly for ruminants but may also be fed to pigs in some regions.

GROWING SUGAR BEET

Sugar beet is a temperate plant grown at latitudes between 30° and 60°N throughout Europe, North America, USSR, China and Japan. They are grown on many different soil types (Draycott, 1972) and have traditionally been used in a crop rotation with cereal grains. Sugar beet is a biennial which stores sugar in its roots during the first year, allowing overwintering and subsequent production of flowers and seed during the second. The seeds are sown in March and the crop is harvested and processed between September and January of the first year. Yields up to 50 tonnes/ha are not uncommon. Annual global production of beet sugar is about 28 million tonnes, which represents 40% of the world's sugar production, with the production of nearly 20 million tonnes of beet pulp available for animal nutrition.

PROCESSING SUGAR BEET

Beets destined for the sugar industry are harvested at the end of their first growing season. The beets (minus leaves and crown) are washed, thinly sliced, and the majority of the sugar is extracted in hot water. The beet slices are then pressed to obtain the remaining sugar liquor. The spent beet slices, known as sugar beet pulp, can be fed to animals either wet or, more usually, after drying.

Sucrose is crystallized from the sugar liquor which is a syrup containing a variety of sugars, minerals, proteins and other biopolymers. The syrup which remains after three successive crystallizations is called sugar beet molasses. This is a palatable product and is frequently combined with dried sugar beet pulp to yield a high quality animal feed.

NUTRIENT CONTENT OF SUGAR BEET

Sugar beet and sugar beet pulp are both principally sources of energy for pigs due to their high percentage of soluble carbohydrates and a highly fermentable fiber fraction. The chemical composition of sugar beet and of its main products are shown in Table 44.1. The values are drawn from a wide variety of sources and in many cases the range of values quoted in the literature is shown. This range is due to differences not only in cultivar, agronomy and climate but also to differences in the operating conditions in sugar beet factories.

The crude protein level is generally low, although dried sugar beet tops and dried sugar beet pulp have a protein content similar to most cereal grains. Although the lysine content of dried plain sugar beet pulp (0.55%) is approximately 50% higher than that of barley or wheat, there is no information about its digestibility or availability. While threonine levels in dried plain sugar beet pulp are similar to those of cereals, sulphur amino acid levels are very low, but again nothing appears to be known of their digestibility or availability. The amino acid content of sugar beet pulp is shown in Table 44.2.

Table 44.1. Chemical Composition of Sugar Beet Products (%)

	Dry Matter	Crude Protein	Crude Fiber	Fat	NFE	D.E. (kcal/kg DM)
Fresh Roots	24	1.1-1.6	1.0-1.5	0.1	21.4	3220
Dried Roots	91	4.2	4.9	0.4	81.0	----
Wet Pulp	13	1.5-1.6	3.1-4.0	0.2	5.3	----
Dried Pulp	91	8.8-10.0	17.4-19.6	0.6	58.7	3038
Dried Molassed Pulp	91	8.7-9.9	12.0-15.2	0.5	61.8	2919
Fresh Tops	17	1.3-2.7	1.0-2.0	0.3	8.2	2210
Dried Tops	90	9.0-14.0	8.5-10.6	2.5	46.0	----
Molasses	78	3.5-10.6	0.0	0.0	62.0	2510

Halnar and Garner, 1953; Kellner, 1915; McDonald et al., 1981; Morrison, 1956; National Research Council, 1988.

Table 44.2. Amino Acid Composition of Plain and Molassed Sugar Beet Pulp (% DM)

	Molassed Sugar Beet Pulp	Plain Sugar Beet Pulp
Arginine	0.29	0.39
Histidine	0.18	0.26
Isoleucine	0.29	0.29
Leucine	0.51	0.57
Lysine	0.44	0.55
Methionine	0.09	0.09
Cysteine	0.07	0.07
Phenylalanine	0.27	0.36
Threonine	0.31	0.41
Valine	0.46	0.48

William et al. (unpublished).

Some information on the mineral content of sugar beet products is provided in Table 44.3. Calcium levels are high and phosphorous levels low in dried plain sugar beet pulp compared with cereals, but information on their availability is lacking.

Table 44.3. Mineral Composition of Sugar Beet Products (%)

	Total Ash	Calcium	Phosphorus	Potassium
Fresh Roots	0.7-1.0	0.04	0.04	0.25
Dried Roots	3.4	0.15	0.15	0.95
Wet Pulp	0.5-0.6	0.09	0.01	0.02
Dried Pulp	3.1-5.9	0.75	0.08	0.2-0.7
Dried Molassed Pulp	5.8-8.3	0.70	0.08	1.6-1.9
Fresh Tops	2.1-3.9	0.18	0.04	1.0
Dried Tops	10-20	0.95	0.21	5.5
Molasses	5.0-9.0	0.08	0.02	4.7

Low (unpublished).

The composition of the non-starch polysaccharide in dried sugar beet pulp and its digestibility is shown in Table 44.4. Data for the energy content and yield of the non-starch polysaccharides are also shown in this table. The non-starch polysaccharides are fermented by bacteria in the small and large intestines to yield volatile fatty acids, which are absorbed and subsequently used like sugars as energy substrates.

Table 44.4. Non-starch Polysaccharide (NSP) Content and Digestibility of Dried Plain Sugar Beet Pulp (SBP) for Growing Pigs (35 kg)[1]

NSP[2] Monomers	g/kg of SBP NSP	Energy (kcal/ kg sugar)	Energy (kcal/ kg SBP)	Apparent Digest-ibility (%)	D.E. (kcal/ kg NSP)
Arabinose	280	3708	1038	99.0	1028
Xylose	20	3780	76	68.0	51
Mannose	10	3732	37	97.0	36
Galactose	70	3732	261	94.0	246
Glucose	290	3660	1061	86.0	913
Uronic acids	330	2871	947	99.0	938
	----				----
Sum	1000				3212

[1]Low (unpublished).
[2]Non-Starch Polysaccharide comprised 60% of the dry weight of the SBP.

UNDESIRABLE CONSTITUENTS IN SUGAR BEET

Sugar beet plants contain about 55 g/kg DM oxalate of which approximately 90% is found in the leaves (Lennon and Tagle, 1973). This high concentration of oxalate in leaves has been implicated in reduced availability of calcium to sows from ensiled sugar beet tops (Salo and Laakso, 1977). Gorb and Maksakov (1962) suggested that sugar beet tops should be introduced to animals gradually, and that a calcium supplement should be given 2 to 3 hours after a beet-top feed. Feeding very high levels (>30%) of dried plain sugar beet pulp to growing pigs frequently leads to reduced digestibility of nitrogen from other dietary components. Such high amounts also increase faecal bulk and water content, with associated problems of slurry disposal in large intensive pig rearing units.

FEEDING SUGAR BEET PRODUCTS

A. Sugar Beet Roots

Starter Pigs

Sugar beet has been widely used in Eastern Europe as a pig feed for many years. Danilenko et al. (1962) suggested that weaners could be introduced to sugar beet at 10 to 15 days old, levels of sugar beet then being gradually increased to at least half of the total nutrients for pigs of 45 kg and over.

Growing-Finishing Pigs

Danilenko et al. (1962) fed fattening pigs a control concentrate diet or diets where 52% of whole, 65% of chopped or 74% of dried sugar beet (all on a DM basis) replaced most of the concentrates. Daily liveweight gains were 604, 551 and 590 g, respectively, all of which were regarded as satisfactory. Carcasses of pigs fed 74% dried sugar beets did not differ from the control, while those given 52% whole sugar beets had the highest carcass yields.

Seidler et al. (1972) replaced up to 67% of a potato/barley fattening ration with dried sugar beet slices with no change in daily gain, feed efficiency or carcass quality. Likewise, Korniewicz (1982) reported that carcass and meat quality were not affected by replacing 50% of the dietary cereals by dried sugar beets, but daily gain, digestibility of crude fiber and retention of nitrogen were significantly improved for the sugar beet diets. Reduced feed costs coupled with improved carcass yield were obtained by Ibragimov and Baljan (1965) when 25 or 40% sugar beets were fed instead of concentrates.

The form in which sugar beet is fed seems to affect animal performance, dried sugar beet appearing to be better than the fresh product, which in turn is used more efficiently when chopped. Treatment of sugar beets may also affect the site of its digestion and hence efficiency of utilization. Kesting and Bolduan (1986), working with sugar beet of various particle sizes, found that coarse particles, either fresh or ensiled, were predominantly digested in the large intestine whereas steamed or mashed sugar beet was digested pre-cecally. It is of note that protein digestion was greatest from mashed sugar beet.

As with cereals, sugar beet has been used to replace traditionally-fed root crops. Sugar beet appears to give better performance than fodder beet (Korniewicz, 1982; Palamaru, 1965) and swedes (Parts, 1978). Comparisons between sugar beet and potatoes did not, however, show clear superiority of one crop over the other.

Breeding Stock

Sows can digest and utilize sugar beet to a high degree. According to Navratil and Zeman (1978) sugar beet is equivalent in feeding value to ground whole maize plants for both pregnant and lactating sows. In studies by Karpus (1968), sows fed diets containing up to 50% of the total nutrients as sugar beet during gestation had high weight gains. During lactation, they were fed up to 44% sugar beet and at this level they lost the most weight but yielded the most milk. However, there was no difference between the groups in subsequent litter size, piglet birthweight or weaning weight (Karpus, 1968).

Pregnant sows can eat large quantities of fresh sugar beet. Wildgrube (1969) recommended feeding 10 kg sugar beet/day with a supplement containing protein, minerals and vitamins to pregnant sows. Even higher levels of sugar beet have been fed with success by Ionescu et al. (1975) who gave pregnant sows 12 kg sugar beet/day and 1.2 kg of mixed feed. During lactation, sugar beet levels were reduced while those of mixed feed were increased to 3.5 kg/day of each. When these sows were compared with those fed mixed feed only, there was no difference in the progress of gestation, lactation or growth of the piglets.

B. Sugar Beet Pulp

Starter Pigs

Dried plain sugar beet pulp has often been regarded as an unsuitable feed for young pigs because of its high non-starch polysaccharide content. However, it has been demonstrated that the microflora of young pigs can digest in excess of 85% of this non-starch polysaccharide to provide energy in the form of volatile fatty acids (Longland and Low, 1988; Longland et al., 1988b). Inclusion of 15% molassed or dried plain sugar beet pulp in a barley-based diet did not reduce performance in 4 to 8 week old piglets (Low et al., 1990).

Growing-Finishing Pigs

It has generally been reported that diets containing up to 15% of molassed or dried plain sugar beet pulp gave similar performance to controls (O'Donovan and Curran, 1965; Longland et al., 1988a; Bulman et al., 1989). However, at higher levels of inclusion, results have been more variable.

Aherne and McAlese (1965) found no deleterious effects of feeding 20 or 30% dried plain sugar beet pulp, and none were seen by Marambio and

Sepulveda (1966) who fed 25% dried plain sugar beet pulp. However, Chauvel et al. (1975) found that inclusion of 20% dried plain sugar beet pulp in a maize-soya diet slightly reduced growth rate. When diets containing 30 or 45% dried sugar beet pulp (molassed or plain) were given to growing pigs by Longland et al. (1988a) and Bulman et al. (1989), somewhat depressed performance and lower killing out percentages were seen. Recently, workers at the Terrington Experimental Husbandry Farm in the U.K. have concluded that inclusion of 20% molassed dried plain sugar beet pulp gave good performance, with a 5% reduction in feed costs.

It seems that an inclusion level of about 20% dried plain sugar beet pulp represents the most economically satisfactory inclusion level, although factors such as type of pig, housing and management could influence the optimum level. Higher levels reduce carcass gain rates. This is the result of less fat deposition and lowered killing out percentage rather than reduced rates of lean tissue growth. If the cost of the product is favorable then these performance and carcass changes may be acceptable (O'Donovan and Curran, 1965; Marambio and Sepulveda, 1966; Longland et al., 1988a; Bulman et al., 1989).

Breeding Stock

There have been few reports of sows being fed dried plain sugar beet pulp. Studies on sows during pregnancy showed that their weight gains were almost identical when they were fed diets containing 0, 22.5 or 45% dried plain sugar beet pulp. However, during lactation, when sows were fed a conventional cereal-based diet, feed intake and weight losses of sows that had been fed 45% dried plain sugar beet pulp during pregnancy were higher than those fed 0 or 22.5% (Close et al., unpublished). In another study, Mentler (1966) showed that sows could be fed 30% dried plain sugar beet pulp in a cereal-based diet without affecting liveweight changes or the general condition of the sows, piglet-foetal development or newborn piglet vigour.

C. Ensiled Sugar Beet Pulp

Sugar beet pulp is increasingly being stored as silage because of rising costs of drying (Courtin and Spoelstra, 1989). Indeed, upwards of 25% of sugar beet pulp in many European countries is being stored in this manner. Ensiled sugar beet pulp, fed to pigs (30-90 kg) at a level of 25% in a cereal-based diet, led to reduced daily gain, but improved feed efficiency so that the ensiled sugar beet pulp diet was economically beneficial (Thielmans and Bodart, 1983). Linder et al. (1984) reported that the metabolizable energy value of ensiled sugar beet pulp was 2637 kcal/kg in pigs. Ensiled sugar beet pulp is likely to become increasingly available as a pig feed in Europe, and may well prove to be a valuable diluent of cereals, possibly at levels similar to those of the dry product (on an equivalent DM basis).

D. Sugar Beet Tops

Sugar beet tops can provide a useful source of digestible nutrients for pigs, as shown by Urakov (1963), who fed pigs (80-87 kg) cereal/silage based diets supplemented with two levels of sugar beet top meal. Levels of 8 to 10% sugar beet top were found to be nutritionally and economically sound. Sugar beet top silage has been shown to be a valuable protein and energy source for pregnant sows, containing 14.2% DM digestible crude protein, and an ME content of 2210 kcal/kg, when fed at 20 or 48% of DM in a barley-based diet (Salo and Laakso, 1977).

E. Sugar Beet Molasses

Sugar beet molasses has been fed at levels of 20 and 40% to pigs of 24 to 50 or 51 to 90 kg, respectively (Karamitros, 1987). Rates of gain of pigs fed 20% sugar beet molasses were similar to those of the controls, while those of the heavier pigs fed 40% sugar beet molasses were reduced, but the feed efficiency of both groups was raised. However, neither carcass yields nor quality were affected by the diets containing sugar beet molasses. Similarly, Poltársky et al. (1973) found that there was no difference in carcass parameters between pigs fed maize-based diets containing 12.5 or 25% of the total digestible nutrients as sugar beet molasses.

SUMMARY

The almost total lack of information on the digestibility and/or availability of the nutrients in sugar beet and its components, other than energy, makes it difficult to offer sound recommendations about how to supplement these materials in complete diets. On the basis of the gross composition of these materials, it appears necessary to supplement with protein and minerals, at least if they form a high proportion of complete diets. However, in practice, a substantial proportion of diets will inevitably contain other ingredients.

Guidelines for the use of sugar beet and its components are shown in Table 44.5. There are insufficient data available for more exact recommendations to be made. However, there is potential for using sugar beet in diets fed to all classes of pig. At high levels of inclusion, lower growth rates are counterbalanced by less carcass fat. The extent to which these feedstuffs are used is clearly related to their cost. If cost is low, then their potential is considerable.

Table 44.5. Guidelines on Maximum Feeding Levels (% of Diet) of Sugar Beet Products to Pigs[1, 2]

	Pig Weight (kg)			Pregnant Sows	Lactating Sows
	<15	15-35	36-100		
Fresh Roots	Introduced	-->	50	10 kg/d	2-3 kg/d
Dried Roots	Introduced	50[3]	70	70[3]	30
Dried Pulp	10-15	15	15	45	20
Dried Molassed Pulp	10-15	15	20	45[3]	20
Tops[4]		*	*	10	20*
Molasses	5[3]	20	20	20[3]	*

[1]Low (unpublished).
[2]The feeding levels given are those quoted in the literature as maintaining pig performance on a par with the controls. Depending on the cost of the product fed, higher levels of some products may be given which, although they may reduce certain aspects of performance, can be economically beneficial.
[3]Levels not reported in the literature but, from values quoted for similar products, the levels given here are likely to be safe.
[4]No Values Found in the Literature = *

REFERENCES

Aherne, F.X. and McAlese, D.M., 1965. Molassed beet pulp in the rations of pigs. Proc. Royal Dubl. Soc. 1: 173-181.

Bulman, J.C., Longland, A.C., Low, A.G., Keal, H.D. and Harland, J.I., 1989. Intake and performance of growing pigs fed diets containing 0, 150, 300 or 450g/kg molassed or plain sugar beet pulp. Anim. Prod. 48: 626 (Abstract).

British Sugar Bureau, 1976. Energy from the Sun. The Story of Sugar. British Sugar Bureau, London.

Chauvel, J., Villain-Guillot, J. and Bourdon, D., 1975. Energy value of dried beet pulp and utilization of the growing finishing pig. Journées de la Recherche Porcine en France, 53-60.

Courtin, M.G. and Spoelstra, S.F., 1989. Counteracting structure loss in pressed sugar beet silage. Anim. Feed Sci. Technol. 24: 97-110.

Danilenko, I.A., Bogdanov, G.A. and Maksakov, V.Ja., 1962. Quality of pork after fattening for meat on beet and maize rations. Zivotnovodstvo 7: 10-18.

Draycott, A.P., 1972. Sugar Beet Nutrition. Applied Science Publishing. Essex, England.

Gorb, T.V. and Maksakov, V. Ja., 1962. Influence of sugar beet tops on animals. Veterinarija Moscow 39: 66-68.

Halnar, E.T. and Garner, F.H., 1953. The Principles and Practise of Feeding Farm Animals. Longmans Green and Company Ltd, London.

Ibragimov, B.G. and Baljan, G.A., 1965. Sugar beet is a valuable feed. Svinovodstvo 8: 27-28.

Ionescu, T., Isar, M. and Colceriu, C., 1975. Sugar beet in feeding pregnant and lactating sows. Lucrarile Stiintifice ale Institutului de Cercetari Pentru Nutritie Animala 5: 103-107.

Karamitros, D., 1987. Sugar beet molasses for growing and fattening pigs. Anim. Feed Sci. Technol. 18: 131-142.

Karpus, N.M., 1968. Sugar beet for pigs. Svinovodstvo 11: 26-28.

Kellner, O., 1915. The Scientific Feeding of Animals. Duckworth and Company, London.

Kesting, U. and Bolduan, G., 1986. Use of sugar beet for feeding pigs. 1. Effect of different methods of preparation on the gastro intestinal tract. Arch. Anim. Nutr. 36 (6): 499-507.

Korniewicz, A., 1982. Cereal-saving methods of fattening pigs. Biuletyn Informacyjny, Zootechniki Instyut Zaktad Informacji Zootechnicznej 20: 45-63.

Lennon, I. and Tagle, M.A., 1973. Total oxalate in different samples of sugar beet (*Beta vulganis* var. saccharata). Archivos Latinoamericanos de Nutricion 23: 243-249.

Linder, J.P., Koch, G. and Burgstaller, G., 1984. Nutrient digestibility and protein and energy contents of silage of pressed sugar beet pulp for pigs. Archiv fur Tierernährung 34: 467-479.

Longland, A.C. and Low, A.G., 1988. The digestion of three sources of dietary fibre by growing pigs. Proc. Nutr. Soc. 47: 104A.

Longland, A.C., Low, A.G., Keal, H.D. and Harland, J.I., 1988a. Dried molassed and plain sugar beet pulp in diets for growing pigs. Proc. Nutr. Soc. 47: 102A.

Longland, A.C., Low, A.G., Keal, H.D. and Harland, J.I., 1988b. The digestibility of growing pig diets containing dried molassed or plain sugar beet pulp. Proc. Nutr. Soc. 47: 103A.

Low, A.G., Carruthers, J.C., Longland, A.C. and Harland, J.I., 1990. Performance and digestibility of non-starch polysaccharides in cereals and sugar beet pulp in pigs of 3 to 8 weeks. Anim. Prod. (In press).

Marambio, J. and Sepulveda, R., 1966. Dried sugar beet pulp for fattening pigs. Nutricion Bromatologia Toxicologia, Santiago 5: 7-13.

McDonald, P., Edwards, R.A. and Greenhalgh, J.F., 1981. Animal Nutrition (3rd ed.). Longman Group Ltd, Harlow, England.

Mentler, L., 1966. Unprocessed sugar beet pulp for breeding sows. Allattenyésztés 15: 43-44.

Morrison, F.B., 1956. Feeds and Feeding. 22nd ed. Morrison Publishing Company, Ithaca, New York, 1165 pp.

National Research Council, 1988. Nutrient Requirements of Domestic Animals, No. 2. Nutrient Requirements of Swine. 9th ed. National Academy of Sciences, Washington, D.C.

Navratil, B. and Zeman, L., 1978. Effect of the daily ration and type of complete feed on changes in liveweight of pregnant sows. Zivocisna Vyroba 23: 27-34.

O'Donovan, P.B. and Curran, S., 1965. Dried beet pulp in pig fattening rations. Irish J. Agric. Res. 4: 179-187.

Palamaru, E., 1965. Sugar beet for dairy cows and for fattening cattle and pigs. Lucrarile Stiintifice ale Institutului de Cercetari Zootechnice Bucharest 22: 5-32.

Parts, J., 1978. Results of production trials conducted feeding bacon pigs sugar beet, semi-sugar beet, fodder potatoes and hybrid swedes. Estonian Research Institute of Animal Breeding and Veterinary Science 46: 98-103.

Poltársky, J., Sommer, A. and Zilla, T., 1973. Sugar beet molasses in feeds for fattening pigs. Pol'nohospodarstvo 19: 210-217.

Salo, M.L. and Laakso, E., 1977. The digestibility and nutritive value of sugar beet top silage for sows. J. Sci. Agric. Soc. Fin. 49: 203-208.

Seidler, S.A., Lubowicki, R. and Niewiarowska, T., 1972. Feeding value of dried sugar beet slices for fattening pigs. Roczniki Nauk Rolniczych. 94: 69-76.

Thielmans, M.F. and Bodart, C., 1983. Pressed ensiled sugar beet pulp in the feeding of fattening pigs. Revue de l'agriculture 36: 1411-1414.

Urakov, V.I., 1963. Meal from sugar beet tops is a valuable pig feed. Svinovodstvo 9: 23-24.

Wildgrube, M., 1969. Sugar beet and turnips as basic feed for sows. Jahrbuch Tierernährung Fütterung, Jena. 7: 204.

CHAPTER 45

Sunflower Meal

W.E. Dinusson

INTRODUCTION

The production of sunflower seed (*Helianthus annus*) has increased dramatically in recent years with world production exceeding twenty-one million metric tons in 1988 (National Sunflower Association, 1988). This interest in the sunflower is based on the fact that the oil contained in the sunflower seed supplies a very high level of polyunsaturated fatty acids and is highly sought after as a vegetable oil for human consumption. The meal, produced as a by-product of this crushing process, has made available to the swine industry an alternative feed resource that has considerable potential for use during all phases of production.

NUTRIENT CONTENT OF SUNFLOWER MEAL

The nutrient composition of the sunflower meal will vary according to the quality of the seed and the method used for oil extraction. These methods vary from plant to plant. The energy content of dehulled sunflower meal is reported to be about 3000 kcal/kg (Table 45.1). This will vary with hull and oil residue in the meal. However, this compares favorably with other protein sources. It can be estimated that a 26% solvent-extracted meal would supply about 2337 kcal/kg.

Most sunflower seeds have a black hull, thus the meal produced is gray in color, the shade varying with amount of hull residue. The crude fiber of 12 to 13% in dehulled meals with 40% or more protein would increase to 21 to 26% for meals where hulls are included. The oil residue varies from less than 1% in high protein meals to 12% or more in poorly extracted meals.

Table 45.1. Proximate Analysis of Sunflower Meal

	Soybean Meal[1]	Dehulled SFM[2]	Partially Dehulled SFM[2]	SFM (with hulls)[2]
Dry Matter (%)	90.0	93.0	89.0	89.0
D.E. (kcal/kg)	3680	3047	2425	2337
Crude Protein (%)	48.5	45.5	32.0	26.0
Crude Fiber (%)	3.4	11.7	21.0	30.0
Ether Extract (%)	0.9	1.6	0.5	0.5

[1]National Research Council, 1982.
[2]Cargill Processing Group, (unpublished).

The protein content of sunflower meals varies with method of processing, season produced and area of production (National Sunflower Association, 1988). Complete removal of hulls (decortication) prior to oil extraction will produce a meal with a protein content in excess of 40% and a crude fiber content of 12% or less. Partial dehulling prior to solvent extraction will produce meals of 30 to 35% protein. Whole sunflower meal supplies about 26% crude protein. In practice, the protein percentage has been one to three units higher than the guarantee on the tag.

The amino acid profile of sunflower meals is presented in Table 45.2. Obviously, lysine is the first limiting amino acid. Although higher in lysine than the protein in barley or corn, sunflower meal contains only about 65% as much as that found in the protein of soybean meal. Literature references vary in the values reported. Values given by the National Research Council in 1982 are 113% of those reported in 1988. It is worthy of note that Smith (1968) reported even lower values. The average of values from nine laboratories was only 0.81% lysine. There was also lack of agreement in values reported for other amino acids, particularly the sulfur containing amino acids. It is not readily apparent whether these differences relect a changing amino acid profile due to newer varieties being grown, the area of production or an improvement of analytical techniques.

Jorgensen and Sauer (1982), using an ileo cecal cannulation, reported the lysine in sunflower meal was only 70.6% available as compared to 80.7% for soybean meal. In the same study, they found the availabilities of methionine to be 81.1%, threonine 68.6% and phenylalanine 73.3%. Apparent availabilities as measured by fecal analysis were higher reflecting some microbial synthesis in the lower tract.

Table 45.2. Essential Amino Acid Content of Sunflower Meal (% as fed)

	Soybean Meal[1]	Dehulled SFM[2]	Partially Dehulled SFM[2]	SFM (with hulls)[2]
Protein	48.5	45.5	32.0	26.0
Arginine	3.67	3.62	2.90	2.10
Histidine	1.20	0.96	0.65	0.65
Isoleucine	2.13	1.97	1.45	1.05
Leucine	3.63	2.77	2.40	1.65
Lysine	3.12	1.68	1.35	0.90
Methionine	0.71	0.82	0.65	0.60
Phenylalanine	2.36	2.12	1.75	1.20
Threonine	1.90	1.63	1.30	1.25
Tryptophan	0.69	0.60	0.35	0.20
Valine	2.47	2.22	1.90	1.30

[1]National Research Council, 1982.
[2]Cargill Processing Group (unpublished).

The ash content of sunflower seed is reported to be about 6%, but varies somewhat by area produced, variety grown and growing season. Calcium is usually about 0.4% and somewhat higher than that of soybean meal. Phosphorus is about 0.9%. Both these minerals need to be supplemented in swine diets. Other minerals vary widely from soybean meal (Table 45.3).

Table 45.3. Mineral Composition of Sunflower Meal

	Soybean Meal[1]	Sunflower Meal[2]	Partially Dehulled Sunflower Meal[2]
Calcium (%)	0.26	0.42	0.35
Magnesium (%)	2.13	1.19	0.62
Phosphorus (%)	0.64	0.94	1.30
Potassium (%)	2.13	1.19	1.00
Sodium (%)	0.01	0.16	0.20
Copper (mg/kg)	20.30	4.00	3.50
Iron (mg/kg)	1.31	31.00	14.00
Selenium (mg/kg)	0.10	2.13	1.50
Zinc (mg/kg)	0.57	98.00	98.00

[1]National Research Council, 1982.
[2]Various sources.

An indication of the vitamin content of sunflower seed is presented in Table 45.4. The values are those of the National Research Council (1988). It is not known how stable these vitamins are during storage and feed mixing.

Table 45.4. Vitamin Composition of Sunflower Meal (mg/kg)

	Soybean Meal	Sunflower Meal (dehulled)
Vitamin E	3.30	11.1
Biotin	0.32	----
Choline	2753	3632
Niacin	22	242
Pantothenic Acid	14.8	40.6
Riboflavin	2.9	3.5
Thiamin	3.1	3.1
Vitamin B_6	4.8	16.0

National Research Council, 1988.

UNDESIRABLE CONSTITUENTS IN SUNFLOWER MEAL

Sunflower meal is not known to contain any anti-nutritional factors. The major factor limiting its use in swine diets is the lack of lysine or the poor availability of the lysine. Additional fiber, when the hulls are not or only partially removed prior to oil extraction, detracts from the usefulness in diets for growing-finishing swine. Fiber content can reduce feed intake and feed efficiency (Seerley et al., 1974).

FEEDING SUNFLOWER MEAL

Starter Pigs

Sunflower meal has not received much attention for use in creep or starter diets. The high fiber and lower energy content would likely reduce gains and feed efficiency. Lysine and threonine supplementation would also be needed in such diets. Moser et al. (1985) found no differences in gains of pigs (initial weight of 10 kg) fed corn-sunflower or soybean meal diets. Wahlstrom et al. (1985) did report a response to tryptophan and threonine additions to a low protein, lysine supplemented corn-sunflower diet with pigs of nine kilogram initial weight.

Growing-Finishing Pigs

Substituting sunflower meal for one-third, two-thirds or all of the soybean meal in pelleted barley or corn based diets for growing-finishing swine (20-95 kg) had little effect on daily gains provided lysine was added to equal that of the soybean meal control diet (Table 45.5). Feed efficiency was slightly reduced, reflecting the additional fiber and lower energy value of the sunflower meal.

Table 45.5. Swine Performance on Rations with Sunflower Meal (SFM) Replacing Soybean Meal (SBM)[1]

	SFM/SBM			
	0/100	33/67	67/33	100/0[2]
Barley-Based Diets (15% Protein)				
Daily Gain (kg)	0.68	0.75	0.67	0.68
Daily Intake (kg)	2.27	2.29	2.19	2.23
Feed Efficiency	3.12	3.08	3.24	3.29
Corn-Based Diets (15% Protein)				
Daily Gain (kg)	0.77	0.80	0.78	0.83
Daily Intake (kg)	2.13	2.22	2.22	2.36
Feed Efficiency	2.77	2.79	2.86	2.84

[1]Dinusson et al., 1980a,b.
[2]Lysine added to SFM diets to equal level in soybean meal (0/100) diet.

Georgia workers (Seerley et al., 1974) using expeller meal found reduced gains and feed efficiency when 50% or more of the soybean meal was replaced by sunflower meal. Lysine supplementation largely overcame the reduced performance on the higher levels of sunflower meal. In comparisons of 28 vs. 44% sunflower meal in swine diets, gains were reduced by 4% and feed for gain was increased by 9% for pigs on the lower protein meal (Table 45.6). Feed intake was increased by 7% to equal the intake in the soybean meal control diet.

Table 45.6. 28 vs. 44% Sunflower Meal Compared to Soybean Meal in Growing-Finishing Swine Diets[1]

	SBM (44%)	SFM (44%)[2]	SFM (28%)[2]
Daily Gain (kg)	0.81	0.75	0.72
Daily Intake (kg)	2.72	2.57	2.75
Feed Efficiency	3.37	3.51	3.81

[1]Dinusson et al., 1981a,b.
[2]Lysine added to SFM diets to equal that in soybean meal diet.

Lysine supplementation is essential if a major portion of the protein addition is sunflower meal. Moser et al. (1985) reported that a 34% protein sunflower meal can replace up to 45% of the lysine in the ration with no detrimental effects on swine performance. In pelleted barley-based diets using sunflower meal (34%) as the protein source, increasing lysine from an estimated 0.57 to 0.70% increased gains by over 15% and reduced feed for gain by 6%. Increasing the sunflower meal to provide 18 instead of 16% protein did increase gain 6%. Adding methionine had no effect on gain. Lysine plus methionine increased gains less than 3% over lysine alone (Table 45.7). Availability of lysine in sunflower meal was assumed to be similar to that of barley in these trials (National Research Council, 1988).

Table 45.7. Barley-Based Swine Diets with Added Protein, Lysine and Methionine (16% protein diets)[1]

	0	L	M	L+M[2]	18% Protein
Daily Gain (kg)	0.64	0.74	0.64	0.76	0.68
Daily Intake (kg)	2.25	2.49	2.35	2.50	2.42
Feed Efficiency	3.52	3.37	3.69	3.29	3.55

[1]Dinusson et al. 1982; 1983.
[2]L=Lysine, M=Methionine.

Wahlstrom et al. (1985) added either tryptophan or threonine to a low protein (12%), lysine fortified corn-sunflower diet. Tryptophan gave a small but non-significant improvement in both gain and feed for gain. Adding both of these together to the 12% protein diet significantly improved both rate of gain and feed efficiency to equal that of a 16% lysine fortified positive control diet. Isoleucine or isoleucine plus methionine had no effect. These results suggest that, on a lysine fortified corn-sunflower meal diet, tryptophan and threonine are limiting amino acids. Compared to corn, barley has higher levels of these amino acids.

Growing pigs gained only two-thirds as rapidly on sunflower meal supplemental diets as did those on soybean meal supplemental diets. It also took 15% more feed for gain on the sunflower meal diets. When lysine was added to the sunflower meal diets to equal that of the soybean meal diets there was no difference in pig performance (Dinusson et al., 1985). Baird (1981) reported that if lysine was added to corn-sunflower meal diets to equal the level in corn-soybean meal diets, the performance of growing pigs was similar. He further reported that up to 50% of the soybean meal could be replaced by sunflower meal, even without added lysine, without deleterious effect on pig performance.

Sunflower meals of lower protein and higher fiber content due to only partial or no dehulling tend to reduce pig performance. Lysine supplementation corrects the poorer performance in most instances, but the higher fiber diets required more feed for gain. Increasing the percent protein in the diets with higher levels of sunflower meal can also be used to maintain gains of growing-finishing pigs (Moser et al., 1985). This method did reduce feed efficiency due to the higher fiber and lower usable energy of the sunflower meal.

The use of sunflower meal to supplement diets based on cereals other than barley or corn has received scant attention. It is likely that more lysine would have to be added to rations based on proso (hog millet) or milo.

Breeding Pigs

Sunflower meal has been used as part of the supplement in gestation rations. Bred gilts, sows and adult boars require only 12% protein and 0.43% lysine in their rations (National Research Council, 1988). Therefore, sunflower meal could be used in rations to meet these requirements. However, energy requirements for these classes of swine are higher than those found in sunflower meal so a greater feed intake would be needed. In lactation rations, sunflower meal could be used as part of the protein supplementation with a meal of higher lysine content or lysine could be added to make up the deficit. However, energy intake is likely to be lower.

SUMMARY

The extent of use of sunflower meal in swine diets will be governed by economics, availability and quality of the meal and lysine availability and cost. In the past decade, the price of sunflower meal has been very competitive with other protein sources in areas where sunflowers are grown and processed. The outlook in the future is for a greater use of this protein source for swine. Research on how best to use this meal in starter diets and for adult swine rations is needed to allow for more efficient use.

REFERENCES

Baird, D.M., 1981. Sunflower meal as partial and total supplement for finishing pigs. J. Anim. Sci. 53 (Suppl. 1): 81.

Dinusson, W.E., Harrold, R.L. and Johnson, J.N., 1980a. Sun oil meal in swine rations. 10th Annual Livestock Research Report, Department of Animal and Range Sciences, North Dakota State University, Fargo.

Dinusson, W.E., Harrold, R.L. and Johnson, J.N., 1980b. Sun oil meal as replacement for soybean meal in corn based rations for swine. Annual Report, Department of Animal and Range Sciences, North Dakota State University, Fargo.

Dinusson, W.E., Harold, R.L., Johnson, J.N. and Zimprich, R.C., 1981a. 28 vs 44% protein sunflower meal for swine. Annual Report, Department of Animal and Range Sciences, North Dakota State University, Fargo.

Dinusson, W.E., Harrold, R.L. and Johnson, J.N., 1981b. Sun oil meal in rations for growing-finishing swine. J. Anim. Sci. 53 (Suppl. 1): 85.

Dinusson, W.E., Harrold, R.L. and Johnson, J.N., 1982. Lysine and protein levels for growing-finishing swine. Annual Report, Department of Animal and Range Sciences, North Dakota State University, Fargo.

Dinusson, W.E., Harrold, R.L., Johnson, J.N. and Zimprich, R.C., 1983. Lysine and methionine additions to barley-sunflower meal rations for swine. Annual Swine Research Report, Department of Animal and Range Sciences, North Dakota State University, Fargo.

Dinusson, W.E., Johnson, J.N. and Zimprich, R.C., 1985. Protein and lysine levels in barley-sunflower meal rations for growing-finishing swine. Swine Progress Report, Department of Animal and Range Sciences, North Dakota State University, Fargo.

Jorgensen, H. and Sauer, W.C., 1982. Amino acid availabilities in protein sources for swine. University of Alberta, Edmonton, 61st Annual Feeder's Day Report, pp. 86-88.

Moser, R.L., Cornelius, S.G., Pettigrew, J.E. and Hanke, H.E., 1985. Efficiency of 34% crude protein sunflower meal for growing pigs. Nutr. Rep. Int. 31: 583-591.

National Academy of Sciences, 1982. United States-Canadian Tables of Feed Composition. 3rd ed. National Research Council, Washington, D.C.

National Research Council, 1988. Nutrient Requirements of Domestic Animals, No. 2. Nutrient Requirements of Swine. 9th ed. National Academy of Sciences, Washington, D.C.

National Sunflower Association, 1988. United States Sunflower Crop Quality Report. Bismarck, North Dakota.

Seerley, R.W., Burdick, D., Russom, W.C., Lowrey, R.S., McCampbell, H.C. and Amos, H.E., 1974. Sunflower meal as a replacement for soybean meal in growing swine and rat diets. J. Anim. Sci. 38: 947-953.

Smith, K.J., 1968. A review of nutritional value of sunflower meal. Feedstuffs 40: 20-23.

Wahlstrom, R.C., Libal, G.W. and Thaler, R.C., 1985. Efficiency of supplemental tryptophan, threonine, isoleucine and methionine for weanling pigs fed a low-protein lysine supplemental corn-sunflower meal diet. J. Anim. Sci. 60: 720-724.

CHAPTER 46

Sunflower Seeds

R.C. Wahlstrom

INTRODUCTION

Sunflower seeds (*Helianthus annus*) are of economic importance in some parts of the world because they are the best-adapted, high oil-protein crop available. World wide production of sunflowers has increased steadily in the past 10 to 20 years with the major sunflower producing areas being the USSR, China, Argentina, the United States and the European Economic Community (McMullen, 1985). These areas produce about 90% of the world's production of sunflower seed.

Although sunflowers are grown primarily for the production of oil, processing plants may not be available in all areas and not all sunflower seeds are suitable for the production of oil. Thus, sunflower seed may be available as an alternate feed resource for use in swine production. Because of their high oil content, they can be used as both an energy and protein source.

GROWING SUNFLOWERS

Sunflowers are basically a temperate zone plant. However, the crop is adaptive to a wide range of environments. A temperature range of 8 to 34°C is tolerated without significant reduction in yield, indicating adaptation to regions with warm days and cool nights. However, temperature does affect seed-oil content. In general, temperatures which remain above 25 or below 16°C at flowering are believed to reduce seed yield and seed-oil content.

Sunflowers use soil moisture efficiently and are considered to be drought resistant since they root deeply and extract water at depths not attained by other crops (Weiss, 1983). In addition, sunflower seedlings are more tolerant of frost than many crops (Robinson, 1978). They grow well on neutral to moderately alkaline soils (pH 6.5-8.0) but perform poorly on acid soils (Weiss, 1983). Average yields of 1400 to 2000 kg/ha are common, although yields may vary from 1,000 to 3,000 kg/ha depending on plant population, fertilizer application, weed control, soil type, temperature and available moisture.

NUTRIENT CONTENT OF SUNFLOWER SEEDS

Sunflower seeds can be considered as high in energy and medium in protein content. They contain on average about 38% oil, 17% crude protein and 15% crude fiber. The additional energy provided by the fat present in sunflower seed is partially offset by the high crude fiber content. The high crude fiber arises because of the thick hull on the sunflower seed. The chemical composition of sunflower seed is presented in Table 46.1.

Table 46.1. Chemical Composition of Sunflower Seed (%)

	Sample A	Sample B
Moisture	8.8	7.4
Crude Protein	17.3	17.4
Ether Extract	37.9	39.2
Crude Fiber	16.2	14.2
Ash	3.8	3.6
Calcium	0.21	---
Phosphorus	0.60	---

Sample A: Kepler et al., 1982.
Sample B: Hartman et al., 1985.

An amino acid profile of sunflower seed is given in Table 46.2. Sunflower seed protein is deficient in lysine and diets containing sunflower seed may need to be combined with feed ingredients that are good sources of lysine or supplemented with synthetic lysine.

Sunflower seeds are a good source of dietary fat for swine. Sunflower seed contains a high amount (85%) of unsaturated fatty acids. Oil from sunflowers produced in cool northern climates normally contains about 70% linoleic acid, whereas oil produced in more southerly latitudes may contain as little as 30 to 40% linoleic acid (Dorrell, 1978). There is a strong inverse relationship with oleic acid.

Table 46.2. Amino Acid Composition of Sunflower Seed

	% of Protein	% of Seed
Arginine	9.5	1.64
Histidine	2.4	0.42
Isoleucine	3.9	0.68
Leucine	6.1	1.05
Lysine	3.8	0.65
Methionine	1.7	0.30
Phenylalanine	4.4	0.76
Threonine	3.6	0.62
Valine	5.0	0.87

Kepler et al., 1982.

UNDESIRABLE CONSTITUENTS IN SUNFLOWER SEEDS

Sunflower seed appears to be relatively free of anti-nutritional factors that reduce animal performance. However, its high crude fiber content (14-17%) must be a concern, as the addition of crude fiber to swine diets decreases the digestible and metabolizable energy concentration of the diet and may also influence the digestibility of other nutrients.

Heating sunflower seed to 100°C for five hours decreased chlorogenic acid, an effective trypsin inhibitor, by 42% (Milic et al., 1968). However, heat treatment causes a reduction in amino acid content and availability (Alexander and Hill, 1952; Bandemer and Evans, 1963). Any benefit obtained from changing the levels of chlorogenic acid content by heating were lost by decreases of lysine, threonine and arginine (Amos et al., 1975). As a consequence, heat treatment of sunflower seed does not appear to be economically beneficial in swine production.

FEEDING SUNFLOWER SEEDS TO SWINE

Starter Pigs

The effect of including sunflower seed in starter diets has received little attention. The oil in sunflower seed is well utilized by young swine (Adams and Jensen, 1985b). However, Fitzner et al. (1989) fed diets containing 15 and 25% sunflower seed to pigs weaned at 21 days and weighing approximately 6 kg. Average daily gain and feed intake decreased as the level of sunflower seed increased in the diet. Because of the high fiber content of sunflower seed, and based on data such as that presented in Table 46.3, it is recommended that sunflower seed be limited to 15% of the diet of young weaned pigs.

Table 46.3. Performance of Starter Pigs (6-19 kg) Fed Sunflower Seed

	Level of Sunflower Seed (%)				
	0	10	15	20	25
Daily Gain (kg)	0.39	0.38	0.38	0.35	0.35
Daily Intake (kg)	0.69	0.68	0.67	0.59	0.59
Feed Efficiency	1.77	1.79	1.76	1.68	1.68

Fitzner and Hines, 1988.

Growing-Finishing Pigs

Increased production and availability of sunflower seed can result in its being economically competitive as an energy and protein source for growing-finishing swine. Sunflower seed has been included at various levels up to 60% of growing-finishing swine diets. Laudert and Allee (1975) included 0, 20, 40 or 60% sunflower seed in diets of 23 kg pigs and reported a linear decrease in feed consumption and an improvement in feed efficiency as the level of sunflower seed increased. Feeding 62 kg pigs diets of 25 or 50% sunflower seed reduced gain and feed consumption but did not affect feed efficiency. Supplementing diets containing 50% sunflower seed with 0.3% lysine resulted in increased gains and feed consumption and an improvement in feed efficiency (Baird, 1980). Rate of gain and feed efficiency were slightly superior to pigs fed the control maize-soybean meal diet.

The results of a feeding trial using lower levels (0, 2.5, 5, 10 and 20%) of sunflower seed in the diets of growing-finishing swine (Hartman et al., 1985) are given in Table 46.4. Rate of gain increased when diets contained 5% and decreased if diets contained 20% sunflower seed. The inclusion of 10% sunflower seed in the diets resulted in a slight reduction in rate of gain in Trial 1 but an increase in Trial 2. Feed efficiency was not affected by any level of sunflower seed. Rate of gain appears to be affected by feed intake.

Adams and Jensen (1985b) added 28 or 56% sunflower seed to pig diets to provide 10 or 25% dietary fat and reported fat digestibility was unaffected. They observed a decrease in the digestibility value for dry matter, gross energy and nitrogen as dietary fat level increased. This probably reflects the increase in dietary crude fiber content as the level of sunflower seeds increased. Noland et al. (1980) replaced 0, 25, 50 or 100% of the soybean meal in pig diets with sunflower seed and reported no differences in digestible energy. However, all diets containing sunflower seed had significantly higher protein digestibility than the diet containing only soybean meal as the protein source. Processing method can affect the availability of fat in sunflower seed. Roasting tended to decrease digestibility of fat in whole seeds and dehulled sunflower seed had a lower fat digestibility than regular sunflower seed (Adams and Jensen, 1985a).

Table 46.4. Performance of Pigs Fed Diets Containing Sunflower Seed[1]

	Sunflower Seed (%)			
	0	5	10	20
Trial 1				
Daily Intake (kg)	2.41	2.41	2.20	2.20
Daily Gain (kg)[2]	0.78	0.82	0.74	0.71
Feed Efficiency	3.06	3.00	2.93	3.08
Trial 2	0	2.5	5	10
Daily Intake (kg)	2.36	2.46	2.53	2.61
Daily Gain (kg)[3]	0.71	0.74	0.76	0.80
Feed Efficiency	3.32	3.34	3.28	3.30

[1]Hartman et al., 1985.
[2]Cubic response ($P<0.05$).
[3]Linear response ($P<0.01$).

Diets containing high levels of unsaturated fatty acids produce soft carcasses. Increasing amounts of sunflower seed in diets of growing-finishing swine result in a linear increase in linoleic acid in the backfat and a deleterious effect on carcass firmness (Marchello et al., 1984; Hartman et al., 1985). Muscles from carcasses of pigs fed diets containing 26 or 39% sunflower seed are soft, pale and exudative (Marchello et al., 1984). Diets containing up to 20% sunflower seed, although producing soft carcasses, have been reported to have no effect on the amount of carcass backfat, length, longissimus muscle area, chemical composition or taste panel scores (Hartman et al., 1985). Panelists evaluated chops for flavor, tenderness, juiciness and overall desirability. The decrease in carcass firmness when over 10% sunflower seed is included in the diet could make carcasses less acceptable for normal processing procedures of packing plants and also less acceptable to the retail trade.

Sunflower seed does not affect performance of growing-finishing pigs at dietary levels of 10% and has only a slight effect when included as 20% of the diet. However, because of the effect of high levels of sunflower seed on carcass quality, it is recommended that diets of growing-finishing swine not contain over 10% sunflower seed.

Breeding Stock

Research on the effect of sunflower seed as a feed ingredient during gestation and lactation is limited. The addition of fat to sow diets during late gestation has been reported to increase pig survivability by increasing energy stores of the newborn pig and increasing the fat content of milk to provide increased energy for the young pig (Pettigrew, 1981). The high fat content of sunflower seed could provide a convenient source of fat in diets of gestating and lactating sows. However, pig survival and body weights at birth and 14 days were not improved when 25 and 50% sunflower seeds were fed to sows from day 100 of gestation and during the first two weeks of lactation (Kepler et al., 1982). Data are presented in Table 46.5. The percentage of fat in sows' milk increased as level of dietary sunflowers increased. Some problems with consumption were encountered during the first few days of feeding when diets contained 50% sunflower seed. However, there were no differences in diet consumption during lactation.

Table 46.5. Sunflower Seed in Sow Diets[1]

	Sunflower Seed (%)		
	0	25	50
Number of Farrowings	35	31	30
Pigs Born Alive/Litter	9.8	10.1	9.3
Survival to 14 Days (%)	74.3	74.8	69.9
Birth Weight (kg)	1.5	1.4	1.5
14-Day Weight (kg)	3.6	3.6	3.8
Fat in Colostrum (%)	5.4	6.4	6.6
Fat in Milk (%)[2]	7.8	10.1	11.4

[1]Kepler et al., 1982.
[2]Linear effect ($P<0.01$).

Replacing part of the energy and protein in sow diets with 25% sunflower seed has no adverse effect on sow and piglet performance. Diets containing 25% sunflower seed will contain approximately 12% fat, a level that will increase percentage of milk fat. Because of inconsistent consumption of diets containing 50% sunflower seed, it is recommended that sunflower seed be limited to 25% in breeding swine diets.

SUMMARY

Sunflower seed is a high energy feed ingredient that can replace cereal grain and protein supplement in the diet in a ratio of approximately four parts of grain to one part of protein supplement if diets are balanced with lysine.

For optimum benefit from this nontraditional food source, it is recommended that sunflower seed should not be included in starter diets at levels greater than 15% of the diet. It should be limited to 10% of diets of growing-finishing swine and should not be included at levels greater than 25% in gestation and lactation diets.

REFERENCES

Adams, K.L. and Jensen, A.H., 1985a. Effect of processing on the utilization by young pigs of the fat in soya beans and sunflower seeds. Anim. Feed Sci. Technol. 12: 267-274.

Adams, K.L. and Jensen, A.H., 1985b. Effect of dietary protein and fat levels on the utilization of the fat in sunflower seeds by the young pig. Anim. Feed Sci. Technol. 13: 159-170.

Alexander, J.C. and Hill, D.C., 1952. The effect of heat on the lysine and methionine in sunflower seed oil meal. J. Nutr. 48: 149-159.

Amos, H.E., Burdick, D. and Seerley, R.W., 1975. Effect of processing temperature and L-lysine supplementation on utilization of sunflower meal by the growing rat. J. Anim. Sci. 40: 90-95.

Baird, D.M., 1980. Sunflower meal and sunflower seed with supplemental lysine for finishing pigs. J. Anim. Sci. 51 (Suppl. 1): 185.

Bandemer, S.L. and Evans, R.J., 1963. The amino acid composition of some seeds. J. Agr. Food Chem. 11: 134-137.

Dorrell, D.G., 1978. Processing and utilization of oilseed sunflower. In: J.F. Carter, ed. Sunflower Science and Technology. American Society of Agronomy, Madison, pp. 407-440.

Fitzner, G.E. and Hines, R.H., 1988. Sunflower oil seeds in nusery pig diets. Kansas State University, Manhattan, Swine Day Report, pp. 49-51.

Fitzner, G.E., Hines, R.H., Goodband, R.D., Thaler, R.C., Stoner, G.R. and Weeden, T.L., 1989. Effect of black sunflower oil seeds on weanling pig performance. J. Anim. Sci. 67 (Suppl. 1): 111.

Hartman, A.D., Costello, W.J., Libal, G.W. and Wahlstrom, R.C., 1985. Effect of sunflower seeds on performance, carcass quality, fatty acids and acceptability of pork. J. Anim. Sci. 60: 212-219.

Kepler, M.A., Libal, G.W. and Wahlstrom, R.C., 1982. Sunflower seeds as a fat source in sow gestation and lactation diets. J. Anim. Sci. 55: 1082-1086.

Laudert, S.B. and Allee, G.L., 1975. Nutritive value of sunflower seed for swine. J. Anim. Sci. 41: 381 (Abstract).

Marchello, M.J., Cook, N.K., Johnson, V.K., Slanger, W.D., Cook, D.K. and Dinusson, W.E., 1984. Carcass quality, digestibility and feedlot performance of swine fed various levels of sunflower seed. J. Anim. Sci. 58: 1205-1210.

McMullen, M.P., 1985. Sunflower production and pest management. North Dakota State University Extension Bulletin 25: 1-76.

Milic, B., Stojanovic, S., Vucurevic, N. and Turcic, M., 1968. Chlorogenic and quinic acids in sunflower meal. J. Sci. Food Agric. 19: 108-113.

Noland, P.R., Campbell, D.R. and Johnson, Z.B., 1980. Use of unextracted sunflower seeds as a protein source. Anim. Feed Sci. Technol. 5: 51-57.

Pettigrew, J.E., 1981. Supplemental dietary fat for peripartal sows: A review. J. Anim. Sci. 53: 107-117.

Robinson, R.G., 1978. Production and culture. In: J. F. Carter, ed. Sunflower Science and Technology. American Society of Agronomy, Madison, pp. 89-143.

Weiss, E.A., 1983. Oilseed Crops. Longman Inc., New York, 660 pp.

CHAPTER 47

Sweet Potato

E. Nwokolo

INTRODUCTION

Tropical, humid regions usually have a high incidence of cloud cover, immense amounts and rates of precipitation, poor soil drainage, very high soil moisture and tremendous plant pest and disease burdens. Cereal crops are affected much more by these factors than root and tuber crops. Yet governments and individuals spend more money and time attempting to cultivate cereal grains than they spend in cultivating root and tuber crops. Part of the reason for this may be the scarcity of information on the chemical and nutritional qualities of tuber and root crops for animal feeding.

The sweet potato (*Ipomoea batatas* L.) has been grown primarily for human food and has been so used for many centuries (Edmond, 1971). Sweet potatoes which are not suitable for human use because of damage, poor size or grade, are culled and may be processed for other industrial uses. Industrial uses of sweet potato include starch production and fermentation into alcohol. However, there is a large excess which is either currently being used as animal feed or has the potential for such utilization. By providing information on the nutritional composition of these root and tuber crops, it is hoped that their utilization can be enhanced in the formulation and production of animal feeds, particularly those fed to swine.

GROWING SWEET POTATOES

The sweet potato belongs to the family Convolvulaceae, which comprises about 50 genera and more than 1200 species of plants. It is an herbaceous perennial, with a trailing, twining stem whose leaves are simple and are arranged alternately around the slender stem (Onwueme, 1978). It is not certain where the sweet potato originated, although it was being used in Central and South America in the 15th century (Edmond, 1971). From such obscure origins, the sweet potato has spread worldwide to all tropical and subtropical regions of the world, being cultivated extensively in Africa, Asia, Central and South America, Oceania and the Pacific islands, and to a lesser extent in the United States and Europe. World production is about 111 million metric tonnes with over 90% of sweet potatoes produced in Asia and 6% in Africa (FAO, 1985).

Yeh and Bouwkamp (1985) reported that the dry matter production of high yield sweet potato cultivars is comparable with the yield from many other roots, tubers and grains. Yields of sweet potato range from an average of 5.7 tonnes/ha in Africa to 15.9 tonnes/ha in Asia. Under intensive irrigation and fertilizer application, productivity ranges from 30.7 tonnes/ha in Hong Kong to 40.0 tonnes/ha in Israel. Walter et al. (1984) observed that protein yield from sweet potato was about 184 kg/ha, a value that compares favorably with yields of 200 kg/ha for wheat, and 168 kg/ha for rice. Harvesting is manual in many parts of Africa, and highly mechanized in Israel, Hong Kong and many countries in Asia.

NUTRIENT CONTENT OF THE SWEET POTATO

The nutrient composition of fresh and dried root and leaf of sweet potato is shown in Table 47.1. Fresh sweet potato is a perishable, high moisture feed containing about 31.5% solids, 28% total carbohydrates, 1.8% protein, 1.1% ash and 0.7% fat, with an energy content of 1244 kcal/kg (O'Hair, 1984). On the other hand, dehydrated sweet potato has been reported to contain 89.6% total carbohydrates, 3.7% protein, 2.5% ash, 1.5% fat and 2.7% crude fiber, with a gross energy content of 4061 kcal/kg and a metabolizable energy content of 2962 kcal/kg (Ravindran and Rajaguru, 1985). Collins and Walter (1985) observed that the carbohydrate fraction contains starch, sugar, cellulose, hemicellulose and water-insoluble pectins. The starch is composed of 60 to 70% amylopectin, the rest being amylose. The sugars are comprised mainly of sucrose, with smaller amounts of glucose and fructose.

The crude protein content of fresh roots varies greatly depending on cultivar. Yeh and Bouwkamp (1985) reported values which ranged from 1.1% to as high as 8.8%. While the content of protein in the root is generally very low, leaf protein content is considerably higher and may be as high as 18 to 24%, on a dry matter basis (O'Hair, 1984; Yeh and Bouwkamp, 1985). Most of the protein in sweet potato exists as the globulin ipomoein (Jones and Gersdorff, 1931) although sweet potato also contains significant quantities (15 to 35% at harvest) of non-protein nitrogen (Purcell et al., 1978).

Table 47.1. Nutrient Composition of Sweet Potato Root and Leaves (% as fed)

	Fresh Weight[1]		Dry Weight[2]	
	Root	Leaf	Root	Leaf
Moisture	68.8-73.3	86.7	---	---
Dry Matter	26.7-31.2	13.3	100	100
Carbohydrate	25.6-28.5	8.0	91-95	60.2
Fiber	0.8-1.0	1.6	3.0-3.2	12.0
Protein	1.0-1.9	3.2	3.7-6.1	24.1
Ash	0.7-1.0	1.4	2.6-3.2	10.5

[1]O'Hair, 1984.
[2]Calculated using the data of O'Hair.

There are indications that sweet potato protein may be deficient in some essential amino acids, particularly the sulfur amino acids and lysine (Table 47.2). Thus, Walter and Catignani (1981) observed that both lysine and the sulfur amino acids are limiting. It is recommended that sweet potato-based diets be supplemented with both lysine and the sulfur amino acids. The addition of small amounts of fish or meat meal may be a convenient way of ensuring lysine and sulfur amino acid sufficiency. Alternatively, synthetic lysine and methionine may be used to supplement sweet potato-based diets for swine.

Table 47.2. Amino Acid Composition of Normal and High Protein Sweet Potato Chips (% air dry sample)

	Normal Sweet Potato	High Protein Sweet Potato	Yellow Corn
Dry Matter	90.50	90.40	89.00
Crude Protein	2.90	6.35	8.90
Cystine	0.03	0.10	0.09
Isoleucine	0.12	0.24	0.45
Lysine	0.13	0.26	0.18
Methionine	0.04	0.12	0.09
Phenylalanine	0.13	0.31	0.45
Threonine	0.11	0.29	0.36
Tryptophan	0.41	----	0.09
Tyrosine	0.05	0.27	0.39

Yeh and Bouwkamp, 1985.

Mineral components of fresh root and leaf, as well as dried root and leaf, of sweet potato are presented in Table 47.3. While dried roots are only fair sources of calcium, phosphorus and potassium, dried leaves are excellent sources of these minerals. Fresh roots are very poor sources of most minerals because of dilution with large quantities of water. However, trace mineral analysis data reported recently (Collins and Walter, 1985; Yeh and Bouwkamp, 1985) indicate that dried tops have a much higher content of trace minerals than dried roots.

Table 47.3. Vitamin and Mineral Content (mg/100 g) of Sweet Potato Root and Leaves

	Fresh Weight[1]		Dry Weight[2]	
	Root	Leaf	Root	Leaf
Calcium	21.0-33.0	86.0	79-106	647
Phosphorus	38.0-50.0	81.0	142-160	609
Iron	0.9-2.0	4.5	3.4-6.4	33.8
Sodium	31.0	5.0	107	37.6
Potassium	210.0	562	724	4225
Thiamin	0.12	0.1	0.35	0.8
Riboflavin	0.04	0.2	0.16	1.6
Niacin	0.70	0.7	2.40	5.3
Ascorbic Acid	21-37	17.0	79-119	127.8
Beta-carotene	35-2400	2215	---	---

[1]O'Hair, 1984.
[2]Calculated using the data of O'Hair.

Sweet potatoes contain large quantities of beta-carotene (Lease, 1941; Ezell and Wilcox, 1946) with the content being very high in yellow cultivars of sweet potato but moderate-to-low in white cultivars. Dried leaves of sweet potato contain extremely high levels of carotene. Fresh leaves and roots of sweet potato have been reported to contain significant amounts of ascorbic acid (Jenkins and Moore, 1954) while the root has also been observed to contain moderate quantities of niacin, riboflavin, thiamin and pantothenic acid (Pearson and Leucke, 1954). Detailed nutrient composition of fresh sweet potato roots and leaves has been provided by O'Hair (1984). The content of these nutrients, expressed on a wet weight basis, has been recalculated on a dry weight basis and is shown in Table 47.3.

UNDESIRABLE CONSTITUENTS IN SWEET POTATO

The existence of two active fractions of a trypsin inhibitor in sweet potato was reported by Sohonie and Honawar (1956) who noted that the presence of these anti-nutritional factors could adversely affect proteolytic digestion. The trypsin inhibitor is heat-labile and is destroyed by exposure to high temperatures for relatively short periods of time. However, it is quite unlikely that the heat used in feed manufacture (milling and pelleting) is sufficient for the destruction of the trypsin inhibitor in sweet potato. Indeed, Yeh et al. (1979) have demonstrated that exposure of sun-cured sweet potato chips to 100°C for 5 minutes did not completely destroy the trypsin inhibitor in the chips.

PROCESSING OF SWEET POTATO FOR SWINE FEEDING

In Africa, Asia and Oceania, sweet potatoes are grown in small lots along side pig fattening operations. The only processing operations performed involve washing of the potatoes, trimming to remove excessive root growth and slicing or grinding of the sweet potatoes prior to feeding. When fresh sweet potatoes are fed, it is important to estimate the dry matter intake of the pigs. This is to ensure that recommended nutrient requirements for the physiological function, age and production rate of the pigs are met.

Sweet potatoes may also be dried to produce a low moisture, high energy feed ingredient for swine and poultry. The roots are usually washed, trimmed and cut up. The cut up material is dried in air that is heated to 80°C to 100°C in a forced-draught dehydration chamber. The cut up sweet potatoes are placed on a conveyor belt which passes through the heated chamber. Products which have attained a pre-set degree of dehydration are separated, allowed to cool to room temperature and bagged. In tropical countries, cut up roots may also be air-dried by spreading them out on mats or they may be dehydrated using solar driers. Under these conditions, drying takes a longer time than in forced-draught dehydration chambers and consequently a slight increase in the deterioration of the final product should be expected.

Table 47.4. The Effect of Trypsin Inhibitor on The Performance of Growing and Finishing Pigs[1]

		Sweet Potato-Based Diets[2]		
	Control	2	3	4
Inhibition of TIA of Feed (%)	23	44	52	49
Days Needed to 90 kg Weight	118	147	163	201
Daily Gain (kg)	0.63	0.48	0.44	0.36
Daily Intake (kg)	1.85	1.83	1.78	1.74
Feed Efficiency	3.08	3.84	4.09	4.99

[1]Yeh and Bouwkamp, 1985.
[2]Control=corn-soybean diet; Diet 2=corn-sweet potato-soybean meal; Diet 3=sweet potato-soybean meal; Diet 4=sweet potato + vine-soybean.

Processing of sweet potato aims to increase carbohydrate and protein digestibility while reducing the content of trypsin inhibitor in the chips to a level which will not interfere with growth or production. Rapid, high temperature drying (390 to 430°C for 1 minute) does not improve nutrient digestibility but popping at 6 to 8 kg/cm^2 pressure at 164 to 178°C, does (Yeh et al., 1979). Yeh and Bouwkamp (1985) quote unpublished work from Taiwan Sugar Corporation which showed that trypsin inhibitor activity (TIA/g dry matter) was 1320 in sun-cured sweet potato. This was reduced to 1100 in chips dry-heated at 390°C for 1 minute, and further decreased to 630 in chips dry-heated at 430°C for 1 minute. Data showing the influence of trypsin inhibitors on performance of growing-finishing pigs is presented in Table 47.4.

FEEDING SWEET POTATO

As early as the beginning of this century, it was recognized that sweet potato roots were a satisfactory feed ingredient for swine. However, some workers indicated that when pigs were fed raw sweet potatoes, they grew more slowly unless the diet included corn. From a review of earlier experiments, conducted in Mississippi, Louisiana and South Carolina in the 1940s, Edmond (1971) suggested that for swine, a 2:1 mixture of corn and sweet potatoes was more efficiently utilized for growth than sweet potatoes alone. There was also an indication from these results that pigs fed blanched sweet potatoes performed better than those fed raw roots and were almost equal in performance to those fed corn meal.

The growth depression in pigs fed raw sweet potatoes could be a result of insufficient energy intake from the fresh potatoes or the presence of trypsin inhibitors. Results from short term feeding trials with pigs showed that growth rate and feed efficiency of pigs fed heat-treated sweet potato chips were significantly better than those fed raw sweet potato chips, the performance being comparable to pigs fed corn-based diets (Lee, 1979; Wu, 1980). Further evidence of improved nutritional quality of cooked versus raw sweet potatoes was provided by Yeh et al. (1979), who reported that pigs fed raw sweet potatoes at the 40% level as a replacement for corn grew significantly slower than those given corn or popped sweet potato diets. Yeh and Bouwkamp (1985) quoted unpublished data from Taiwan workers which indicated that 25% might be the optimum level of substitution of sweet potato for corn in swine diets. However, there is evidence that with 0.2% L-Lysine supplementation, 50% of the corn could be replaced by sweet potato chips without any detrimental effects on growth or productivity (Chen et al., 1980). Supplementation of sweet potato diets with L-lysine has been shown to be highly beneficial, implying that in high sweet potato diets a lysine deficiency exists. As Table 47.5 shows, acceptable rates of growth can be achieved with levels of sweet potato inclusion of up to 60%, when such diets are supplemented with sulfur amino acids and lysine.

Table 47.5. The Performance of Fattening Pigs Fed Different Proportions of Corn and Sweet Potato Chips

Corn in Diet (%)	Potato in Diet (%)	Daily Gain (kg)	Feed Efficiency
63-85	0	0.53	3.93
30-39	30-39	0.48	3.83
0	56-72	0.37	4.79
63-81	0	0.65	3.38
45-58	15-20	0.66	3.37
29-37	29-39	0.62	3.54
14-18	42-54	0.58	3.74
0	54-68	0.56	3.81
72-84	0	0.60	3.08
35-41	35-41	0.48	3.84
0	68-81	0.44	4.08
69-75	0	0.69	2.95
33-36	33-36	0.66	3.13
0	63-68	0.60	3.37

Yeh and Bouwkamp, 1985.

Carcass quality is affected by the incorporation of sweet potato into swine diets. There is evidence that in growing-fattening pigs fed sweet potatoes (raw or cooked), there is a decrease in back fat thickness (Tai and Lei, 1970). This is probably because the bulkiness of diets containing high levels of sweet potatoes does not permit sufficient food intake to encourage excessive back fat deposition. Thus, these workers noted less back fat and a higher percentage of lean cuts when sweet potato substituted for half or all of the corn in swine diets. There is also practical knowledge among hog farmers feeding sweet potatoes, that feeding sweet potatoes to hogs causes a hardening of the fat, resulting in a firmer carcass. This hardening of the carcass fat implies that the proportion of saturated fats in the carcass is increased. The carcass measurements of growing-fattening pigs fed sweet potato-based diets are shown in Table 47.6.

Studies to determine the digestibility of crude protein, fiber and energy in sweet potato have been conducted by various workers using unimproved village pigs (Rose and White, 1980) as well as improved strains of pigs (Lee et al., 1977; Lee and Yang, 1981). Village pigs (*Sus scrofa papuensis*), from Papua New Guinea, fed raw sliced sweet potatoes had digestibility coefficients of 95.3% for dry matter, 96.1% for organic matter, 94.2% for energy, 72.4% for acid detergent fiber and 57.2% for crude protein. These digestibility coefficients are not significantly different from literature values reported for cassava and other root crops when fed to pigs. However, they are lower than nutrient digestibility coefficients for corn and wheat, implying a slight inferiority of sweet potato to corn and wheat when fed to growing pigs. In an experiment in which 30% of a corn-soybean diet was replaced by sweet potato chips, digestibility coefficients for most nutrients were uniformly high (Lee et al., 1977).

Table 47.6. Carcass Measurements of Growing-Finishing Pigs Fed Different Proportions of Corn and Sweet Potato Chips

Corn in Diet (%)	Potato in Diet (%)	Backfat Thickness (cm)	Lean Cuts (%)
63-81	0	4.47	----
45-58	15-20	4.34	----
29-37	29-37	3.81	----
14-18	42-54	3.53	----
0	54-68	3.84	----
72-84	0	3.57	47.19
35-41	35-41	3.20	47.35
0	69-81	3.30	47.43
69-75	0	3.25	52.77
33-36	33-36	3.10	54.89
0	63-68	3.22	54.12

Yeh and Bouwkamp, 1985.

FEEDING SWEET POTATO SILAGE TO SWINE

Whole sweet potatoes including stem, leaves and roots have been ensiled and fed to swine. Jung et al. (1977) reported studies in which pigs were fed either a commercial cereal-soybean diet or one in which sweet potato silage replaced 15, 30, 45 or 60% of the cereal in the diet, all diets being isocaloric and isonitrogenous. Average daily gain and feed intake were similar on all diets, with back fat thickness decreasing as the level of sweet potato silage in the diet increased. However, sweet potato silage may be associated with a depression in crude protein digestibility in diets in which the silage is fed. This has been demonstrated by Tomita et al. (1985) who reported that in an experiment in which barrows were fed (i) a basal diet, (ii) 66.5% of basal diet + 33.5% of sweet potato silage, or (iii) 70.4% of basal diet + 29.6% of sweet potato silage, digestibility of dry matter, organic matter and energy was high (89-91%), but digestibility of protein was low (32%).

Sweet potato silage can be fed to breeding stock with excellent results. The moderately high nutrient content of sweet potato leaves, coupled with a fiber content that is higher than in the root, ensures success in feeding sweet potato silage to breeding sows.

SUMMARY

Fresh sweet potato roots need to be heat treated in order to destroy trypsin inhibitors present in the raw root, as well as to improve nutrient digestibility. Cutting and drying the fresh root to chips results in a product with better digestibility and utilization. It is evident that up to 50% of the grain in corn-soybean diets may be replaced with sweet potato chips, without a significant depression in growth or production. With such high levels of replacement, it may be necessary to supplement such sweet potato-based diets with 0.2% to 0.5% of lysine and with additional sulfur amino acids.

Sweet potato silage may be fed to growing, fattening or breeding pigs as a replacement for up to 40% of the cereal in the diet, provided that a good protein supplement like soybean meal is used.

REFERENCES

Chen, Y., Yeh, T.P., Yang, Y.S., Chang, T.C., Wu, M.C., Siao, C.M. and Yang, W.L., 1980. Studies on the formula feeds with sweet potatoes as diets for growing-finishing pigs. Taiwan Sugar Corporation Annual Research Report.

Collins, W.W. and Walter, W.M. Jr., 1985. Fresh roots for human consumption. In: J.C. Bouwkamp, ed. Sweet Potato Products: A Natural Resource For The Tropics, CRC Press, Boca Raton, Florida, pp. 153-173.

Edmond, J.B., 1971. Physiology, Biochemistry and Ecology. In: J.B. Edmond and G.R. Ammerman, eds. Sweet Potatoes: Production, Processing, Marketing. The AVI Publishing Company, Westport, Conneticut, pp. 30-57.

Ezell, B.B. and Wilcox, M.S., 1946. The ratio of carotene to carotenoid pigments in sweet potato varieties. Science 103: 193-194.

FAO, 1985. Production Year Book. Food and Agriculture Organization of the United Nations, Rome.

Jenkins, W.F. and Moore, E.L., 1954. The distribution of ascorbic acid and latex vessels in three tissue regions of sweet potatoes. Proc. Amer. Soc. Hort. Sci. 63: 389-392.

Jones, D.B. and Gersdorff, C.E.F., 1931. Ipomein, a globulin from sweet potatoes, *Ipomoea batatas*. J. Biol. Chem. 93: 119-126

Jung, C.Y., Park, B.H. and Lee, K.H., 1977. Studies on feeding value of sweet potato silage for growing-finishing pigs. Research Reports of the Office of Rural Development (Livestock) in South Korea. 19: 87-92.

Lease, E.J., 1941. Sweet potato as a source of vitamin A for man and domestic animals. Proc. Assoc. South. Agric. Workers 42: 162-163.

Lee, P.K., 1979. Study on hog feed formula of using high protein sweet potato chips and dehydrated sweet potato vines as the main ingredient. J. Taiwan Livest. Res. 12: 49-71.

Lee, P.K. and Yang, Y.F., 1981. Studies on digestibility of crude protein and energy with pigs fed on diets containing locally produced maize meal, sorghum grains, sweet potato chips or cassava meal. J. Taiwan Livest. Res. 14: 65-74.

Lee, P.K., Yang, Y.F. and Chen, F.N., 1977. Comparative study on the nutrient digestibility of sweet potato chips, cassava pomace and dried banana chips by pigs. J. Taiwan Livest. Res. 10: 215-225.

O'Hair, S.K., 1984. Farinaceous Crops. In: F.W. Martin, ed. Handbook of Tropical Food Crops. CRC Press, Boca Raton, Florida, pp. 109-137.

Onwueme, I.C., 1978. The Tropical Tuber Crops: Yam, Cassava, Sweet Potato and Cocoyam. John Wiley and Sons, Chichester, New York.

Pearson, P.R. and Leucke, R.W., 1954. The B-vitamin content of raw and cooked sweet potatoes. Food Research 10: 325-329.

Purcell, A.E., Walter, W.M. Jr. and Giesbrecht, F.G., 1978. Changes in dry matter, protein and non-protein nitrogen during storage of sweet potatoes. J. Amer. Soc. Hort. Sci. 103: 190-192.

Ravindran, V. and Rajaguru, A.S.B., 1985. Nutrient contents of some unconventional poultry feeds. Indian J. Anim. Sci. 55: 58-61.

Rose, C.J. and White, G.A., 1980. Apparent digestibilities of dry matter, organic matter, crude protein, energy and acid detergent fiber of chopped raw sweet potato (*Ipomoea batatas* L.) by village pigs in Papua New Guinea. Papua New Guinea Agric. J. 31: 69-72.

Sohonie, K. and Honawar, P.M., 1956. Trypsin inhibitors of sweet potato (*Ipomoea batatas*). Science and Culture 21: 538.

Tai, N.L. and Lei, T.S., 1970. Determination of proper amount of yellow corn and dried sweet potatoes in swine feed. J. Agric. Assoc. China 70: 71-76.

Tomita, Y., Hayashi, K. and Hashizume, T., 1985. Palatability of pigs to sweet potato silage and digestion trial by them. Bulletin of Faculty of Agriculture, Kagoshima University, Japan, 35: 75-80.

Walter, W.M. Jr. and Catignani, G.L., 1981. Biological quality and composition of sweet potato fractions. J. Agric. Food Chem. 29: 797-799.

Wu, J.F., 1980. Energy value of sweet potato chips for young swine. J. Anim. Sci. 51: 1261-1265.

Yeh, T.P. and Bouwkamp, J.C., 1985. Sweet potato roots and vines as animal feed. In: J.C. Bouwkamp, ed. Sweet Potato Products: A Natural Resource For the Tropics. CRC Press, Boca Raton, Florida, pp. 235-253.

Yeh, T.P., Wong, S.C., Lin, H.K. and Kuo, C.C., 1979. Popping sweet potato chips for pigs. J. Anim. Sci. 49 (Suppl. 1): 257.

CHAPTER 48

Triticale

S.V. Radecki and E.R. Miller

INTRODUCTION

Triticale (*Tricicale hexaploide*) is a relatively new, synthetic small-grain crop that was produced by crossing Durum wheat with rye. The goal of plant breeders in developing triticale was to combine the high crude protein and digestible energy content of wheat with the high yields and protein quality of rye. Its name was derived by combining Triticum, the botanical name for wheat, with Secale, the botanical name for rye.

Although rye-wheat hybrids have been researched for over 100 years, it was not until the mid 1960s that triticale varieties of practical agronomic and agricultural importance were developed. The ability of this grain to grow in acidic soils and extreme climates, as well as having yields that can surpass those of rye, make triticale a practical and economical feedstuff in northern climates where temperatures may limit corn production, and in milder regions as a winter crop.

Tritcale is not a major crop in North America and therefore, large quantities of triticale have not been fed to livestock. However, from time to time, significant quantities of triticale become available and can be successfully utilized as an energy source in swine diets.

NUTRIENT CONTENT OF TRITICALE

Much variation exists between cultivars of triticale with respect to nutrient composition (Erickson, 1979; Hale and Utley, 1985; Adeola et al., 1986a; Coffey and Gerrits, 1988). Compared to the more traditional sources of energy such as corn and sorghum, triticale is much higher in crude protein content (Table 48.1). However, it has a lower ether extract content and higher crude fiber and as a consequence, it provides a lower level of digestible energy than does corn or sorghum.

Table 48.1. Nutrient Composition of Triticale, Corn and Sorghum

	Corn	Sorghum	Triticale
Dry Matter (%)	88.0	89.0	90.0
D.E. (kcal/kg)	3530	3415	3299
Crude Protein (%)	8.5	8.9	15.8
Crude Fiber (%)	2.3	2.2	4.0
Ether Extract (%)	3.6	2.8	1.5
Calcium (%)	0.03	0.03	0.05
Phosphorus (%)	0.28	0.28	0.30

National Research Council, 1988.

Triticale has a more favorable balance of amino acids (including a higher lysine content) compared with sorghum or corn (Table 48.2). In triticale-soybean meal diets, lysine is the first and threonine the second limiting amino acid (Shimada and Cline, 1974). Overall ileal digestibility of amino acids in triticale is less than corn (Adeola et al., 1986a), with the amino acids arginine, histidine, leucine, phenylalanine and threonine, specifically, having lower availabilities (Adeola et al., 1986b). The other indispensible amino acids have availabilities similar to that of corn and net protein utilization is similar for corn and triticale (Adeola et al., 1986a).

The mineral composition of triticale is also of interest, especially with regard to phosphorus. Some cultivars of triticale, such as Beagle 82, may have twice the phosphorus concentration of corn (Erickson, 1979). The Wintri cultivar also has a greater phosphorus concentration than corn (Adeola et al., 1986a). This additional phosphorus allows for a reduction in the amount of supplemental phosphorus that needs to be included in the diet, and thus can be of economic importance.

There is little information on the vitamin composition of triticale. However, some estimates indicate that triticale may be significantly lower in choline and riboflavin than are corn or sorghum (National Research Council, 1988).

Table 48.2. Amino Acid Content of Common Cultivars of Triticale, Corn and Sorghum

	Corn[1]	Sorghum[1]	Triticale B858[2]	Triticale Beagle 82[3]	Triticale Wintri[4]
Arginine	0.43	0.37	0.83	0.65	0.79
Histidine	0.27	0.24	0.38	0.23	0.37
Isoleucine	0.35	0.44	0.61	0.47	0.52
Leucine	1.19	1.32	1.07	0.88	0.76
Lysine	0.25	0.23	0.46	0.43	0.56
Methionine	0.18	0.16	0.18	0.22	0.20
Cystine	0.22	0.13	0.32	0.25	0.14
Phenylalanine	0.46	0.49	0.80	0.58	0.51
Threonine	0.36	0.27	0.36	0.41	0.36
Tryptophan	0.09	0.10	----	0.11	----
Valine	0.48	0.53	0.82	0.52	0.41

[1]National Research Council, 1988.
[2]Coffey and Gerrits, 1988.
[3]Hale and Utley, 1985.
[4]Adeola et al., 1986b.

UNDESIRABLE CONSTITUENTS IN TRITICALE

Various cultivars of triticale, especially varieties developed prior to 1975, may contain levels of trypsin and chymotrypsin inhibitors that may limit the use of this cereal in swine diets. Differences in the performance of pigs fed triticale, as well as variations in protein availability of triticale, may be partially explained by the variation in trypsin and chymotrypsin concentrations of the various triticale varieties. Palatability of these varieties also can be of concern. Susceptibility to ergot and other diseases may also limit the production of triticale. However, recent hybrids have produced triticale varieties with acceptable levels of anti-nutritional factors, especially the Beagle 82 cultivar. Therefore, these factors should not limit the use of triticale in swine diets. Palatability trials have indicated that these new triticale varieties are as palatable as corn (Erickson, 1979).

USE OF TRITICALE IN SWINE DIETS

Starter Pigs

If the lysine content of the triticale is not considered when balancing the diet and triticale is simply substituted into starter diets on an equal weight basis for corn, daily gain is depressed as the percent triticale reaches 60% of the total corn (Table 48.3).

Table 48.3. Influence of Triticale in Starter Diets (7.5 kg) on Pig Growth Performance

	Percent of Total Grain as Triticale					
	0	20	40	60	80	100
Daily Gain (kg)	0.33	0.37	0.33	0.32	0.27	0.22
Daily Intake (kg)	0.75	0.76	0.73	0.74	0.70	0.54
Feed Efficiency	2.29	2.03	2.20	2.30	2.60	2.47

Erickson et al., 1979.

In diets balanced for lysine, replacement of all the corn and 9.6% of the soybean meal with triticale does not influence daily gain or feed intake by nursery pigs (Table 48.4). However, pigs fed high levels of triticale tend to be less efficient.

Table 48.4. Influence of Substituting Triticale for Corn in Starter Pig Diets Balanced for Lysine.

	Percent of Total Grain as Triticale		
	0	50	100
Daily Gain (kg)	0.47	0.44	0.44
Daily Intake (kg)	0.93	0.88	0.96
Feed Efficiency	1.99	2.01	2.19

Hale and Utley, 1979.

Growing-Finishing Pigs

Feeding pigs a fortified triticale-soybean meal diet, balanced for lysine, from weaning to market weight, has no adverse effect on daily gain, as compared to a corn-soybean meal diet (Table 48.5). This may indicate that pigs in the finishing stage of production are better able to use this feedstuff and the balance of amino acids in triticale more closely meets their requirements than it does for starter pigs. However, feed efficiency is lower for pigs fed triticale, as the energy content of triticale is lower than that of corn. Addition of lysine or methionine to triticale diets improved performance but did not improve rate of gain to the level of pigs consuming a corn-based diet (Erickson et al., 1979; Coffey and Gerrits, 1988). The lack of an improvement in gain from pigs fed triticale based diets and supplemented with lysine and methionine may be partially explained by an apparent threonine imbalance, as threonine is likely the second limiting amino acid in triticale-soybean meal diets (Shimada and Cline, 1974).

Table 48.5. Influence of Triticale in Swine Diets Balanced for Lysine

	Corn	Triticale	Triticale + Lysine	Triticale + L + M
Starter Phase				
Daily Gain (kg)	0.48	0.47	0.45	0.46
Daily Intake (kg)	0.91	0.94	0.95	0.95
Feed Efficiency	1.91	2.02	2.13	2.05
Grower Phase				
Daily Gain (kg)	0.69	0.64	0.67	0.69
Daily Intake (kg)	1.92	1.85	1.79	1.87
Feed Efficiency	2.75	2.91	2.71	2.71
Finisher Phase				
Daily Gain (kg)	0.77	0.79	0.78	0.82
Daily Intake (kg)	2.64	2.91	2.74	2.88
Feed Efficiency	3.48	3.74	3.52	3.50
Overall				
Daily Gain (kg)	0.66	0.64	0.64	0.66
Daily Intake (kg)	1.88	1.94	1.85	1.91
Feed Efficiency	2.87	3.06	2.91	2.87

Coffey and Gerrits, 1988.

It appears that, in a completely mixed ration when the diet is balanced for lysine and triticale is used to replace all the corn and some of the soybean meal, there is little influence on the digestibility of the nutrients in that diet. Total diet dry matter and energy digestion coefficients are similar when diets are balanced for lysine using triticale in place of corn (Coffey and Gerrits, 1988). A similar response is seen when triticale is substituted for corn on an equal percentage basis, and no adjustment is made for lysine content (Hale et al., 1985).

Pigs consuming a corn-soybean meal diet retained more nitrogen as a percent of the nitrogen consumed than pigs consuming a diet in which triticale was substituted for the corn (Hale et al., 1985). But, when diets are formulated on an isolysinic basis, triticale-soybean meal diets have greater crude protein digestibilities than corn-soybean meal diets (Coffey and Gerrits, 1988).

Breeding Stock

Research with triticale in breeding herd diets has not been reported. Until more research is conducted to determine the nutritive value of triticale for breeding stock, a limit of 25% is suggested.

SUMMARY

Due to the variation in nutrient composition of triticale varieties, laboratory analysis of this feedstuff is important prior to use in swine diets. New triticale varieties, which are low in trypsin inhibitors and more disease resistant, make the use of this feedstuff in swine diets quite practical. To take advantage of the lysine content of this feedstuff, all diets should be balanced for lysine, and the corn not simply replaced by the triticale. Under these conditions, triticale can be used to replace 100% of the corn in the diet of growing pigs. Without feed analysis, and when substituting triticale for corn on an equal weight basis, no more than 60% of the grain should be triticale. The greater amino acid content, and especially the greater lysine concentration in triticale, may allow for a reduction in the amount of soybean meal, or other sources of amino acids in the diet, which may have important economic implications.

REFERENCES

Adeola, O., Young, L.G., McMillan, E.G. and Moran, E.T., 1986a. Comparative protein and energy value of OAC wintri triticale and corn for pigs. J. Anim. Sci. 63: 1854-1861.

Adeola, O., Young, L.G., McMillan, E.G. and Moran, E.T. Jr., 1986b. Comparative availability of amino acid in OAC wintri triticale and corn for pigs. J. Anim. Sci. 63: 1862-1869.

Coffey, M.T. and Gerrits, M.J., 1988. Digestibility and feeding value of B858 triticale for swine. J. Anim. Sci. 66: 2728-2735.

Erickson, J.P., 1979. Triticale as a replacement for other grains in swine diets. Michigan Agricultural Experiment Station, East Lansing, Report No. 383: 86-91.

Erickson, J.P., Miller, E.R., Elliott, F.C., Ku, P.K. and Ullrey, D.E., 1979. Nutritional evaluation of triticale in swine starter and grower diets. J. Anim. Sci. 48: 547-553.

Hale, O.M., Morey, D.D. and Myer, R.O., 1985. Nutritive value of Beagle 82 triticale for swine. J. Anim. Sci. 60: 503-510.

Hale, O.M. and Utley, P.R., 1985. Value of Beagle 82 triticale as a substitute for corn and soybean meal in the diet of pigs. J. Anim. Sci. 60: 1272-1279.

National Research Council, 1988. Nutrient Requirements of Domestic Animals, No. 2. Nutrient Requirements of Swine. 9th ed. National Academy of Sciences, Washington, D.C.

Shimada, A. and Cline, T.R., 1974. Limiting amino acids of triticale for the growing rat and pig. J. Anim. Sci. 38: 941-946.

CHAPTER 49

Wheats: Soft and Dwarf

W.I. Magowan

INTRODUCTION

Wheat (*Triticum aestiuum*), is an important grain crop both in regard to its antiquity and its use as a source of human food. The pre-eminence of wheat as a staple food is due to many factors. It has adapted to a wide range of soil and climatic conditions and can be grown extensively throughout the world. It is economical in production and gives good yields with excellent storage properties. Its popularity as a human food is due to its mild, acceptable flavor and to the unique ability of its principle proteins to form gluten when mixed with water. Until recently, whole grain has not been used extensively as a livestock feed on account of the high costs involved in growing wheat and the availability of cheaper feedstuffs. However, the development of specific feed grade wheats in the early 1970s has made wheat more competitive for use in many livestock feeds.

Soft wheat varieties, both red and white, are characterized as having starchy or soft grains with little resistance to grinding and crushing and with lower flour yields than the hard wheats. They are used mostly for cake, cracker, pastry and family flours while the hard wheats are used chiefly for bread flours. Periodically, off-grade soft wheat is available for animal feed and appears to support similar performance in growing pigs to both hard feed wheats.

Dwarf varieties of wheat have also been developed. The development of these shorter stemmed varieties has reduced lodging problems and produced dramatic increases in yields. Dwarf wheats would also appear to have a place in swine production.

NUTRIENT CONTENT OF WHEATS

Wheat cultivars vary greatly in chemical composition and therefore in nutritive value (Ivan and Farrell, 1975). In the World Collection of Wheat, protein levels range from 6% to 22% with the highest frequency distribution between 13% and 14% (Johnson et al., 1970). The proximate analysis of hard, soft and dwarf wheats shown in Table 49.1 indicates that hard and dwarf wheats tend to be higher in crude protein than the soft wheats. The crude fiber content of soft wheats is slightly lower than the hard and dwarf wheats but their energy content is usually similar. Dry matter, energy and overall nitrogen digestibilities are not significantly different between diets based on hard or soft wheat (Ivan and Farrell, 1976).

Table 49.1. Nutrient Composition of Hard, Soft and Dwarf Wheats (as fed)

	Hard Wheat[1]	Soft Wheat[1]	Dwarf Wheat[2]
Crude Protein (%)	12.6	11.4	12.6
Crude Fiber (%)	2.6	2.3	2.6
Ether Extract (%)	1.6	1.6	1.9
Ash (%)	1.9	1.8	1.3
D.E. (kcal/kg)	3402	3402	----

[1]National Research Council, 1988.
[2]Bell and Keith, 1988.

The amino acid content of hard, dwarf and soft wheats is compared in Table 49.2. Although there are not large differences between wheat cultivars in amino acid content, the soft and dwarf wheats tend to have a superior amino acid profile in comparison with the hard wheats. However, hard wheat showed a better availability of some essential amino acids (lysine, arginine, isoleucine and tyrosine) than soft wheat (Ivan and Farrell, 1976). As lysine is the first limiting amino acid in wheat for growing pigs, its higher digestibility (80% in hard vs. 71% in soft wheat), is of practical importance. Arginine, isoleucine and tyrosine are not usually limiting in wheat-based diets.

Table 49.2. Amino Acid Content of Hard, Soft and Dwarf Wheats (% as fed)

	Hard Wheat[1]	Soft Wheat[1]	Dwarf Wheat[2]
Arginine	0.65	0.65	0.68
Histidine	0.30	0.32	0.36
Isoleucine	0.53	0.45	0.41
Leucine	0.87	0.90	1.08
Lysine	0.40	0.40	0.43
Methionine	0.22	0.22	0.28
Cystine	0.30	0.36	0.39
Phenylalanine	0.71	0.64	0.76
Threonine	0.37	0.39	0.47
Tryptophan	0.17	0.27	----
Valine	0.58	0.58	0.39

[1]National Research Council, 1988.
[2]Stothers et al., 1986.

Mineral levels in both hard and soft wheat are presented in Table 49.3. Values for both wheats are similar and in most cases average values for local grain are used in diet formulation with macro and trace minerals added to supply the required levels.

Table 49.3. Mineral Composition of Hard and Soft Wheat (as fed)

	Hard Wheat	Soft Wheat
Calcium (%)	0.04	0.05
Phosphorus (%)	0.37	0.36
Sodium (%)	0.02	0.01
Magnesium (%)	0.12	0.10
Potassium (%)	0.43	0.41
Copper (mg/kg)	5.10	7.00
Iron (mg/kg)	35.00	29.00
Manganese (mg/kg)	30.40	33.40
Selenium (mg/kg)	0.29	0.04
Zinc (mg/kg)	35.00	42.00

National Research Council, 1988.

The vitamin content of soft and hard wheats are compared in Table 49.4. There would appear to be little difference in vitamin content between these wheats and therefore it should not be necessary to use a specially formulated premix when formulating diets based on soft or hard wheats.

Table 49.4. Vitamin Content (mg/kg) of Hard and Soft Wheat (as fed)

	Hard Wheat	Soft Wheat
Vitamin E	11.10	15.60
Biotin	0.11	---
Choline	1004	892
Folic Acid	0.40	0.40
Niacin	53	53
Pantothenic Acid	10.10	10.10
Riboflavin	1.30	1.50
Thiamine	4.50	4.70
Vitamin B_6	3.00	3.30

National Research Council, 1988.

UTILIZATION OF SOFT WHEATS

Starter Pigs

Including soft white spring wheat in starter diets for weanling pigs up to a level of 60% was found to have no significant effect on pig performance (Magowan and Aherne, 1987). Daily feed intake, daily gain and feed efficiency were similar for diets with 0% to 60% soft wheat (Table 49.5). Protein digestibility was greater for the diet with 60% hard wheat but this was not reflected in pig performance.

Table 49.5. Replacement of Hard Red Spring Wheat with Soft Spring Wheat in Pig Starter Diets (7-21 kg)

Soft Wheat (%)	0	20	40	60
Hard Wheat (%)	60	40	20	0
Daily Intake (kg)	0.67	0.68	0.71	0.68
Daily Gain (kg)	0.46	0.47	0.51	0.48
Feed Efficiency	1.45	1.45	1.39	1.41

Magowan and Aherne, 1987.

In a two year study with milling wheats and feed wheats, Bowland (1974) concluded that both types of wheat were similar in feeding value when used as the sole cereal component in diets for weanling pigs. Protein digestibility data indicated a superiority for some of the diets based on feed wheats compared to those based on milling wheats. However, this difference was not reflected in performance.

Growing-Finishing Pigs

In each of three feeding trials conducted by Hines (1982), comparing soft winter wheat with hard winter wheat, there was no difference in daily gain and feed efficiency (Table 49.6). Pigs were fed from an average initial weight of 60 kg to an average final weight of 120 kg.

Table 49.6. Effect of Hard Versus Soft Wheat on Performance of Finishing Swine

	Hard Wheat	Soft Wheat
Daily Gain (kg)	0.81	0.84
Daily Intake (kg)	2.96	3.10
Feed Efficiency	3.67	3.69

Adapted from Hines, 1982.

UTILIZATION OF DWARF WHEATS

Stothers et al. (1986) investigated the nutrient content and the performance of growing pigs (Table 49.7) on diets where a feed wheat

Table 49.7. Performance of Pigs Fed Starter, Grower and Finisher Diets Containing Glenlea, HY320 or Marshall Wheat

	Glenlea	HY320	Marshall
Starter Diets (18-40 kg)			
Daily Intake (kg)	1.35	1.43	1.47
Daily Gain (kg)	0.52	0.59	0.56
Feed Efficiency	2.60	2.42	2.62
Grower Diets (40-62 kg)			
Daily Intake (kg)	2.20	2.34	2.22
Daily Gain (kg)	0.69	0.70	0.73
Feed Efficiency	3.19	3.34	3.04
Finisher Diets (62-98 kg)			
Daily Intake (kg)	3.12	3.55	3.12
Daily Gain (kg)	0.97	1.06	0.96
Feed Efficiency	3.20	3.34	3.25

Stothers et al., 1986.

(Glenlea) and semi-dwarf wheats (HY320 and Marshall) were the sole cereal grain. There was no significant difference in pig performance during the starter, grower and finisher phase. The only consistent effect reported was a higher feed consumption of pigs fed the HY320 diet, whether starter, grower or finisher.

Bell and Keith (1988) compared the dwarf wheat HY-320 with hulless barley and corn as energy sources for pigs fed from 23 to 100 kg (Table 49.8). There was no significant difference in pig performance as a result of the energy source used.

Table 49.8. Performance of Growing-Finishing Pigs Fed Diets Containing HY-320 Wheat, Hulless Barley or Corn

	Dwarf Wheat	Hulless Barley	Corn
Daily Gain (kg)	0.81	0.81	0.83
Daily Intake (kg)	2.45	2.46	2.46
Feed Efficiency	3.06	3.05	2.97
Dressing Percentage (%)	77.3	76.3	76.2
Backfat Thickness (mm)	24.9	26.2	25.1

Bell and Keith, 1988.

SUMMARY

Performance data from growing pigs fed soft wheat and dwarf wheat indicates that they can be included in swine diets according to their nutrient profiles. As for wheat in general, when diets are formulated correctly, these wheats can be used to produce high quality feed for pigs. For animals that require high energy diets, wheat can often be the most cost effective way of supplying the necessary energy. Due to the higher energy content of wheat over barley, care must be taken when a direct substitution is made. If the energy content is increased through the addition of wheat, the amino acid/calorie ratio will change and in growing pigs this may lead to a reduction in carcass quality. Therefore, when wheat is used to replace a lower energy grain, the amino acid levels should be adjusted to maintain the correct amino acid to energy balance.

When formulating diets with high protein wheat, care must be taken to ensure that the essential amino acid levels are adequate. As the crude protein level increases, the essential amino acids do not increase at the same rate and thus if diets are formulated on the basis of crude protein alone, a diet which is deficient in essential amino acids may result. The increase in protein in the wheat mainly reflects changes in the amount of gluten present and gluten is a poor source of essential amino acids. Thus, when formulating diets with high protein grains, it is important to formulate on the basis of amino acids to ensure that requirements are met. As the lysine digestibility in hard wheat is higher than in soft wheat, diets which are based on soft wheat require more supplementary lysine in order to achieve similar growth rates to hard wheat diets.

When wheat is ground too finely palatability may be reduced and the incidence of stomach ulcers may also increase. It is recommended that wheat should be ground through a 4.5 to 6.4 mm screen to achieve a medium grind and reduce potential problems.

REFERENCES

Bell, J.M. and Keith, M.O., 1988. Comparisons of HY-320 wheat, tupper hulless barley and yellow corn in feeding growing-finishing pigs. University of Saskatchewan, Saskatoon, Department of Animal Science Research Report No. 559, pp. 215-219.

Bowland, J.P., 1974. Comparison of several wheat cultivars and a barley cultivar in diets for young pigs. Can. J. Anim. Sci. 54: 629-638

Hines, R.H., 1982. Hard wheat compared to soft wheat for finishing swine. Kansas State University, Manhattan, Swine Day Report, pp. 104-107.

Ivan, M. and Farrell, D.J., 1975. Nutritional evaluation of wheat. 2. The sequence of limiting amino acids in wheats of different protein content as determined with growing rats. Anim. Prod. 20: 77-91.

Ivan, M. and Farrell, D.J., 1976. Nutritional evaluation of wheat. 5. Disappearance of components in digesta of pigs prepared with two re-entrant canulae. Anim. Prod. 23: 111-119.

Johnson, V.A., Mattern, P.J. and Schmidt, J.W., 1970. The breeding of wheat and maize with improved nutritional value. Proc. Nutr. Soc. 29: 20-31.

Magowan, W.I. and Aherne, F.X., 1987. An evaluation of the nutritive value of soft white spring wheat in starter diets for weanling pigs. Proc. 8th Western Nutrition Conference, Edmonton, Alberta pp. 125-127.

National Research Council, 1988. Nutrient Requirements of Domestic Animals, No. 2. Nutrient Requirements of Swine. 9th ed. National Academy of Sciences, Washington, D.C.

Stothers, S.C., Campbell, L.D. and Guenter, W., 1986. Proc. 7th Western Nutrition Conference, Saskatoon, Saskatchewan, pp. 251-260.

CHAPTER 50

Wild Oat Groats

P.A. Thacker

INTRODUCTION

Despite intensive efforts at chemical and cultural control, wild oats continue to contaminate a significant proportion of the world's cereal grain production. Approximately 40% of the dockage assessed on cereal grains and as much as 1% of the total harvest of grain is comprised of wild oats. Due to their black-colored hull, these wild oats must be removed before the grain can be exported or processed for human consumption and large quantities of wild oat seeds are separated annually at commercial seed cleaning plants.

Wild oats have traditionally been marketed as Mixed Feed Oats and have been utilized almost exclusively in diets fed to ruminants. The presence of a highly fibrous hull limits their usefulness in diets formulated for monogastrics. However, a technique for dehulling wild oats has been recently developed (Sosulski and Sosulski, 1985). The dehulled kernels of wild oats, commonly called groats, may have the potential to replace domestic oat groats in rations fed to poultry and swine.

PROCESSING OF WILD OAT GROATS

A typical batch of unprocessed screenings obtained from a seed cleaning plant contains approximately 38.5% wild oats, 24.7% barley, 13.8% wheat and

20.6% straw (Sosulski and Sosulski, 1985). In order to produce wild oat groats, these screenings are first purified on a rotary thickness grader followed by indent disc and air separation to eliminate the barley, wheat and chaff. This process results in the production of a product comprising approximately 95% wild oats (Sosulski and Sosulski, 1985). The wild oat hulls are then separated from the groat by impact on a Buhler-Maig huller followed by air aspiration to remove the loose hulls and chaff. The groats are then passed over a gravity table to remove any light-weight kernels which may have missed being dehulled on the first pass. These light kernels are recycled through the huller. The green groats are then treated with live steam (98°C) before being dry kilned at 92°C for 5 h to inactivate the lipase enzyme in the bran. Yields of groats from this process range from 50 to 60%. Further processing with a groat cutter and roller mill results in the production of the wild oat flakes used in the food industry as a breakfast cereal.

NUTRIENT CONTENT OF WILD OAT GROATS

A comparison of the chemical composition of wild and domestic oat groats and oat flakes is presented in Table 50.1. Wild oat groats contain significantly higher levels of crude protein in comparison with domestic oat groats (19.9 vs. 14.5%). Therefore, use of wild oat groats in a ration may reduce the amount of protein supplementation required in order to provide a balanced swine ration.

Table 50.1. Chemical Composition of Wild and Domestic Oat Products (% as fed)

	Wheat	Oat Groat	Feed Oats	Wild Oat Groat	Wild Oat Flake
Moisture	12.9	12.7	10.7	11.4	10.2
Crude Protein	15.0	14.5	12.2	19.9	19.1
Ether Extract	1.8	6.3	7.4	7.8	8.7
Crude Fiber	3.0	3.3	1.5	3.8	1.3
Ash	1.6	1.9	1.7	2.2	2.0

Thacker et al., 1986.

Wild oat groats contain a higher level of ether extract than domestic oat groats (7.8 vs. 6.3%). Chemical analysis has shown that almost 90% of the lipid in wild oats is in the form of triglyceride (Sosulski, 1984). These triglycerides are highly unsaturated, with the oleic, linoleic and linolenic acid levels being 46, 35 and 2%, respectively, of the total fatty acids (Campbell et al., 1987). The polyunsaturated fatty acids are desirable nutritionally, but they are also subject to oxidative instability during storage, especially after seed grinding.

There is an active lipase in the pericarp of the wild oat (Sosulski, 1984). The lipase is not in contact with the lipids in the intact kernel, but crushing or milling the seed will result in hydrolysis of the triglycerides into free fatty acids, even at relatively low seed moisture levels. The free fatty acids are much more susceptible to oxidation into rancid, bitter breakdown products than are intact triglycerides. Therefore, it is essential to steam the groats thoroughly to destroy the lipase in the bran before proceeding to process the groat into feed products. Once the enzyme is inactivated, the dry milled products can be stored for several months without a significant breakdown of lipids, especially if cool temperatures are maintained.

The crude fiber content of the wild oat groat is slightly higher than that of the common oat groat (3.8 vs. 3.3%). From the standpoint of animal performance, this may be a negative factor. However, in terms of human nutrition, the higher crude fiber content of the wild oat groat may be a positive factor due to the proposed link between dietary fiber and the development of coronary heart disease. Therefore, research is being conducted into the possibility of utilizing wild oat products in human breakfast cereals (Chang and Sosulski, 1985).

The amino acid content of wild and domestic oat groats and oat flakes is presented in Table 50.2. Despite the large difference in protein content between the wild and domestic oat products, there would appear to be little difference in amino acid composition between common and wild oat groat protein (Thacker et al., 1986). The protein contained in wild oat groats would appear to be of fairly high quality. All of the essential amino acids, with the exception of lysine and threonine, are present in sufficient quantity to meet the requirements of the starter pig (10-20 kg).

Table 50.2. Amino Acid Composition of Wild and Domestic Oat Products (% of protein)[1]

	Wheat	Oat Groat	Feed Oats	Wild Oat Groat	Wild Oat Flake	Starter[2] Pig
Arginine	4.1	6.7	6.7	6.5	6.6	1.3
Histidine	1.9	2.2	2.4	2.4	2.3	1.1
Isoleucine	3.0	3.5	3.6	3.7	3.4	3.1
Leucine	6.1	7.6	7.7	7.4	7.4	3.7
Lysine	2.3	3.9	4.1	4.0	3.6	4.4
Methionine	1.3	1.2	1.4	1.5	1.4	2.8
Phenylalanine	4.3	5.4	5.9	5.7	5.6	4.4
Threonine	2.2	3.2	2.7	1.7	2.6	2.8
Valine	3.7	5.6	5.6	5.0	5.4	3.1

[1]Thacker et al., 1986.
[2]National Research Council, 1988.

A comparison of the micronutrient composition of wild and domestic oat groats is presented in Table 50.3. Of particular interest is the relatively high level of phosphorus contained in wild oat groats, which is reported to be of fairly high availability. Therefore, inclusion of wild oat groats in a ration may be useful as a means of providing additional organic phosphorus. The high vitamin A content of the wild oat cannot be substantiated on the basis of literature values and further research will be necessary in order to confirm this finding.

Table 50.3. Micronutrient Composition of Domestic and Wild Oat Groats[1,2]

	Domestic Groat	Wild Groat
Calcium (%)	0.11	0.08
Phosphorus (%)	0.44	0.57
Copper (mg/kg)	4.44	7.00
Iron (mg/kg)	50.50	59.80
Manganese (mg/kg)	31.90	28.00
Zinc (mg/kg)	40.00	44.20
Vitamin A (IU/g)	1.08	2.26
Niacin (mg/100g)	1.02	1.03
Riboflavin (mg/100g)	0.18	0.21
Thiamin (mg/100g)	0.56	0.69

[1]Dry Matter Basis.
[2]Sosulski and Sosulski, 1985.

UNDESIRABLE CONSTITUENTS IN WILD OAT GROATS

Wild oat groats contain a moderately high level of beta-glucan (3.3%; Campbell et al., 1987) which is a cell wall component consisting of glucose units linked by beta-1,4 and beta-1,3 linkages (Fleming and Kawakami, 1977). Beta-glucans have been shown to reduce the nutritive value of several feeds for poultry by increasing the viscosity of the intestinal fluid (Burnett, 1966). Although there has been little research conducted with swine, there are several reports which suggest that pig performance may also be adversely affected by the presence of beta-glucan (Newman et al., 1980; Newman et al., 1983).

FEEDING WILD OAT GROATS

Domestic oat groats are widely utilized in creep and starter rations for swine because they are palatable, highly digestible and contain a relatively low crude fiber content. If wild oat groats are shown to be comparable in nutritional value to domestic oat groats, it could make available to the swine industry a valuable feed resource from a previously unused product.

There has been very little work conducted on the nutritive value of wild oat groats for swine. However, Thacker et al. (1986) did test their potential for use in starter rations. In this experiment, the control diet was based on wheat and soybean meal while the remaining diets contained 25% domestic oat groats, wild oat groats, toasted oat flakes (Feed oats) or toasted wild oat flakes added at the expense of wheat. All diets were formulated to contain approximately 20% crude protein and synthetic lysine was added so that all diets supplied approximately 0.9% lysine.

The performance of starter pigs fed diets containing wild and domestic oat groats and flakes is presented in Table 50.4. The growth rate of pigs fed diets containing either wild oat groats or wild oat flakes was similar to that obtained with domestic oat groats and flakes. Toasting and rolling of oat groats or wild oat groats did not appear to improve their nutritional value. The results of this experiment indicate that it may be possible to include up to 25% wild oat groats in starter diets without any adverse effects on performance.

Table 50.4. Performance of Starter Pigs (7-25 kg) Fed Various Oat Products

	Wheat/ SBM	Oat Groat	Feed Oats	Wild Oat Groat	Wild Oat Flake
Daily Gain (kg)	0.39	0.39	0.38	0.37	0.35
Daily Intake (kg)	0.64	0.65	0.65	0.61	0.60
Feed Efficiency	1.64	1.62	1.70	1.64	1.70

Thacker et al., 1986.

Digestibility coefficients for dry matter, crude protein and energy obtained from starter pigs fed the same diets used in the growing trial are presented in Table 50.5. In general, digestibility coefficients were highest for the control diet and lowest for the diet containing wild oat groats. The presence of beta-glucan may account for the decline in digestibility with the diets containing

Table 50.5. Digestibility Coefficients For Starter Pigs Fed Various Oat Products

	Wheat/ SBM	Oat Groat	Feed Oats	Wild Oat Groat	Wild Oat Flake
Dry Matter (%)	88.1	87.1	87.3	82.2	85.4
Crude Protein (%)	85.8	83.0	83.6	78.4	81.3
Energy (%)	87.3	86.1	86.2	79.9	83.9

Thacker et al., 1986.

wild oat products. Toasting and rolling produced only marginal improvements in digestibility over that obtained with the unprocessed groat.

Wild oat groats are likely to find limited use in rations fed to feeder pigs and breeding stock. It takes almost 160 kg of wild oats to produce 100 kg of wild oat groats. Therefore, when the initial purchase price of the mixed feed oats is combined with the cost of dehulling, the selling price demanded by manufacturers of wild oat groats is generally higher than can be justified on the basis of their nutrient content. However, if they can be purchased at a sufficiently low price, there does not appear to be any reason why wild oat groats could not be used to provide a significant proportion of the protein and energy required during the growing-finishing phase or in rations fed to breeding stock.

SUMMARY

Wild oat groats contain significantly higher levels of crude protein than domestic oat groats (19.9 vs. 15.5%). Therefore, use of wild oat groats in a ration may reduce the amount of protein supplementation required in order to provide a balanced swine ration. For starter rations, wild oat groats would appear to be equal to domestic oat groats in nutritive value. Therefore, it may be possible to include up to 25% wild oat groats in starter diets without any adverse effects on performance. However, as a consequence of the current high cost of wild oat groats, there would appear to be limited potential for including them in grower diets or in diets fed to breeding stock.

REFERENCES

Burnett, G.S., 1966. Studies of viscosity as the probable factor involved in the improvement of certain barleys for chickens by enzyme supplementation. Brit. Poult. Sci. 7: 55-75.

Campbell, G.L., Sosulski, F.W. and Classen, H.L., 1987. Evaluation of wild oat groats as a feed ingredient for broiler chicks. Anim. Feed Sci. Technol. 16: 243-252.

Chang, P.R. and Sosulski, F.W., 1985. Functional properties of dry milled fractions from wild oats (*Avena fatua* L.). J. Food. Sci., 50: 1143-1147.

Fleming, M. and Kawakami, K., 1977. Studies of the fine structure of β-D-glucans of barleys extracted at different temperatures. Carbohydrate Research 57: 15-23.

National Research Council, 1988. Nutrient Requirements of Domestic Animals, No.2. Nutrient Requirements of Swine. 9th ed. National Academy of Sciences, Washington D.C.

Newman, C.W., Eslick, R.F. and El-Negoumy, A.M., 1983. Bacterial diastase effect on the feed value of two hulless barleys for pigs. Nutr. Rep. Int. 28: 139-145.

Newman, C.W., Eslick, R.F., Pepper, J.W. and El-Negoumy, A.M., 1980. Performance of pigs fed hulled and covered barleys supplemented with or without a bacterial diastase. Nutr. Rep. Int. 22: 833-837.

Sosulski, F.W., 1984. Wild oats: A new food commodity. University of Alberta Agriculture and Forestry Bulletin 7: 60-61.

Sosulski, F.W. and Sosulski, K., 1985. Processing and composition of wild oat groats (*Avena fatua* L.). J. Food Engineering 4: 189-203.

Thacker, P.A., Christison, G.I. and Sosulski, F.W., 1986. Use of wild oat groats in starter rations for swine. University of Saskatchewan, Department of Animal Science Research Reports No. 460: 124-129.

Milton Keynes UK
Ingram Content Group UK Ltd.
UKHW030902141024
449569UK00025B/1264